Geology

Volume 1
Alluvial Systems – Magmas
1-388

edited by
James A. Woodhead

SALEM PRESS, INC.
Pasadena, California   Hackensack, New Jersey

Essays originally appeared in *Magill's Survey of Science: Earth Science Series* (1990), consulting editor James A. Woodhead, and *Magill's Survey of Science: Earth Science Series, Supplement,* edited by Roger Smith. New material has been added.

Library of Congress Cataloging-in-Publication Data
Geology / edited by James A. Woodhead
    p.   cm. — (Magill's choice)
Includes bibliographical references and index.
    ISBN 0-89356-522-9 (set : alk. paper). — ISBN 0-89356-523-7 (v. 1 : alk. paper). — ISBN 0-89356-524-5 (v. 2 : alk. paper)
    1. Geology—Encyclopedias. I. Woodhead, James A. II. Series.
QE5.G465   1999
550'.3—dc21   98-53009
                                                                            CIP

Second Printing

PRINTED IN THE UNITED STATES OF AMERICA

Contents

Publisher's Note

Magill's Choice is a Salem Press imprint designed to meet the basic reference needs of libraries on limited budgets. The two volumes of *Geology* collect 87 essential articles and the glossary from Salem's popular *Magill's Survey of Science: Earth Science Series* and *Magill's Survey of Science: Earth Science Series, Supplement,* a six-volume set containing 432 essays drawing on the expertise and insights of nearly two hundred scholars. The majority of the articles have been updated with new material to reflect advances in research and scholarship. In selecting these articles, special attention has been given to the curricula of high school and college undergraduate earth science courses; the topics chosen thus emphasize core materials—the fundamentals of geology and its related disciplines.

Averaging thirty-five hundred words each, the articles in *Geology* are organized to supply information in a quickly retrievable format. Each article begins by identifying the subject and the fields in which it is studied. This information is followed by a concise italicized description of the subject and definitions of five to ten principal terms used in the essay. The body of each essay is divided into three sections: *Summary, Methods of Study,* and *Context. Summary* sections introduce and explain the parameters of the subject; *Methods* sections explain how the subject is investigated by geologists and other earth scientists, offering examples; and *Context* sections place the subject within broader fields of earth science. The essays conclude with updated annotated bibliographies and cross-references to closely related essays in *Geology.* Dozens of photographs and other illustrations supplement and enliven the text.

In keeping with the Magill's Choice emphasis on basics, the set emphasizes the fundamentals of the various subdisciplines of geology. The set includes four essays on topics in geochemistry; eight on geochronology and paleontology; five on geomorphology; nine on geophysics; three on glacial geology; two on mineralogy and crystallography; two on petroleum geology and engineering; eleven on petrology; ten on sedimentology; four on stratigraphy; eight on structural geology; thirteen on tectonics; and eight on volcanology.

Readers can locate information in these volumes in several ways. The articles themselves have simple, straightforward titles and are arranged alphabetically, making it easy to go directly to topics such as "Calderas," "Lakes," and "Sand Dunes." Titles of articles are also listed under broader subject categories in an appendix at the end of volume 2. The cross-reference lists appended to individual essays direct readers to related topics, and the general index in volume 2 directs readers to more specific information.

Salem Press once again wishes to thank the many scholars who contributed the essays presented in *Geology*. A listing of their names and affiliations is included at the beginning of volume 1.

List of Contributors

John L. Berkley
State University of New York College at Fredonia

Danita Brandt
Eastern Michigan University

Michael Broyles
Collin County Community College

David S. Brumbaugh
Northern Arizona University

James A. Burbank, Jr.
Western Oklahoma State College

Robert E. Carver
University of Georgia

Habte Giorgis Churnet
University of Tennessee at Chattanooga

John H. Corbet
Memphis State University

William W. Craig
University of New Orleans

Robert L. Cullers
Kansas State University

E. Julius Dasch
NASA Headquarters

Dennis R. Dean
University of Wisconsin—Parkside

René De Hon
Northeast Louisiana University

Albert B. Dickas
University of Wisconsin—Superior

Bruce D. Dod
Mercer University

Dean A. Dunn
University of Southern Mississippi

Steven I. Dutch
University of Wisconsin—Green Bay

Richard H. Fluegeman, Jr.
Ball State University

John W. Foster
Illinois State University

A. Kem Fronabarger
College of Charleston

Charles I. Frye
Northwestern Missouri State University

Karl Giberson
Eastern Nazarene College

Billy P. Glass
University of Delaware

Pamela J. W. Gore
De Kalb Community College

William R. Hackett
Idaho State University

Charles E. Herdendorf
Ohio State University

David F. Hess
Western Illinois University

Louise D. Hose
Westminster College

Pamela Jansma
Jet Propulsion Laboratory, National Research Council

Richard C. Jones
Texas Woman's University

Pamela R. Justice
Collin County Community College

Christopher Keating
Angelo State University

Diann S. Kiesel
University of Wisconsin Center— Baraboo/Sauk County

W. David Liddell
Utah State University

J. Lipman-Boon
Independent Scholar

Donald W. Lovejoy
Palm Beach Atlantic College

Spencer G. Lucas
New Mexico Museum of Natural History

Michael L. McKinney
University of Tennessee, Knoxville

David W. Maguire
C. S. Mott Community College

Michael W. Mayfield
Appalachian State University

Randall L. Milstein
Oregon State University

Otto H. Muller
Alfred University

John E. Mylroie
Mississippi State University

Bruce W. Nocita
University of Southern Florida

Edward B. Nuhfer
University of Wisconsin—Platteville

Steven C. Okulewicz
City University of New York—Hunter College

Michael R. Owen
St. Lawrence University

Robert J. Paradowski
Rochester Institute of Technology

C. Nicholas Raphael
Eastern Michigan University

Donald F. Reaser
University of Texas at Arlington

Jeffrey C. Reid
North Carolina Geological Survey

J. A. Rial
University of North Carolina at Chapel Hill

James L. Sadd
Occidental College

Cory Samia
Oregon Museum of Science and Industry

David M. Schlom
California State University, Chico

Kenneth J. Schoon
Indiana University Northwest

John F. Shroder, Jr.
University of Nebraska at Omaha

Stephen J. Shulik
Clarion University of Pennsylvania

Paul P. Sipiera
William Rainey Harper College

Joseph L. Spradley
Wheaton College, Illinois

Anthony N. Stranges
Texas A&M University

Eric R. Swanson
University of Texas at San Antonio

Stephen M. Testa
California State University, Fullerton

Keith J. Tinkler
Brock University

James L. Whitford-Stark
Sul Ross State University

Grant R. Woodwell
Mary Washington College

Geology

Geology

Mount Everest and the other peaks of the Himalayas are composed in part of alluvial sedimentary rocks. *(Archive Photos)*

the fan's entire surface or predominantly on the outer regions, away from the region where the stream leaves the confines of the mountain valley. Arid fans are also built by deposition from mudflows and debris flows. These flows differ from braided stream flows in that they contain less water and much more debris. Humid fans have braided stream systems operating continuously on their surface. Their overall deposits are similar to braided stream deposits formed in other sedimentary environments.

Meandering rivers have a greater sinuosity than braided rivers and are usually confined to a single channel. They have a lower gradient and therefore are typically located downstream from braided rivers. Sediment in meandering rivers is moved mostly as fine-grained suspended load. Several different types of sedimentary deposit result from the activity of meandering rivers. The coarsest material available to the river is moved and deposited within the deepest part of the channel. These gravelly deposits are thin and discontinuous. Point bars develop on the inside curve of a meander bend and are a major site of sand deposition, although silt and gravel may also be components of point-bar deposits. Deposition takes place on point bars because of flow conditions in the river as water travels around the bend. Erosion takes place on the bank opposite the point bar, and in this way, meanders migrate. When a river floods and tops its banks, water and finer-grained sediment spill out of the channel and onto the surrounding valley floor, which is called the floodplain. On the bank directly adjacent to the river channel, large amounts of sediment are deposited to form natural levees, elongate narrow ridges parallel to the channel; levees help to confine the river but are still topped during major floods. As the river spreads out across the floodplain, silt and clay are deposited.

Meandering rivers are constantly shifting their location within the river valley. In this way, very thick alluvial deposits may accumulate through time. New channels may be created between meanders such that old meander bends are cut off and isolated. These isolated meander bends are termed "oxbow lakes" and tend to fill quickly with sediment.

A special type of meandering river is an anastomosing river. It is characterized by a system of channels which do not migrate as much as meandering river channels and which are separated by large, permanent islands.

Alluvial sediments require running water for transport and deposition. This water may be available year-round, such that rivers and streams are constantly active. Sporadic stream discharge produces alluvial sediments in arid and semiarid environments. Alluvial sediments are also associated with glaciers. Large amounts of sediment with a wide range in grain size are deposited directly by glaciers. Streams fed by glacial meltwaters are important and effective agents of transport and deposition of this sediment. Most streams associated with glaciers are bedload streams and therefore have a braided pattern. Their sediment is usually quite coarse-grained.

Alluvial systems form broad, interconnected networks of drainage that feed water and sediment from highlands or mountainous regions to lowlands and eventually to the sea. These drainages form recognizable patterns that are controlled by the type of rock; the type of deformation, if any, that the earth's crust in the region has undergone; and the region's climate. As river systems age, the landscape changes and evolves. In this way, the surface of the land is sculpted by rivers, in concert with other surface processes.

Methods of Study

The study of alluvial systems can be divided roughly into two categories: the study of modern river systems and the study of sedimentary rocks interpreted to have formed in some type of alluvial system. The study of modern alluvial processes and the sedimentary deposits that they generate is crucial to the understanding of such deposits in the rock record. By understanding modern rivers—the ways in which sediment is moved and deposited, the changes through time in channel shape and location, and the characteristics of the deposits in relation to specific physical conditions—geologists can begin to interpret ancient alluvial sediments.

Modern alluvial systems are studied in a number of ways. It is important to know as much as possible about the flow of water itself, so measurements are made of flow velocity, depth, and width. The channel shape and configuration are also measured. Samples of river sediment, both suspended load and bedload, are collected, and estimates are made of how much sediment is moved by a river. It is also important to look at recent alluvial sediments not now directly associated with the river system. That may be done by digging trenches or collecting samples from floodplains or other alluvial sediments. This type of study looks at the last few hundred or few thousand years of river activity and is critically important to the understanding of these systems. The geologist must study not only what is happening today but also how the system has evolved. In this way, knowledge becomes a predictive tool which greatly enhances the overall understanding of the phenomenon.

It was not until the 1960's that geologists began to realize how large has been the contribution of alluvial systems to sedimentary rocks. This realization came about as a direct result of studies of modern alluvial systems which provided the necessary information to allow the correct identification of ancient alluvial deposits. The study of ancient alluvial deposits takes many forms and provides some information that cannot be gathered from modern deposits. For example, modern deposits, for the most part, have only their uppermost surface exposed, except along the bank and in erosion gullies. Ancient accumulations, in contrast, have been transformed into rock such as shale, sandstone, and conglomerate and commonly are parts of uplifted and dissected terrains, including mountain ranges. These exposures of alluvial deposits allow the three-dimensional architecture of alluvial systems to be studied. Geologists look carefully at vertical changes and associations in the types and abundances of sediment which form as a result of alluvial processes.

It has been found that both meandering and braided rivers commonly form cycles of sedimentation that begin with relatively coarse-grained debris and progress to fine-grained debris. These cycles result mainly from the shifting and migration of the channel system. The coarser-grained base represents the channel and bar deposits, and the finer-grained top represents the overbank floodplain deposits.

The study of ancient alluvial deposits also provides clues to the geologic evolution of a region. The specific mineral composition of sedimentary rocks can indicate the nature of the terrain from which the sediment was derived. That, in turn, contributes to an understanding of the history that led to the generation of the ancient river system that produced the alluvial deposit.

Context

Alluvial systems operate over much of the earth's surface. They move and deposit enormous quantities of sediment each year and are both a boon and a hindrance to humankind. River valleys and floodplains are desirable regions for agricultural development because of the fertile soil often found there. Rivers, however, naturally flood about every 1.5 years. This flooding causes huge losses in property, crops, and sometimes even lives. Flooding associated with alluvial fans can be highly energetic and can occur almost without warning, usually in response to heavy precipitation over a short period. Water levels can build very quickly in narrow valleys with little vegetation and produce a rushing wall of water. Such a flood occurred in 1976 in Big Thompson Canyon, near Rocky Mountain National Park, Colorado.

Many different economically valuable materials are found in alluvial deposits. Because many of these deposits are relatively coarse-grained, they have spaces between the grains which may contain usable fluids, such as water, oil, and natural gas. Many important aquifers are found in alluvial deposits. Petroleum typically originates in marine deposits, but it commonly migrates and may form reservoirs in alluvial deposits. Most sandy alluvial deposits are composed predominantly of quartz grains; however, concentrations of a number of different minerals and ores, including gold and diamonds, occur in alluvial deposits. Another very important economic resource in alluvial deposits is the sand and gravel itself. This material is used for road construction and in the manufacture of cement and concrete.

The deposition of sediment from river systems represents the wearing-away of the land. Much alluvial sediment eventually makes its way to the ocean, where it undergoes reworking by marine processes and deposition on continental shelves or perhaps in the deep sea. Through time, as continents move relative to one another, these sediments may be compressed, folded, and uplifted to become parts of major mountain chains. The Appalachian Mountains in the eastern United States and the Himalayas in northern India and China are only two examples of mountains composed in part of sedimentary rocks, some of which are alluvial in origin.

Bibliography

Davis, Richard A., Jr. *Depositional Systems: A Genetic Approach to Sedimentary Geology.* Englewood Cliffs, N.J.: Prentice-Hall, 1983. This college-level textbook has good introductory chapters on alluvial systems and on related subjects, such as sediment transport.

Reading, H. G., ed. *Sedimentary Environments and Facies.* 2d ed. London: Blackwell, 1986. Probably the most comprehensive text available on sedimentary environments. Much of the material is technical, but the text has excellent figures and photographs and will not overwhelm the careful reader.

Schumm, Stanley A. *The Fluvial System.* New York: John Wiley & Sons, 1977. Suitable for college-level readers, this source is packed with information on river systems.

Smith, David G., ed. *The Cambridge Encyclopedia of Earth Sciences.* Cambridge, England: Cambridge University Press, 1981. This easy-to-read and thorough source contains a chapter on sedimentation. It covers fluvial, desert, and glacial sediments and includes diagrams and photographs of braided and meandering rivers and an alluvial fan. The production and transportation of sediments is also discussed. For general readers.

Tarbuck, Edward J., and Frederick K. Lutgens. *The Earth: An Introduction to Physical Geology.* 2d ed. Columbus, Ohio: Merrill, 1987. An introductory textbook suitable for high school students. Chapter 9, "Running Water," contains sections on channel deposits (bars), floodplain deposits, alluvial fans, and various types of river. Includes many diagrams and photographs. Review questions and a list of key terms conclude the chapter.

Walker, R. G., ed. *Facies Models.* 2d ed. Toronto: Geological Association of Canada, 1984. An excellent compilation of nineteen chapters on sedimentary systems. Two chapters are devoted to alluvial systems. Suitable for college students.

Bruce W. Nocita

Cross-References

Deltas, 126; Drainage Basins, 141; Floodplains, 241; River Flow, 526; River Valleys, 534; Stratigraphic Correlation, 593; Weathering and Erosion, 701.

Alpine Glaciers

Alpine glaciers are masses of ice and snow that move slowly down from the peaks to produce the spectacular landforms that are associated with high mountain scenery. Steep horn-shaped or pyramidal peaks, rushing meltwater rivers, cliff-sided valleys with waterfalls—one associates these features with alpine mountains that have been eroded by glaciers. Where such glaciers are still active, they may threaten lives and property through catastrophic forward surges and floodwaters, or they may be essential sources of meltwater in dry areas.

Field of study: Glacial geology

Principal terms

ABLATION: the result of processes, mainly melting (evaporation is also involved), that waste ice and snow from a glacier

CIRQUE: a steep-sided, gentle-floored, semicircular hollow produced by glacial erosion of bedrock high on mountain peaks

EQUILIBRIUM LINE: the boundary between areas of mass balance gain and loss on a glacier's surface for any one year

LITTLE ICE AGE: a short-term cooling trend that lasted from about A.D. 1450 to 1850, during which mountain glaciers all over the world advanced considerably beyond their present limits

MASS BALANCE: the summation of the net gain and loss of ice and snow mass on a glacier in a year

MORAINE: a ridge of glacial-ice-deposited till

TILL: an ill-sorted mixture of fine and coarse rock debris deposited directly by glacial ice

Summary

Alpine, or valley, glaciers are long, narrow streams of ice that originate in the snowfields and cirque basins of high mountain ranges and flow down preexisting stream valleys. They range from a few hundred meters to more than a hundred kilometers in length. In many ways, they resemble river systems. They receive an input of water in the form of snow in the high parts of the mountains. They have a system of tributaries leading to a main trunk system. The flow direction is controlled by the valley that the glacier occupies, and, as the ice moves, it erodes and modifies the landscape over which it flows.

The essential parts of the mass balance of the alpine glacier system are the zone of accumulation, where there is a net gain of ice, and the zone of ablation, where the ice leaves the system by melting and evaporating. In the zone of accumulation, snow is transformed into glacial ice through a process of metamorphism, or change of form. Freshly fallen snow consists of delicate hexagonal (six-sided) ice crystals, or needles, with as much as 90 percent of the total volume as empty air space. As snow accumulates, the ice at the points of the snowflakes melts from the pressure of the snow buildup and migrates toward the center of the flake, where it refreezes. Eventually, many small, elliptical grains about the size of BB shot (about 0.45 centimeter in diameter) are formed of recrystallized ice. The accumulation of masses of these ice pellets is called firn, or névé. With repeated deposits, each year, the loosely packed firn granules are compressed by the weight of the overlying snow. Meltwater, which results from daily temperature fluctuations and the pressure exerted by the overlying snow, seeps through the open pore spaces between the granules; when it refreezes, it adds to the recrystallization process. Air in the pore spaces is forced out. When the ice reaches a thickness of about 30-40 meters, it can no longer support its own weight and yields to slow plastic flow. The upper part of a glacier is thus rigid and tends to fracture, but the ice beneath moves by plastic deformation and flow.

On the surface of the alpine glacier, the boundary between the zone of accumulation and the zone of ablation is approximated by the equilibrium line. Above this line, the surface of the glacier tends to be smooth and white because more new snow accumulates than is lost by melting and all the irregularities are soon covered and smoothed with snow. These areas are dangerous to mountain climbers, who can fall into buried fractures or crevasses there. Below the equilibrium line, melting and evaporation exceed snowfall. There, the surface of the ice is rough and pitted and is commonly broken by open crevasses or streaked with rock debris.

The glacier ice flows and slides from the positive accumulation area to the negative wastage zone; the amount of gain or loss in each area constitutes the mass balance. Should a glacier have an excess of one over the other for a significant time interval of tens to hundreds of years, it will advance or retreat in consequence. As glacial ice flows over the land beneath, it erodes, transports, and deposits vast amounts of rock and soil material. Glaciers erode in two ways: by glacial plucking, in which meltwater beneath the ice freezes blocks onto the passing ice, and by abrasion of the substrate by the rock blocks held fast in the overlying ice. Eroded material can be carried throughout the glacier, though especially at the bottom and the top, and is later mainly deposited as moraines composed of till near the margins of the glacier, where melting dominates.

Alpine glacial landforms of the ice itself include a variety of features attributable to the characteristic fracture and flow of the movement. Fracturing, or breaking, of the brittle ice occurs close the surface of the glacier, where the ice is under lower pressure. Flow of the ice occurs through recrystallization of ice crystals at depth in the glacier, where the greater pressure causes slow change and consequent movement.

At the head of an alpine glacier, where the ice pulls away from the wall of the mountain, a bergschrund (from the German word for "mountain crack") crevasse develops in summer; during winter, it is filled with avalanche snow. Wherever a glacier moves over an irregular rock bed below, the ice on the surface fractures into a variety of other crevasses. Above the equilibrium line, these features are usually covered with snow and thus are dangerous to traverse, but below they are uncovered and of far less hazard to mountain travelers. When crevasses, such as those described, create tumbled cliffs and icefalls in causing flow over a submerged cliff, the (originally French) term "serac" is used to describe them. Below such icefalls, where compressive flow occurs as the ice piles up, the glacier surface is commonly formed into a series of semicircular waves, or bands, called ogives. Ogives are formed when the broken crevasses collect dust and dirt during the summer melt season. In winter, only snow accumulates in the crevasses. Thus, when the dirty seracs close up and move away from the icefall, they form a dark band, whereas the snow-filled crevasses form a light band. Monitoring of the formation and movement of these light and dark ogives assists in investigations of ice velocity.

At the front of an alpine glacier, some of the ice usually stagnates, and other ice may be forced to override it. The result commonly is sheafs of overthrust ice slabs that carry the debris forward into the terminal moraines. Ice that melts away in these regions produces a considerable amount of meltwater that itself is capable of eroding much sediment from the mo-

An alpine glacier in Alaska. *(Lynn Abigail)*

raines. Because water is an efficient fluid, it sorts the sediment into different sizes: gravel, sand, silt, and clay. Redeposition of these sediments into layered deposits provides valuable sand and gravel supplies for construction in mountain areas.

Various landforms are produced by glacier meltwater processes. Kame terraces (mounds of sediment dropped from the meltwater) are formed between the glacier ice and the valley wall by the streams and lakes impounded there. Similarly, out in front of the ice, the valley train, or outwash, sediments are spread out into plains of sorted and stratified sediment. Commonly, blocks of ice are stranded in these sediments to melt away later and leave kettle holes and kettle lakes to pockmark the plain. Beneath the flowing ice of the alpine glacier, the bedrock will be abraded by the sediment carried along in the ice. This "glacial rasp" will groove, striate, and finally polish the bedrock over which it glides. Where blocks of bedrock are frozen onto the base of the ice and plucked out, a roche moutonnée (from the French for sheepskin-wig-shaped, or "curly," rock) is formed, smoothed by abrasion on the up-ice side and rough and broken on the down-ice portion.

After retreat of an alpine glacier upvalley for a long time and wasting away of the ice, the eroded bedrock surface of roche moutonnée and polished and striated bedrock will finally be exposed. In the case of the surface beneath a former icefall, for example, a smoothed "cyclopean," or glacial, stairway will emerge. The many intervening troughs eroded into the bedrock of the alpine valleys commonly become filled with small lakes that are linked by small overflowing streams. The resulting "paternoster" lakes are so named for their resemblance to beads on a chain. The valley walls themselves, having been undercut and eroded deeply by the ice, become exposed to show the characteristic U-shaped cross-sectional profile of glaciated valleys.

As alpine glaciers melt away entirely, finally the upper cirque basins high on the mountainsides will be exposed. These cirques commonly have steep headwalls, where the old bergschrunds were originally, and an over-deepened floor, where the flowing ice scooped out a basin. Many over-deepened cirque floors fill with water to form a small, steep-sided lake referred to as a tarn. The steep mountain peaks above the cirques are also steepened and undercut by the ice around them and form sharp glacial horns, the characteristic pyramidal peaks of alpine glacial regions. The famous Matterhorn of Switzerland is an example of this landform.

Where cirques have formed on opposite sides of mountain ridges, their headwalls may have merged through back-to-back or headward erosion of the glaciers on opposite sides. If not far advanced when the glaciers melt away, a knife-edged ridge may be the only result. On the other hand, if the glaciation has continued for a long time, the cirques may merge and remove much of the intervening rock mass. After the ice has melted away, a low point, or col, will result. Many of the world's most famous mountain passes are cols that were formed in this fashion.

Methods of Study

Alpine glaciers occur in some of the most scenic mountain ranges of the world and are relatively accessible for direct observation. As a result, they have been studied for many years, and their general characteristics are fairly well understood. In the mid-nineteenth century, European scientists were among the earliest to traverse alpine glaciers and note their apparent movement features. At first, it was necessary to prove that the ice actually moved, although imperceptibly. They accomplished this goal by installing rows of stakes across glacier surfaces between known locations on the valley walls. This experiment revealed that the center moved faster than did the sides, where friction retarded flow. Also noted was the movement of particular stones from higher to lower locations. These became known as erratics (Latin for "wanderer") because they were stones out of place. Little by little, the theory of glaciation began to be accepted.

By the late twentieth century, studies of glacier movement had evolved to include drilling through alpine glaciers with "hot-point" electric arc and other drills to install movement and temperature sensors. These activities led to the discovery that many alpine glaciers in midlatitudes are warm-based; that is, they have basal meltwater and slide over their beds. In the highest and coldest altitudes and latitudes, however, many alpine glaciers are instead cold-based; the ice is frozen to its bed, and all movement is through fracture and flow up within the main body of the ice itself. Such glaciers do not interact much with the substrate over which they ride, and plucking and abrasion are at a minimum. Thus, in what seems an anomalous situation, alpine glaciers in the cold and snowy Arctic may actually affect the landscape far less than they do in the warmer south.

Mass balance studies have been another major component of alpine glacier studies. Measurements of amounts and types of annual precipitation at the heads of glaciers and amounts and types of ablation, or melting and evaporation, at glacier snouts allow determination of glacier behavior and response time. If incoming precipitation is greater than ablation for some period of time, in what is then a positive mass balance, eventually the front of the glacier will respond by moving forward at its front. The overall surface of the ice can rise as well. Conversely, if ablation exceeds precipitation in a negative mass balance, after a certain response time, depending upon the size of the glacier, the glacier front will retreat even though ice is still moving downvalley within the glacier itself. A balance between precipitation and wastage for a number of years will result in a stationary glacier snout. Large alpine glaciers have long response times to these mass-balance changes that are measurable in hundreds to thousands of years, whereas the smaller glaciers may respond in but a few years.

A major type of study of alpine glaciers has long been the mapping of past ice-front positions and the sedimentary deposits produced. The dating of such deposits is accomplished by using a variety of means, including historical records, old paintings, photographs, dendrochronology (counting rings of growth) in trees affected by ice movement, and radiometric age dating. Establishment of a chronology of past alpine glaciation helps in understanding the often-dramatic changes that have occurred in so many alpine areas. The sediments and landforms produced by alpine glaciers are prime evidence of past glacial fluctuations. The ice itself has no capacity to sort rock sediment into different sizes or to deposit these sediments as layers, as flowing water does. The result is the deposition of extensive glacial till, which is an unsorted, unstratified jumble of huge rock fragments to clay-sized particles all mixed together. Glacial moraines of different types are the landform expression of till sediments.

Alpine glaciers produce lateral moraines, or linear ridges, of jumbled rock debris along their sides and terminal moraines of till at the front of their farthest advance. As a glacier moves into a negative mass balance for some time, it will begin to retreat back upvalley. Then, if several years of equality, or balance, between precipitation and ablation occur, the front of the ice will remain in the same place for some time, but internally, the ice will still carry debris forward into the front to pile up a new mass of till as a recessional moraine. Here and there, scattered over the landscape, irregular masses, or sheets, of ground moraine will occur.

Because alpine glaciers are climatically sensitive and have response times to climate fluctuations that are measurable in terms of generally less than human life spans, study of them is justified in part to monitor ice fluctuations to understand climate change. Also, more than 70 percent of the world's fresh water occurs in ice and snow, so monitoring of the fluctuation of the world's glaciers provides important information about long-term water storage. With this object in mind, more than two decades ago, a world survey of ice and snow was begun. Much of the work has involved surveys on the ground, in which the scientists must travel the remote and difficult terrain of the great alpine glaciers of the world. Most recently, however, the inventory and monitoring of ice fluctuation is being done by remote sensing through spaceborne satellites.

Context

Alpine glaciers are of great importance to the people who live near them because the masses of ice often threaten villages at the same time that they provide vital meltwater to irrigate fields and maintain local water supplies. The great scenic beauty of the alpine environment is also a source of considerable tourist revenue: Such places as the Alps of Europe, the Southern Alps of New Zealand, the Himalayas of Asia, and the Rocky Mountains and Andes of North and South America draw thousands of visitors every year. In the United States and Canada, many of the national parks and wilderness reserves are the products of alpine glaciation. Rocky Mountain, Yosemite, Glacier, and Mount McKinley national parks and Glacier Bay National Monument all display active glaciers of great natural beauty.

Glacier ice advances are usually slow and of little consequence unless a village happens to be situated near the front. In the Middle Ages in Europe, for example, some small villages were overwhelmed during ice advances of the colder climate, or Little Ice Age, of the time. More recently, however, glacier surges have been recognized as an unusual and sometimes hazardous phenomenon. Surges are sudden rapid movements that may last for several years, during which the glacier may travel at rates of more than 6 kilometers per year, which is quite vigorous for a glacier. Scientists have discovered that a buildup of high water pressure in the basal meltwater passageways beneath such glaciers causes extreme enhancement of basal sliding. A surging glacier can be visualized as almost partially afloat. As a result, the glacier flow rate can speed up to as much as one hundred times the normal rate. Such glaciers can surge forward over roads, fields, or villages, but the most common hazard is with the large quantities of floodwaters that are released from beneath the ice. Such surge floods have been a problem to development in the western Himalaya in recent years.

In another hazardous situation associated with alpine glaciers, anywhere that ice advances across and blocks a river valley, the impounded ice-dammed lakes are a particular danger to those who live downvalley. Because ice floats in water, as an ice-dammed lake grows behind the glacier barrier, eventually the dam will lift upward, fracture, and fail catastrophically. Such self-dumping lakes can be reestablished and drained many times over the years. Some of the largest floods ever recorded are of this sort.

Bibliography

Andrews, J. T. *Glacial Systems: An Approach to Glaciers and Their Environments.* North Scituate, Mass.: Duxbury Press, 1975. Glacial classification, formation, mechanics, mass balance, and landforms are features of this book. Although Andrews is primarily a specialist on continental glaciers, this work is useful because it is not overly technical.

Bailey, R. H. *Glacier.* Alexandria, Va.: Time-Life Books, 1982. This book, one of several in the Planet Earth series, is a finely written, edited, and illustrated volume. Colin Bull, a well-known glacier specialist, was the main consultant. Because of its exceptionally good presentation, this book is one of the best available on alpine glaciers for the nonspecialist, even though much on continental glaciers is included as well.

Davies, J. L. *Landforms of Cold Climates.* Cambridge, Mass.: MIT Press, 1969. This small and reasonably nontechnical work describes a great many features of cold areas, including those of continental glaciers and those of the periglacial, or "not-quite," glacial environments of the tundra and similar regimes. The book is thorough and is designed for the high school and lower-level university student.

Embleton, C., and C. A. M. King. *Glacial Geomorphology.* New York: John Wiley & Sons, 1975. This book is one of the chief sources on all things glacial, but the dominant coverage is of landforms and effects of past glaciation.

Imbrie, J., and K. P. Imbrie. *Ice Ages: Solving the Mystery.* Cambridge, Mass.: Harvard University Press, 1986. This volume is a most readable book that explains the development of the theory of glaciation and examines predictions of future ice ages. While much of the book involves an explanation of past continental glaciers, the orderly exposition of the development of new ideas on general glaciation is most useful. Also, the personalities and inner feelings of many of the scientists involved are described, which adds a human dimension to what can be a dry subject to the nonenthusiast.

Paterson, W. S. B. *The Physics of Glaciers.* Elmsford, N.Y.: Pergamon Press, 1981. This work is one of the best-available references dealing with the basic mechanics of glacier formation, nourishment, structure, flow, and behavior. Although the work contains many pages of rigorous mathematical analyses, a discerning nonprofessional reader can glean a world of useful and reliable information on glacier behavior from the intervening parts of the book.

Schultz, G. *Glaciers and the Ice Age.* New York: Holt, Rinehart and Winston, 1963. A small book written for the layperson that provides a good review of basic aspects of glaciers and emphasizes their impact upon, and relationships to, human interests and activities.

Sharp, R. P. *Living Ice: Understanding Glaciers and Glaciation.* New York: Cambridge University Press, 1988. This book is one of the most readable and detailed works on glaciers. It is very well illustrated and includes a comprehensive glossary with more than three hundred entries.

John F. Shroder, Jr.

Cross-References

Continental Glaciers, 93; Glacial Landforms, 281.

Andesitic Rocks

Andesite is an intermediate extrusive igneous rock. It is porphyritic and contains phenocrysts of plagioclase, little or no quartz, and no sanidine or feldspathoid. Active volcanoes on the earth erupt andesite more than any other rock type. Andesites are primarily associated with subduction zones along convergent tectonic plate boundaries.

Field of study: Petrology

Principal terms

BOWEN'S REACTION PRINCIPLE: a principle by which a series of minerals forming early in a melt react with the remaining melt to yield a new mineral in an established sequence

EXTRUSIVE ROCK: igneous rock that has been erupted onto the surface of the earth

GROUNDMASS: the fine-grained material between phenocrysts of a porphyritic igneous rock

INTERMEDIATE ROCK: an igneous rock that is transitional between a basic and a silicic rock, having a silica content between 54 and 64 percent

PHENOCRYST: a large, conspicuous crystal in a porphyritic rock

PLUTONIC ROCK: igneous rock formed at great depth within the earth

PORPHYRY: an igneous rock in which phenocrysts are set in a finer-grained groundmass

STRATOVOLCANO: a volcano composed of alternating layers of lava flows and ash; also called a composite volcano

SUBDUCTION ZONE: a convergent plate boundary

VISCOSITY: a substance's ability to flow; the lower the viscosity, the greater the ability to flow

Summary

Andesite takes its name from lavas in the Andes mountains of South America. To most geologists, andesites are light gray porphyritic volcanic rocks containing phenocrysts of plagioclase but little or no quartz and no sanidine or feldspathoid. Despite their lackluster appearance, andesites are of great interest to geologists for several reasons. First, active volcanoes on the earth erupt andesite more than any other rock type; andesite is the main rock type at 61 percent of the world's active volcanoes. Second, andesites have a distinctive tectonic setting. They are primarily associated with convergent plate boundaries and occur elsewhere only in limited amounts. Of the active volcanoes that occur within 500 kilometers of a subduction zone, 78 percent include andesite; only three active volcanoes not near a destructive plate boundary include it. Third, andesites have bulk compositions similar to estimates of the composition of continental crust. This similarity, in association with the tectonic setting of andesites, suggests that they may play an important role in the development of terrestrial crust. Fourth, the development and movement of andesitic magma seem to be closely related to the formation of many ore deposits. It appears that andesite genesis is the source of these metals.

Rocks are classified chemically according to how much silica they contain; rocks rich in silica (more than 64 percent) are called silicic. They consist mostly of quartz and feldspars, with minor amounts of mica and amphibole. Rocks low in silica (less than 54 percent), with no free quartz but high in feldspar, pyroxene, olivine, and oxides, are called basic. Basic rocks, free of quartz, tend to be dark, while silicic rocks are lighter and contain only isolated flecks of dark minerals. Basalts are examples of silicic volcanic rocks. Andesites, having a silica content of about 60 percent, are volcanic rocks termed intermediate. Andesite's plutonic equivalent is diorite. There is no cut-and-dried difference between basalt and andesite or between andesite and rhyolite. Instead, there is a broad transitional group of rocks that carry names such as "basaltic andesite" or "andesitic rhyolite."

Nevertheless, some generalizations can be applied to the lavas and magmas that form these rocks. One generalization has to do with viscosity. Andesite lavas are more viscous than are basalt lavas and less viscous than are rhyolite lavas. This difference is primarily an effect of the lava's composition, and to a certain extent it is a result of the high portion of phenocrysts present in the more viscous lavas.

Different minerals crystallize at different temperatures. As a basalt magma cools, a sequence of minerals appears. The first mineral to crystallize is usually olivine, which continues to crystallize as the magma cools until a temperature is reached at which a second mineral, pyroxene, begins to crystallize. As the temperature continues to drop, these two continue to crystallize. Cooling continues until a temperature is reached when a third mineral, feldspar, crystallizes. This chain of cooling and mineral crystallization is known as Bowen's reaction principle. Often, olivine and pyroxene crystallize out early in the process, so they may be present in the final rock as large crystals, up to a centimeter across. These crystals are called phenocrysts. The size of these phenocrysts is in direct contrast to the fine-grained crystals of the groundmass. Igneous rocks with phenocrysts in a fine-grained groundmass are known as porphyries. Most volcanic rocks contain some phenocrysts. The groundmass crystals form when the lava cools on reaching the surface. If the lava has a low viscosity, reaches the surface, spreads out, and cools quickly, individual crystals do not have enough time to grow. The overall rock remains fine-grained. The phenocrysts crystallized out much earlier, while the magma was still underground. There, they had plenty of time to grow and were then carried to the surface with the magma during eruption.

Basalts, as a result of their low viscosity, tend to produce thin lava flows that readily spread over large areas. They rarely exceed 30 meters in thickness. Andesite flows, by contrast, are massive and may be as much as 55 meters thick. The largest single andesite flow described, which is in northern Chile, has an approximate volume of 24 cubic kilometers. Because of their low viscosity, basalt flows can advance at considerable rates; speeds up to 8 kilometers per hour have been measured. Andesite flows often move only a matter of tens of meters over several hours. As a result of higher viscosity, andesite flows show none of the surface features of more "liquid" lavas. Flow features such as wave forms, swirls, or ropy textures often associated with basalt flows never occur in andesite flows. Andesite flows tend to be blocky, with large, angular, smooth-sided chunks of solid lava. The flow tends to behave as a plastic rather than a liquid. An outer, chilled surface develops on the slow-moving flow, with the interior still molten. Plasticity within the flow increases toward the still-molten inner portion. As the flow slowly shifts, moves, and cools, the hard, brittle outer layer breaks into the large angular blocks characteristic of andesite lava flows. As the flow slowly moves, the blocks collide and override one another to form piles of angular andesite blocks.

The differences between basalts and andesites reflect their differences in composition, which is a function of the environment in which their source lavas occur. Basalts are typically formed at mid-ocean ridges and form oceanic crust. The generation of andesite magma is characteristic of destructive plate margins. Here, oceanic plates are being subducted below continental plates. Destructive plate boundaries tend to produce a greater variety of lavas than do spreading zones (ocean ridges). The close association of andesite with convergent plate boundaries is the significance of the "Andesite Line" often drawn around the Pacific Ocean basin (see figure) This fairly well-defined line separates two major petrographic regions. Inside this line and inside the main ocean basin, no andesites occur. All active volcanoes inside the line erupt basaltic magma, and all volcanic rocks associated with dormant volcanoes within this region are basaltic. Outside the line, andesite is common. The Andesite Line is the western and northern boundary of the Pacific plate and the eastern boundary of the Juan de Fuca, Cocos, and Nazca plates. The Andesite Line parallels the major island arc systems, the subducting edges of the tectonic plates listed above, and a chain of prominent and infamous stratovolcanoes known as the Ring of Fire. These stratovolcanoes are exemplified by Mount Rainier and other volcanoes of the western United States, El Chichón of Mexico, San Pedro of Chile, Mounts Egmont and Taupo of New Zealand, Krakatoa and Tambora of the Indonesian Arc, Fujiyama of

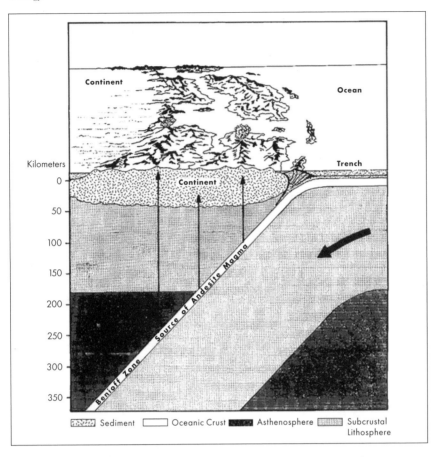

Japan, Bezymianny of Kamchatka, and the Valley of Ten Thousand Smokes in Alaska.

The viscous nature of andesite lavas is responsible for many classic volcanic features. Most notable is the symmetrical cone shape of the strato-volcanoes of the Circum-Pacific region. Short, viscous andesitic flows pouring down the flanks of these volcanoes are alternately covered by pyroclastic material and work to build the steep central cone characteristic of composite stratovolcanoes. When an andesite's silica content rises to a point that it approaches rhyolite composition, it is termed a dacite. Dacite is often so viscous that it cannot flow and blocks the vent of the volcano. This dacite plug is called a lava dome. Such a dome can be seen in pictures of Mount St. Helens. If the plugged volcano becomes active again, the lava dome does not allow for a release of accumulating pressure and explosive gases. Pressure builds until the volcano finally erupts with great force and violence.

The Benioff zone is a plane dipping at an angle of 45 degrees below a continent, marking the path of a subducting oceanic plate at the destructive plate margin. It is a region of intense earthquake activity. At the Benioff zone, the subducting plate melts, producing a chemically complex, destructive margin magma (andesitic). As andesitic magma is formed at the Benioff zone and begins its slow rise toward the surface, it passes through and comes in contact with regions of the asthenosphere, lithosphere, and continental crust. During its rise, the magma also comes in contact with circulating meteoric water, and new magmatically derived fluids are formed. Because andesitic magma rises through a variety of host rocks, the variety of minerals that enrich the new fluids is increased. The rising magma also stresses the surrounding host rocks, causing them to fracture. These fractures and similar open spaces are filled with mineral-rich water and magmatically derived fluids. As the fluids cool, mineral ores precipitate from the solutions and fill the open spaces. These filled spaces often become mineral-rich dikes and sills.

Methods of Study

The study and interpretation of andesitic rocks is accomplished in three basic ways: fieldwork, in which researchers travel to locations to assess and interpret a specific region of the earth that is known or suspected of being andesitic terrain; petrological studies, which use all available methods of study to ascertain the history, origin, conditions, alterations, and decay of the rocks collected during fieldwork; and mineralogical studies, which through intensive laboratory investigations identify the specific mineral characteristic of a sample.

If a scientist is interested in andesitic rocks and knows, or suspects, that a region of the earth is andesitic terrain, travel to this location to do fieldwork is likely. There, the researcher will most likely collect a large number of samples. The location from which each sample is taken is carefully recorded on a map. Additional data describing the stratigraphic interval from which the sample is taken are recorded. These data include thickness, areal extent, weathering, and strike and dip of the location. Photographs are also commonly used to record the surrounding environment of the sample location. Samples are usually given preliminary study at the researcher's field camp.

When the fieldwork is over, the researcher returns with the samples to a laboratory setting. In the laboratory, petrological and mineralogical investigations begin. A petrological study of an andesitic sample will include both a petrographic and petrogenetic analysis. The petrographic analysis will describe the sample and attempt to place it within the standard systematic classification of igneous rocks. Identification of the sample is accomplished by means of examining a thinly sliced and polished portion of the sample under a petrographic microscope—a process known as thin-section analysis. The information obtained by microscopic examination gives a breakdown of the type and amount of mineral composition within the sample. Knowing the conditions under which these minerals form allows the researcher to make a petrogenetic assessment. Petrogenesis deals with the origin and formation of rocks. If a mineral is known to form only at certain depths, temperatures, or pressures, its presence within a sample allows the researcher to draw some specific conclusions concerning the rock's formation and history.

Mineralogical studies of field-gathered samples aid in the petrogenetic portion of the analysis. Mineralogy involves the study of how a mineral forms, its physical properties, chemical composition, and occurrence. Minerals can exist in a stable form over only a narrow range of pressure and temperature. Experimental confirmation of this range enables the researcher to make a correlation between the occurrence of a mineral in a rock and the conditions under which the rock was formed. In this sense, mineralogical and petrological analyses complement each other and work to formulate a concise history of a given sample.

When data gathered from the field and laboratory are combined, the geological history of a given field area can begin to be interpreted. When areas of similar igneous rock types, ages, and mineral compositions are plotted on a map, they form a petrographic province. A petrographic province indicates an area of similar rocks that formed during the same period of igneous activity. On a global scale, the Andesite Line marks the boundary between two great provinces: the basaltic oceanic crust and the andesitic continental crust. Both crustal forms have distinctly different geological histories and mineral compositions.

Context

Andesitic magmas are responsible for the formation of new continental crust. The crust grows by progressive accumulation from the eruptions of andesitic volcanoes on the surface and from dioritic intrusions below ground. In addition, the understanding of andesitic volcanism and its relationship to convergent plate boundaries has aided in the exploration of economic minerals. Certain economically important ores are associated with andesitic rocks and zones of andesitic volcanism. Plotting on a map the locations of the world's major molybdenum and porphyry copper deposits shows that they are situated along the Andesite Line. The deposits are in direct correlation with convergent plate boundaries and with andesitic volcanism. These ores are the result of hydrothermal activity associated with igneous bodies emplaced at high levels in the earth's crust.

Bibliography

Bowen, N. L. *The Evolution of Igneous Rocks.* Mineola, N.Y.: Dover, 1956. An unmatched source and reference book on igneous rocks, written by the father of modern petrology. This book is the basis of Bowen's reaction principle. Written for graduate students and professional scientists.

Carmichael, I. S. E., F. J. Turner, and J. Verhoogen. *Igneous Petrology.* New York: McGraw-Hill, 1974. A college-level textbook on the formation and development of igneous rocks. Very detailed and complete.

Gill, J. B. *Orogenic Andesite and Plate Tectonics.* New York: Springer-Verlag, 1981. A well-documented summary of the entire field of andesite genesis. Written for graduate students and professional earth scientists.

Klein, Cornelis, and Cornelius S. Hurlbut, Jr. *Manual of Mineralogy.* 18th ed. New York: John Wiley & Sons, 1971. A comprehensive physical and chemical description of andesite. For high school and general readers.

Williams, H., and A. R. McBirney. *Volcanology.* San Francisco: Freeman, Cooper, 1979. A classic textbook on volcanoes and volcanology. Well illustrated and very descriptive. Written for the undergraduate or graduate student.

Windley, B. F. *The Evolving Continents.* New York: John Wiley & Sons, 1977. An excellent reference book on plate tectonics and tectonic processes. Written for the college-level reader.

Randall L. Milstein

Cross-References

Archaeological Geology

Archaeology is rapidly becoming a markedly more scientifically based field, a trend started with the "New Archaeology" of the 1970's. Archaeological geology covers the wide range of geological sciences that are applied to archaeology during excavation and postseason, or postexcavation, sorting, classifying, and analyzing.

Field of study: Geochronology and paleontology

Principal terms

ABSOLUTE DATE: a date that gives an actual age, though it may be approximate, of an artifact

CURIE POINT, OR TEMPERATURE: the temperature at which materials containing iron oxides lose their magnetic pattern and align with the earth's magnetic field

RELATIVE DATE: a date that places an artifact as older or younger than another object, without specifically giving an age for it

REMOTE SENSING: any of a wide variety of techniques, such as aerial photography, for collecting data about the earth's surface from a distance

STRATIGRAPHY: the deposition of artifacts in layers, or strata

Summary

Archaeological geology is the application of geological methods and techniques to archaeology. The two disciplines have become so closely intertwined at times that some have spoken for a new term to describe their partnership: "geoarchaeology." A term aimed more specifically at the contributions of the physical and chemical sciences associated with archaeology (such as potassium-argon dating) is "archaeometry." Without such scientific methods, archaeology becomes guesswork at best; dating of finds, for example, should be derived from empirical data, or information that can be proven through experiment and observation, rather than deduced from theory without corroboration. Archaeology is, basically, deductions made about an artifact from the context in which it is found; geology helps to define and date that context, thereby providing the empirical information from which speculation about the artifact can be derived.

The principle of superposition was probably one of the first geological methods that archaeology utilized. This law states that a layer superimposed on another layer should be younger, having been laid down after the lower, or older, layer. This study of stratigraphy is a keystone to archaeological dating but constitutes relative rather than absolute dating. One of the better-known methods of absolute dating is that of carbon 14 dating. Carbon 12 and carbon 14, an isotope of carbon 12, are elements that exist in all living organisms. Once the organism dies, whether it is plant or animal, the input of carbon 14 from the environment stops, and the remaining carbon 14 begins to decay. The amount of carbon 14 left relative to the amount of carbon 12 is used to calculate the amount of time that the organism has been dead, using the known half-life of carbon 14 (5,730 years).

Archaeomagnetism (the term is the archaeological equivalent of geology's "paleomagnetism") is another dating method, but one that uses changes in the earth's magnetic field as recorded by archaeological artifacts such as kilns. The earth experiences continual changes in the intensity and polarity of its magnetic field. If a clay artifact has been heated to its Curie point, or temperature, its magnetic particles will align in the direction of the polarity of the earth's magnetic field in which they cool. Geological identification of the polarity pinpoints the time of firing. Thermoluminescence dating is also a method used on clay. A piece of pottery is heated to 500 degrees Celsius and the ensuing emission of light measured as an indication of the length of time since its firing, as the energy of the thermoluminescence increases as it is stored up over time. Another method for dating inorganic objects, in this case volcanic rock, is potassium-argon dating. When volcanic rock is newly solidified, it contains no argon, but it does contain potassium 40. As the potassium 40 decays, it becomes argon 40. Because the half-life of potassium 40 is known (1 billion years), the amount of argon 40 gives the absolute age of the rock.

Geological techniques are also used for locating and recording archaeological sites. Remote sensing is a technique by which data are collected on a site in a "remote" rather than a hands-on way. Photographic images are a major part of remote sensing. Aerial photography, for example, is used to photograph the landscape from airplanes or satellites. Images and patterns that record electromagnetic radiation provide another source of remote-sensing data, as does soil resistivity surveying. Soil resistivity surveying is a method used to map buried features by finding electrical conductivity differences between the features and the soil around them. The ease or resistance with which the current penetrates the soil is the basic principle. To test the resistivity, four electrodes are inserted into the ground. Two generate the current, and two measure the drops in voltage; an equation taking into account the distance between the electrodes, the amperage, and the drop in voltage then gives the total resistance to conductivity.

Petrology, or the study of different aspects of rocks, is a standard feature of archaeological geology. Most techniques focus on the study of thin sections or powdered samples of rock—for example, through the use of scanning electron microscopes and X-ray diffraction, respectively. In addition, archaeologists also utilize geological studies of cryoturbation (freeze-thaw cycles in soil), argilliturbation (shrinking and swelling cycles in clays), aeroturbation (disturbances by gas, wind, and air), aquaturbation (disturbances from the movement of water), and seismiturbation (disturbances by earthquakes). Study of these different types of environmental disturbance helps identify the different site-disturbing processes at work.

Applications of the Method

Many aspects of the geological sciences can be applied to archaeological sites. It is the nature of the site that determines the appropriate method to use. For example, a historical archaeological site, or one that is dated to within the parameters of recorded history, would not probably be a site at which radiocarbon dating would be useful: The artifacts would not be old enough for a dating method that yields figures in thousands of years. On the other hand, a prehistoric site such as the possibly Iron Age Caer Cadwgan, a probable hillfort in Wales, would benefit primarily from radiocarbon dating: The site is probably about three thousand years old, and charcoal and bone, excellent types of samples for carbon 14 testing, are the main elements found in excavation at this site. This site has also yielded small glass beads, items that would perhaps be datable by thermoluminescence.

Historical sites in general, however, are not the prime candidates for archaeological geology that prehistoric sites are. Prehistoric archaeology, on the other hand, depends completely on geological analysis for some conclusions because it predates any written records; there are no fortuitously preserved documents to fall back on for verification. Archaeological geology can accomplish much, but it is most useful for four principal processes during an excavation: locating a site, recording and analyzing the features of a site, and dating a site.

The archaeological use of soil resistivity for locating sites or individual features was first applied to locate prehistoric stone monuments just after World War II. The differences in ground conductivity are used to locate anomalies, which can range from buried ditches to stone walls. The method was first developed in geology to locate ore deposits, faults, and sinks (sunken land where water can collect). Remote sensing can be a useful method for several different goals. Remote sensing can be used to locate sites, monitor changes in the archaeological record, reveal the distribution of archaeological sites, or map sites. Aerial photography can be used to locate sites, and it can then be used to map a site and record its features for planning an excavation. On some sites, the camera has been sent up in a balloon to take the photographs, as the site may have been too small or the budget of the excavation too limited for airplane-carried camera work.

The techniques of geoarcheology are used to date finds such as this ancient stone "face" discovered in a cave in Spain. *(Leslie G. Freeman/University of Chicago)*

Archaeology is a destructive process; once a site is excavated or even surface collected, it cannot be restored. Remote sensing can, sometimes, take the place of excavation in what is termed nondestructive archaeology. It can also be a useful substitute for excavation when there is not time for a full-scale dig to be mounted, as during times when a formation may be temporarily visible (for example, winter snows revealing significant gradations in the land), but conditions may not be right for excavation. Remote sensing can also preserve at least an image of a site that must be destroyed, as during construction of a road. With this use, it is an invaluable tool in rescue archaeology.

Petrology is a useful tool in general for archaeology. Petrology can be used on many different types of sites, from prehistoric to historic. It can be used to give a provenance, or origin, for building and sculptural stones as old as Stonehenge or as young as the classical Greek marbles. Archaeomagnetism can be used not only to date artifacts but also to distinguish sources for substances. Obsidian sources, for example, have been found to each have different magnetization strengths. It is possible then that the source of ore for coins, which may contain trace amounts of iron, may also be able to be discovered. Archaeomagnetic dating of lava has been used to date the end of the Minoan civilization (about 1500 B.C.), which was destroyed by the volcanic eruptions at Santorin (or Thera) in Greece. It is used mainly for early humankind sites, or up to 10,000 years old.

Thermoluminescence can also be used to date lava as well as burnt flint (implements heated by accident or on purpose to improve certain qualities), burnt stones (heated on a fire and then placed in a food container as a "pot-boiler"), glass (volcanic glass especially), sediments (buried soils), and ceramics. This method is popular because of its absolute dates for a wide range of ages: 50,000-500,000 years old. Potassium-argon dating is also used for inorganic samples but only ones that predate humankind. Samples can only be as recent as 100,000 years old. Another limitation, however, is that because it is used to date lava, in order to be of any use the site must be connected to a particular volcanic eruption, and sites such as Santorin and Pompeii are not common (and too young anyway). The usefulness of potassium-argon dating to archaeology is its ability to fix dates for reversals of the earth's magnetic field, which in turn are used to date archaeological sites through archaeomagnetism.

Using petrology to establish the provenance of a rock artifact is of great importance in prehistoric archaeology. The evaluation of rock types found can help to identify the mining or quarrying skills of a culture, its determination of the usefulness of various kinds of rocks, and even some of the places to which people may have traveled. If a certain type of rock is not native to the area in which it is found yet occurs in large quantities, it may be inferred that perhaps significant trade or travel was taking place. The value that a culture may have placed on a particular rock may also be determined, judging by its use (whether ceremonial or practical, for example). Use-wear as opposed to earth-moving processes that have changed the shape of the rock are another object of study. Petrological methods such as X-ray diffraction can be used to identify the origin of rocks on a site. Studying the rocks from a number of different sites can be a way of tracing the trade routes of a culture.

Context

The main point of archaeology is to investigate humankind's past and to be able to draw some conclusions about what it was like. Yet, one of the criticisms of archaeology has been that it is too heavy on description and too light on substantiated conclusions. Science in general, and geology in particular, helps to give a factual basis to archaeological theorizing. While logic, reasoning, and even perhaps intuition do have a place in archaeology, piecing together the abstract parts of a culture such as religion and philosophy is not as straightforward as piecing together the sherds of a broken pot.

Science thus provides archaeology with the backbone to present its conclusions with some degree of surety. The usefulness of any research is limited unless there is some certainty about the conclusions reached.

As more and more archaeological sites are threatened by construction (as the Rose Theatre in London), environmental pollution (as marble structures such as the Parthenon on the Acropolis in Athens), and other manmade hazards, definitive scientific methods must be developed and utilized while a site is still available for study.

Bibliography

Aitken, M. J. *Thermoluminescence Dating.* San Diego, Calif.: Academic Press, 1985. Aitken has prepared a comprehensive work on thermoluminescence dating. He includes the theoretical and mathematical basis for this method, as well as its applications. This work is specifically aimed at readers without knowledge of physics, but it is still fairly technical and most suitable for college-level students.

Brothwell, Don, and Eric Higgs, eds. *Science in Archaeology.* With a foreword by Grahame Clark. 2d ed. New York: Praeger, 1970. Despite the date of the book, it is helpful because it has chapters on many of the scientific methods that are in use, including thermoluminescence, potassium-argon, and radiocarbon (carbon 14) dating. This volume is useful for seeing the wide range of scientific methods that are applied in archaeology and the beginning of New Archaeology. Suitable for undergraduate college-level students.

Butzer, Karl W. *Archaeology as Human Ecology: Method and Theory for a Contextual Approach.* New York: Cambridge University Press, 1982. Butzer sets out the principles and objectives of geoarchaeology as a subdiscipline of archaeology. His emphasis is on the environmental sciences in general as necessary for the best empirical data in archaeology.

Kelley, Jane H., and Marsha P. Hanen. *Archaeology and the Methodology of Science.* Albuquerque: University of New Mexico Press, 1988. Kelley and Hanen have prepared this volume as a way to reconcile the philosophies of science and archaeology. They include case histories that target the methodological problems of interpreting even sound empirical data. Useful for college-level students seeking to discover how the sciences fit into archaeology.

Kempe, D. R. C., and Anthony P. Harvey, eds. *The Petrology of Archaeological Artefacts.* New York: Clarendon Press, 1983. A very useful book for different techniques of archaeological geology and their various applications, including specific examples. As the title indicates, petrology is the main focus. Technical, so best suited to advanced college-level readers.

Kraft, John C. "Archaeological Geology." *Geotimes* 39, no. 2 (February, 1994): 12-14.

Parkes, P. A. *Current Scientific Techniques in Archaeology.* New York: St. Martin's Press, 1986. An up-to-date work that includes archaeometry in great detail. Technical but well written. Suitable for college-level readers.

Schiffer, Michael. *Advances in Archaeological Method and Theory.* 11 vols. San Diego, Calif.: Academic Press, 1978-1987. This series is a comprehensive review of changes in scientific archaeological methods and new applications of those methods. Each chapter is by a different contributor, and many include specific examples of studied sites.

_____, ed. *Formation Processes of the Archaeological Record.* Albuquerque: University of New Mexico Press, 1987. This book includes chapters on the earth processes that can disturb and change the archaeological record, such as earthquakes and freeze-thaw cycles. Contains photographs and is suitable for the general reader.

J. Lipman-Boon

Cross-References

The Geologic Time Scale, 272.

Basaltic Rocks

"Basalt" is the term applied to dark, iron-rich volcanic rocks that occur everywhere on the ocean floors, as oceanic islands, and in certain areas on continents. It is the parent material from which nearly any other igneous rock can be generated by various natural processes.

Field of study: Petrology

Principal terms

AUGITE: an essential mineral in most basalts, a member of the pyroxene group of silicates

CRUST: the upper layer of the earth and the other "rocky" planets; it is composed mostly of relatively low-density silicate rocks

LITHOSPHERIC PLATES: giant slabs composed of crust and upper mantle; they move about laterally to produce volcanism, mountain building, and earthquakes

MAGMA: molten silicate liquid, including any crystals, rock inclusions, or gases trapped in that liquid

MANTLE: a layer beginning at about 5 to 50 kilometers below the crust and extending to the earth's metallic core

OCEANIC RIDGES: a system of mostly underwater rift mountains that bisect all the ocean basins; basalt is extruded along their central axes

OLIVINE: a silicate mineral found in mantle and some basalts, particularly the alkaline varieties

PERIDOTITE: the most common rock type in the upper mantle, where basalt magma is produced

PLAGIOCLASE: one of the principal silicate minerals in basalt, a member of the feldspar group

SUBDUCTION ZONES: areas marginal to continents where lithospheric plates collide

Summary

Basalt is a dark, commonly black, volcanic rock. It is sometimes called "trap" or "traprock," from the Swedish term *trapp*, which means "steplike." On cooling, basalt tends to form hexagonal columns, which in turn form steplike structures after erosion. Excellent examples can be seen at Devils Postpile, in the Sierra Nevada in central California, and the Giant's Causeway, in County Antrim, Ireland. The term "trap," however, is used mostly by miners and nonspecialists; scientists prefer the word "basalt" for the fine-grained, volcanic rock that forms by the solidification of lava flows. Basaltic magma that crystallizes more slowly below the earth's surface, thus making larger mineral grains, is called "gabbro."

The importance of basalt to the evolution of planets such as the earth cannot be overemphasized. Basalt is considered to be a "primary silicate liquid," in that the first liquids to form by the melting of the original minerals that made up all the so-called rocky planets (those composed mostly of silicates) were basaltic in composition. In turn, basalt contains all the necessary ingredients to make all the other rocks that may eventually form in a planet's crust. Furthermore, many meteorites (which are believed to represent fragments of planetoids or asteroids) are basaltic or contain basalt fragments, and the surfaces of the moon and the planets Mercury, Venus, and Mars are known to be covered to various degrees by basalt lava flows.

Like most rocks in the earth's crust, basalt is composed of silicate minerals, substances whose principal component is the silica molecule. Compared with other silicate rocks, basalts contain large amounts of iron and magnesium and small amounts of silicon. This characteristic is reflected in the minerals in basalts, which are mostly pyroxene minerals (dark-colored, calcium-iron-magnesium silicates, such as augite) and certain feldspar minerals called plagioclase (light-colored, calcium-sodium-aluminum silicates). Pyroxenes and plagioclase are essential minerals in basalts, but some types of basalt also contain the mineral olivine (a green, iron-magnesium silicate). The high abundance of dark green or black pyroxene and, in some cases, olivine gives basalts their characteristic dark color. This color is mainly attributable to the high iron content in pyroxene and olivine.

Basalt is the most common type of igneous rock (rock formed by the crystallization of magma) on or near the surface of the earth. Relatively rare on continents, it is the principal rock in the ocean floors. Drilling into the sea floor by specially designed oceanographic research ships reveals that basalt invariably lies just below a thin cover of fine, sedimentary mud. Ocean floor basalt flows out of mid-ocean ridges, a system of underwater mountain ranges that spans the globe. These ridges commonly trend roughly down the middle of ocean basins, and they represent places where the earth's lithospheric plates are being literally split apart. In this process, basalt magma is generated below the "rift mountains"; it flows onto the cold ocean floor and solidifies. Although oceanic ridges are normally hidden from view under the oceans, a segment of the Mid-Atlantic Ridge emerges above the waves as the island of Iceland.

Oceanic islands are also composed of basalt. The islands of Hawaii, Fiji, Mariana, Tonga, and Samoa, among others, are large volcanoes or groups of volcanoes that rise above water from the ocean floor. Unlike the basalts that cover the ocean floors, however, these volcanoes do not occur at oceanic ridges but instead rise directly from the sea floor. Basalt is also a fairly

The steplike hexagonal basalt columns of the Giant's Causeway in Northern Ireland extend from the headland (above) into the sea (below). *(Robert McClenaghan)*

common rock type on island arcs, volcanic islands that occur near continental margins. These curvilinear island chains arise from melting along subduction zones, areas where the earth's lithospheric plates are colliding. This process generally involves material from the ocean basins diving under the more massive continents; andesite (a light-colored rock) volcanoes are the main result, but some basalt erupts there as well. The Japanese, Philippine, and Aleutian island chains are examples of island arcs, as are the island countries of New Zealand and Indonesia.

Basalt lava flows are not nearly as common on continents as in oceanic areas. Andesites, rhyolites, and related igneous rocks are far more abundant than basalt in continental settings. In North America, basalt lava flows and volcanoes are best exposed to view in the western United States and Canadian provinces, western Mexico, Central America, and western South America. The greatest accumulations of basalt lava flows in the United States are in the Columbia Plateau of Washington, Oregon, and Idaho. In this large area, a series of basalt lava flows have built up hundreds of feet of nearly flat-lying basalt flows over a few million years. These "fissure flows" result from lava's pouring out of long cracks, or fissures, in the crust. They are similar in many respects to the basalt flows produced at oceanic ridges, because no actual volcanic cones are produced—only layer after layer of black basalt. The Snake River Plain in southern Idaho has a similar origin, and other extensive basalt plateaus occur in the Deccan area of southern India, the Karroo area of South Africa, and Paraná State in Brazil.

Not all basalt is erupted as lava flows. If the lava is particularly rich in volatiles such as water and carbon dioxide, it will be explosively ejected from the volcano as glowing fountains of incandescent particles that rain down on the surrounding area. Conical volcanoes composed almost exclusively of basalt ejecta particles are called cinder cones. Good examples of cinder cones are Sunset Crater in northern Arizona, the numerous cinder cones in Hawaii and Iceland, and the very active volcano in Italy known as Stromboli. In fact, the rather violent eruptions that produce cinder cones are called "strombolian eruptions."

Although basaltic rocks may all look alike to the nonspecialist, there are actually many different kinds of basalt. They are arranged by scientists into a generally accepted classification scheme based on chemistry and, to some extent, mineralogy. To begin with, basalt can be distinguished from the other major silicate igneous rocks by its relatively low (about 50 percent) silica content. Within the basalt clan itself, however, other means of classification are used. Basalts are divided into two major groups, the alkaline basalts and the subalkaline basalts. Alkaline basalts contain high amounts of the alkali metal ions potassium and sodium but relatively low amounts of silica. In contrast, subalkaline basalts contain less potassium and sodium and more silica. As might be expected, this chemical difference translates into differences in the mineral content of the basalt types as well. For example, all alkaline basalts contain one or more minerals called "feldspathoids" in addition to plagioclase feldspar. They also commonly contain significant olivine. Subalkaline basalts, on the other hand, do not contain feldspathoids, although some contain olivine, and they may be capable of crystallizing very tiny amounts of the mineral quartz. The presence of this very silica-rich mineral reflects the relatively high silica content of subalkaline basalt magmas versus alkaline basalt magmas. Within these two major groups are many subtypes, too numerous to discuss here.

Methods of Study

Like other igneous rocks, basalts are analyzed and studied by many techniques. Individual studies may include extensive field mapping, in which the distribution of various types of basalt is plotted on maps. Especially if geologic maps of basalt types are correlated with absolute ages determined through radiometric dating techniques, the history of magma generation and its relation to tectonic history (earth movements) can be reconstructed for a particular area. Good examples of such studies are those conducted in recent years on the Hawaiian Islands. These studies indicate that the alkaline basalts on any given island are generally older than the subalkaline basalts, showing that magma production has moved upward, to lower pressure areas, in the mantle with time. This finding supports the idea that oceanic island basalts such as those in Hawaii are generated within so-called mantle plumes, roughly balloon-shaped, slowly rising masses of mantle material made buoyant by localized "hot spots."

Samples of basalt are also analyzed in the laboratory. The age of crystallization of basalt is obtained by means of radiometric dating techniques that involve the use of mass spectrometers to determine the abundances of critical isotopes, such as potassium 40 and argon 40 or rubidium 87 and strontium 86 and 87. To obtain information on how basalt magma is generated and how it subsequently changes in composition before extrusion as a lava flow, scientists place finely powdered samples in metallic, graphite, or ceramic capsules and subject them to heating and cooling under various conditions of pressure. Such procedures are known as experimental petrology. It has been proved that nearly any other igneous magma composition can be derived from basalt magma by the process of crystal fractionation. Widely believed to be the major factor influencing chemical variation among igneous rocks, this process results in ever-changing liquid compositions as the various silicate minerals crystallize, and are thus removed from the liquid, over time. Basalt's parental role gives it enormous importance in the discipline of igneous petrology.

Another fruitful avenue of research is trace element analysis of basalts. Trace elements occur in such low abundances in rocks that their concentrations must usually be expressed in terms of parts per million or even parts per billion. Among the most useful substances for tracing the history of basalt are the rare earth elements. Chromium, vanadium, nickel, phosphorus, strontium, zirconium, scandium, and hafnium are also used. There are many methods for measuring these elements, but the most common, and most accurate, is neutron activation. This method involves irradiating samples in a small nuclear reactor and then electronically counting the gamma-ray pulses generated by the samples. Since different elements tend to emit gamma rays at characteristic energies, these specific energies can be measured and the intensity of gamma pulses translated into elemental concentrations.

Once determined for a particular basalt sample, trace element abundances are sensitive indicators of events that have transpired during the evolution of the basalt. There are two reasons for this sensitivity. First, trace elements are present in such low concentrations as compared with major elements (iron, aluminum, calcium, silicon, and the like) that any small change in abundance caused by changes in the environment of basalt production will be readily noticed. Second, different minerals, including those crystallizing in the magma and those in the source peridotite, incorporate a given trace element into their structures or reject it to the surrounding liquid to widely varying degrees. Therefore, trace element concentrations can be used to show which minerals were involved in producing certain observed chemical signatures in basalts and which were likely not involved.

For example, it is well known that the mineral garnet readily accepts the rare earth element lutecium into its structure but tends to reject most lanthanum to any adjacent liquid. Basalts with very little lutecium but much lanthanum were therefore probably derived by the melting of garnet-bearing mantle rocks. Since garnet-bearing mantle rocks can exist only at great depth, basalts with such trace element patterns must have originated by melting at these depths in the mantle. In fact, that is one of the most important lines of evidence that alkaline basalts originate at high pressure regions in the mantle.

Context

Basaltic islands, particularly in the Pacific basin, are some of the most popular tourist stops in the world. More important, however, basalt magma contains low concentrations of valuable metals that, when concentrated by various natural processes, provide the source for many important ores. Copper, nickel, lead, zinc, gold, silver, and other metals have been recovered from ore bodies centered in basaltic terrains. Some of the richest mines of metallic ores in the world are located in Canada, where ores are found associated with extemely old basaltic rocks, called "green-stones," from long-vanished oceans. The richest of these mines is Kidd Creek, in northern Ontario. These ore-bearing basalts were first extruded more than 2 billion years ago, during what geologists call Precambrian times (the period from 4.6 billion to about 600 million years ago). Other notable ore deposits include the native, or metallic, copper in late Precambrian basalts that was mined for many years in the Keweenaw Peninsula of northern Michigan. In addition, the island of Crete in the Mediterranean Sea has copper mines that were mined thousands of years ago during the "copper" and "bronze" ages of human history. The basalt enclosing these ores is believed to have erupted from an ancient mid-ocean ridge trending between Africa and Europe.

Basalt can also be used as a building stone or raw material for sculptures, but its high iron content makes it susceptible to rust stains. It is also ground up to make road gravel, especially in the western United States, and it is used as decorative stone in yards and gardens.

Bibliography

Ballard, Robert D. *Exploring Our Living Planet.* Washington, D.C.: National Geographic Society, 1983. This book covers every aspect of the earth's volcanic and tectonic features and is lavishly illustrated with color photographs, illustrations, and diagrams. The sections on "spreading" and "hotspots" largely deal with basalt volcanism and its relationship to plate tectonic theory. Well written and indexed, the text will be easily understood and appreciated by specialists and laypersons alike.
Decker, Robert, and Barbara Decker. *Volcanoes.* New York: W. H. Free-man, 1981. This brief book gives a comprehensive treatment of vol-

canic phenomena. It is illustrated with numerous black-and-white photographs and diagrams. Chapters 1, 2, 3, and 6 deal almost exclusively with basalt volcanism. The last four chapters deal with human aspects of volcanic phenomena, such as the obtaining of energy and raw materials, and the effect of volcanic eruptions on weather. Includes an excellent chapter-by-chapter bibliography. Suitable for high school and college students.

Lewis, Thomas A., ed. *Volcano.* Alexandria, Va.: Time-Life Books, 1982. One of the volumes of the Planet Earth series, this book is written with the nonspecialist in mind. Wonderful color photographs, well-conceived color diagrams, and a readable narrative guide the reader through the world of volcanism. The book is especially good for its descriptions of past eruptions and their effects on humankind. Basalt is covered mainly in the chapter on Hawaii and the chapter on Heimaey, Iceland. Has a surprisingly extensive bibliography and index for a book of this kind.

Macdonald, Gordon A. *Volcanoes.* Englewood Cliffs, N.J.: Prentice-Hall, 1972. Written by one of the premier volcanologists in the world, this book is ideal for those desiring a serious but not overly technical treatment. Every conceivable aspect of volcanic phenomena is covered, but the sections on basalt (particularly as it occurs in Hawaii) are particularly good. Includes suggested readings, a comprehensive list of references, a very good index, and an appendix that lists the active volcanoes of the world. Somewhat lengthy.

Putnam, William C. *Geology.* 2d ed. New York: Oxford University Press, 1971. A comprehensive and accessible text. Chapter 4, "Igneous Rocks and Igneous Processes," uses a vivid description of the 1883 eruption of Krakatoa as a way of introducing the formation processes of igneous rocks. Other famous volcanic eruptions are also discussed. The rocks' classification and composition are described in detail. The chapter concludes with a list of references. Illustrated.

Tarbuck, Edward J., and Frederick K. Lutgens. *The Earth: An Introduction to Physical Geology.* 2d ed. Columbus, Ohio: Merrill, 1987. Aimed at the reader with little or no college-level science experience, this textbook includes a chapter devoted to igneous rocks and their textures, mineral compositions, classification, and formation. Illustrated with photographs and diagrams. Includes review questions and list of key terms.

John L. Berkley

Cross-References

Igneous Rock Bodies, 307; Igneous Rocks, 315; Magmas, 383; Rocks: Physical Properties, 550; Shield Volcanoes, 577; Volcanoes: Recent Eruptions, 675; Volcanoes: Types of Eruption, 685.

Biostratigraphy

Biostratigraphy is that branch of the study of layered rocks—stratigraphy—that focuses on fossils. Its goals are the identification and organization of strata based on their fossil content. Biostratigraphy thus investigates one of the principal bases of the geologic time scale of earth history.

Field of study: Stratigraphy

Principal terms

CORRELATION: the determination of the equivalence of age or stratigraphic position of two strata in separate areas, or, more broadly, determination of the geological contemporaneity of events in the geologic histories of two areas

FOSSILS: remains or traces of animals and plants preserved by natural causes in the earth's crust, excluding organisms buried since the beginning of historical time

INDEX FOSSIL: a fossil that can be used to identify and determine the age of the stratum in which it is found

SEDIMENTARY ROCKS: rocks formed by the accumulation of particles of other rocks or of organic skeletons or of chemical precipitates or some combination of these

STRATIGRAPHY: the study of layered rocks, especially of their sequence and correlation

STRATUM (PL. STRATA): a single bed or layer of sedimentary rock

Summary

Biostratigraphy is the method of identifying and differentiating layers of sedimentary rock (strata) by their fossil content. Strata with distinctive fossil content are termed biostratigraphic units, or zones. Zones vary greatly in thickness and in lateral extent. A zone may be a single layer a few centimeters thick and of very local extent, or it may encompass thousands of meters of rocks extending worldwide. The defining feature of a zone is its fossil content: The fossils of a given zone must differ in some specific way from the fossils of other zones.

Zones are usually recognized after fossils have been collected extensively over the lateral and vertical extent of a rock sequence or at many sequences over a broad region. The positions of the fossils in the strata are carefully recorded in the field. Fossils that co-occur in a single layer are noted, as are fossils found isolated in the strata. In the laboratory, the biostratigrapher, usually a paleontologist, then tabulates the vertical and lateral ranges of the fossils collected. It is from these ranges that the paleontologist recognizes zones. Different types of zones are recognized depending on the way in which the fossils in the strata prove to be distinctive. Assemblage zones are strata distinguished by an association (assemblage) of fossils. Thus, not one type but many types of fossils are used to define an assemblage zone. All dinosaur fossils, for example, can be thought of as defining an assemblage zone that encompasses earth history from about 220 to 66 million years ago.

Range zones are strata that encompass the vertical distribution, or range, of a particular type of fossil. Thus, one fossil type, not many, is used to define a range zone. In contrast to the example just given, one type of dinosaur, *Tyrannosaurus rex*, lived only between 68 and 66 million years ago. Its fossils thus define a range zone that corresponds temporally to this two-million-year interval.

Acme zones are rock layers recognized by the abundance, or acme, of a type (or types) of fossil (or fossils) regardless of association or range. Horned dinosaurs (*Triceratops* and its allies) reached an acme between 70 and 66 million years ago; that is, during this period they were most diverse and most numerous. This acme zone thus overlaps the *Tyrannosaurus rex* range zone and represents a small portion of the dinosaur assemblage zone.

Finally, interval zones are recognized as strata between layers where a significant change in fossil content takes place. For example, the mass extinctions that took place 250 and 66 million years ago bound a 184-million-year-long interval zone that is popularly referred to as the "age of reptiles."

Biostratigraphy developed independently in England and France just after 1800. In England, William Smith, a civil engineer, worked in land surveying throughout the country. From his vast field experience, he recognized that a given stratum usually contains distinctive fossils and that the fossils (and the stratum) could often be recognized across a large area. Smith's work culminated in his geological map of England (1815), based on his tracing of rock-fossil layers across much of the country.

Meanwhile, in France, Georges Cuvier and Alexandre Brongniart studied the succession of rocks and fossils around Paris. They too discovered a definite relationship between strata and fossils and used it to interpret the geological history of the rocks exposed near Paris. In this history, Cuvier saw successive extinctions of many organisms coinciding with remarkable changes in the strata. To him, these represented vast "revolutions" in geological history, which Cuvier argued were of worldwide significance. It is now known that Cuvier was mistaken, but the discovery that a particular fossil type (or types) was confined to a particular stratum became the basis for biostratigraphy. This allowed geologists to identify strata from their fossil content and to trace these strata across broad regions of the earth's crust.

Biostratigraphy is the process of identifying and differentiating rock strata according to their fossil content. *(Ben Klaffke)*

Almost simultaneous with the development of biostratigraphy was the development of biochronology. Biochronology is the recognition of intervals of geologic time by fossils. It stemmed from the realization that during earth's history, different types of organisms lived during different intervals of time. Thus, the fossils of any organism represent a particular interval of geologic time. (Such fossils are called index fossils because they act as an "index" to a geologic time interval.) Biochronology thus identifies intervals of geologic time based on fossils. These time-distinctive fossils are the fossils by which zones are defined, which is to say that each zone represents, or is equivalent to, some interval of geologic time.

The time value of zones made them more useful in tracing strata and deciphering local geological histories. Biostratigraphy now became one of the central methods of stratigraphic correlation. With the aid of fossils, it became possible to determine the ages of strata and thus demonstrate the synchrony or diachrony of these strata in different areas. Through its use in stratigraphic correlation, biostratigraphy became one of the bases for constructing what is called the relative geological time scale of earth history composed of eons, eras, periods, epochs, and ages. This time scale is the "calendar" by which all geologists temporally order their understanding of the history of the earth.

Applications of the Method

Biostratigraphy is generally used as a method of stratigraphic correlation, the process of determining the equivalence of age or stratigraphic position of layered rocks in different areas. Stratigraphic correlation by biostratigraphy is extremely important in deciphering geological history; it reveals the sequence of geological events in one or more regions. Understanding geological history is of interest for its own sake to scientists and laypersons alike. It is crucial to the discovery of mineral deposits and energy resources within the earth's crust. In addition, it provides insight into the biological events that have taken place on this planet for the last 3.9 billion years.

A good example of the use of biostratigraphy in this last regard comes from the study of dinosaur extinction. When dinosaurs were first discovered in England in 1824, and when the term "dinosaur" was coined by the British anatomist Sir Richard Owen in 1841, nobody realized that dinosaurs had lived on earth for only 150 million years and that their extinction had taken place rather rapidly about 66 million years ago. By 1862, however, enough dinosaur fossils had been collected around the globe that a biostratigraphic pattern was beginning to emerge. In that year, the American geologist James Dwight Dana, in his classic *Manual of Geology*, noted that all dinosaurs disappeared before the end of the Mesozoic era, which is now considered as the interval of earth history between 250 and 66 million years ago. This biostratigraphic generalization was possible because geologists noticed that many Mesozoic rocks (but no older or younger rocks) were full of dinosaur fossils, and thus the Mesozoic came to be termed "the age of reptiles." It might just as well be referred to as the "dinosaur zone," except of course for the first 30 or so million years of the Mesozoic, during which dinosaurs apparently did not exist.

More than a century of research has confirmed Dana's biostratigraphic generalization and considerably refined it. Scientists now generally agree that the last dinosaurs disappeared worldwide 66 million years ago, give or take one or two million years. It is also known that dinosaurs first appeared about 220 million years ago. Thus, scientists are able to recognize a dinosaur zone and erect many types of zones based on the ranges and acmes of specific types of dinosaurs. This biostratigraphy of dinosaurs is the basis for informed discussion of the sequence and timing of events during the evolution of the dinosaurs. For example, scientists are now confident that *Stegosaurus* lived long before *Tyrannosaurus* and that stegosaurs as a group of dinosaurs became extinct long before the end of the Mesozoic.

Although discussion here has relied heavily on dinosaurs for examples of biostratigraphy at work, the fossils of these giant reptiles are not ideal for use in biostratigraphy, because it is not easy to identify most dinosaur fossils precisely and because most dinosaurs were not animals with broad geographic ranges. Indeed, the fossils of most use in biostratigraphy, index fossils, are those that are easy to identify precisely and that represent organisms that had wide geographic ranges, enjoyed broad environmental tolerances, and lived only for a brief period of geologic time.

Usually an entire skull or skeleton is needed to identify a dinosaur fossil precisely; the isolated bones most often found are not enough, although they do indicate the fossil is that of a dinosaur. Most dinosaurs (there are some notable exceptions) seem to have lived in one portion of one continent; indeed, fossils of the horned dinosaur *Pentaceratops* (a cousin of *Triceratops*) have been found only in New Mexico. There is strong evidence that some dinosaurs preferred coastlines, whereas others preferred dry areas. Thus, many, if not most, dinosaurs did not live in a wide range of environments. Finally, although many dinosaurs apparently lived for only brief intervals of geologic time, the fossil record of most of these giant reptiles is not extensive enough to pin down their exact interval of existence.

The factors that mitigate the use of most dinosaur fossils in biostratigraphy are quite different for microscopic fossils of pollen grains and the shelled protozoans known as foraminiferans. These microscopic fossils fit well the four criteria listed above that identify fossils most useful in biostratigraphy. Indeed, such "microfossils" (studied by micropaleontologists) are some of the mainstays of biostratigraphy.

Context

Biostratigraphy, the recognition of strata by their fossil content, is a cornerstone of stratigraphic correlation. By using fossils to identify bodies of rock, they can be traced over broad areas, and their sequence in distant areas can often be determined. Stratigraphic correlation by biostratigraphy is critical to deciphering geological history; without it, the search for mineral deposits and energy resources would be considerably more difficult. Furthermore, understanding the history of geological disasters—earthquakes, volcanic eruptions, meteorite impacts, and the like—and thereby being able to predict future disasters, relies on knowledge of the sequence and timing of geological events, knowledge often derived from biostratigraphy. Deciphering the history of life on this planet, including the myriad appearances, changes, and extinctions of earth's biota during the last 3.9 billion years, largely depends on the sequence and timing established by biostratigraphy.

Biostratigraphy has also given rise to biochronology, the recognition of intervals of geologic time based on fossils. As a result, scientists have been able to construct a relative global geologic time scale, and it is within the context of this time scale that all geological and biological events in earth history have been placed.

Bibliography

Ager, Derek V. *The Nature of the Stratigraphical Record.* 2d ed. New York: Halsted Press, 1981. A witty and unabashed look at stratigraphy; some of the discussion centers on biostratigraphy. An extensive bibliography, index, and a few well-chosen illustrations illuminate the text.

Barry, W. B. N. *Growth of a Prehistoric Time Scale.* Rev. ed. Palo Alto, Calif.: Blackwell Scientific Publications, 1987. Largely devoted to the history of how the global geologic time scale was formulated, much of this book is a history of biostratigraphy. Well illustrated, with a good bibliography and an index.

Brenner, R. L., and T. R. McHargue. *Integrative Stratigraphy Concepts and Applications.* Englewood Cliffs, N.J.: Prentice-Hall, 1988. Chapter 11 of this college-level textbook provides a detailed look at biostratigraphic concepts, methods, and applications. Well illustrated, with extensive reference lists and an index.

Eicher, D. L. *Geologic Time.* 2d ed. Englewood Cliffs, N.J.: Prentice-Hall, 1976. Chapter 4 provides a less technical look at biostratigraphy than do Brenner and McHargue. Very readable, well illustrated, with some references and an index.

Hedberg, H. D., ed. *International Stratigraphic Guide.* New York: John Wiley & Sons, 1976. The international "rule book" for stratigraphy. It sets procedures and standards to be met when naming stratigraphic units. It also defines many terms used in stratigraphy and has an extensive bibliography. Chapter 6 is devoted to biostratigraphy.

Stanley, S. M. *Earth and Life Through Time.* 2d ed. New York: W. H. Freeman, 1989. An excellent introductory-level college textbook on historical geology. It reviews the history of life and the many fossil forms found in strata in the earth's crust. Chapter 5 includes a discussion of biostratigraphy. Lavishly illustrated, with extensive references, glossaries, appendices on fossil groups, and an index.

Spencer G. Lucas

Cross-References

The Cretaceous-Tertiary Boundary, 118; The Fossil Record, 257; The Geologic Time Scale, 272.

Calderas

With the exception of impacts by asteroid-sized meteorites, the largest caldera-forming eruptions represent the most catastrophic geologic events known. Ancient calderas are sites of many of the earth's ore deposits, and recently formed calderas are important resources of geothermal energy.

Field of study: Volcanology

Principal terms

CALDERA: a large collapse depression, more or less circular in form, the diameter of which is many times greater than the size of any included vents

HOT SPOT: a volcanic center that has persisted for tens of millions of years and that is thought to be the surface expression of a rising plume of mantle material

IGNIMBRITE: an igneous rock deposited from a hot, mobile, ground-hugging cloud of ash and pumice

PLATE TECTONICS: a theory which describes the earth's outer layer as consisting of large, independently moving fragments

PYROCLASTIC: pertaining to volcanic material formed by explosion

SHIELD VOLCANO: a volcano in the shape of a flattened dome, broad and low, built by flows of very fluid basaltic lava

STRATOVOLCANO: a volcano constructed of layers of lava and pyroclastic rock; also called a composite volcano

Summary

Caldera is Portuguese for "kettle" or "cauldron" and was used by inhabitants of the Canary Islands to refer to all natural depressions, including the island's volcanic craters and calderas. The term was introduced into the geologic literature in the 1800's to describe volcanic depressions. There remains, however, a debate over the difference between craters and calderas.

In general, craters are caused by the explosive removal of material, while calderas form by the subsidence of the surface during or immediately after explosive volcanism. Because subsidence structures are usually larger than craters, many geologists consider all volcanic depressions larger than 1 mile (or 1 kilometer, to some) to be calderas. Other geologists prefer to emphasize origin; they use "craters" for depressions produced by explosions and "calderas" for all collapse depressions. It is only when explosion structures are extremely large or calderas small that a problem exists. Geologists, for example, are about evenly divided on whether the large depression produced by the 1980 eruption of Mount St. Helens is, by origin, a crater or, by size, a caldera. One of the small ironies of science is that all volcanologists agree that one of the world's best examples of a caldera is Oregon's Crater Lake.

In general, then, calderas form when the support provided by the underlying molten rock or magma is removed, either by eruption or by withdrawal to a lower level. Three classes of calderas are common: those located at the summits of basaltic shield volcanoes, such as Hawaii's Mauna Loa; those that "behead" andesitic stratovolcanoes, such as Crater Lake; and those that contain the source vents for widespread layers of rhyolitic ash, such as Wyoming's Yellowstone caldera. In addition, planetary geologists have discovered shieldlike volcanoes with summit calderas on Mars, Venus, and Jupiter's satellite Io.

The smallest terrestrial calderas are associated with basaltic shield volcanoes, such as those in the Hawaiian Islands. Shield volcanoes are composed mostly of layers of basalt, a dark-colored volcanic rock rich in magnesium and iron but relatively poor in silica. The Hawaiian Islands are the exposed southeastern end of a largely submarine mountain range of volcanic origin. Volcanic islands and submerged seamounts can be traced for nearly 6,000 kilometers to where they disappear into a deep oceanic trench off Alaska's Aleutian Peninsula. Plate tectonic theory and age determinations performed on volcanic rock suggest that the entire chain took nearly 80 million years to be produced as the Pacific Ocean floor moved at an average rate of about 8.6 centimeters per year over a subcrustal magma source, or hot spot. Hawaii is now located over the hot spot and contains two active shield volcanoes, Mauna Loa and Kilauea. Each shield volcano has a summit caldera from which emanate radial rift zones marked by recent lava flows, minor vents, and lines of craters. A magma chamber is located at a relatively high level within each volcano.

Although basaltic shield volcanoes erupt frequently, their eruption style is the mildest known. A typical eruptive sequence begins with the magma reservoir within the volcano gradually filling and producing a measurable inflation of the volcano's summit. Swarms of small earthquakes caused by magma movement occur below the impending vent site. The eruption commonly begins with lava fountaining at the summit. As magma works its way along the rift zones to erupt at lower elevations, activity ceases at the summit. The continued removal of magmatic support causes a part of the summit area to subside along arcuate faults, forming a caldera.

Summit calderas are slightly elliptical in outline, with flat floors and steep walls. Because of the frequency with which basaltic shield volcanoes erupt, summit calderas have a complex history of collapse, uplift, and infilling by later lava eruptions. Kilauea's summit contains several collapse features,

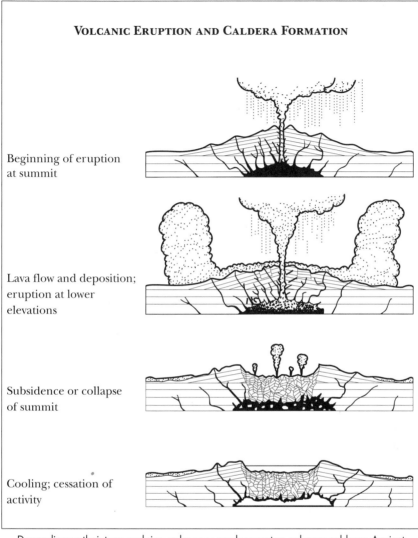

VOLCANIC ERUPTION AND CALDERA FORMATION

Beginning of eruption
at summit

Lava flow and deposition;
eruption at lower
elevations

Subsidence or collapse
of summit

Cooling; cessation of
activity

Depending on their type and size, volcanoes produce craters or larger calderas. Ancient calderas are the sites of many ore deposits; some more recent calderas are regions of geothermal energy.

including Kilauea caldera, the major structure. Its approximate dimensions are 4 by 3 kilometers, and its average depth is about 100 meters. A second, smaller caldera, Kilauea Iki, is within Kilauea caldera, and both structures are surrounded by arcuate faults along which minor collapse has occurred. Mauna Loa's summit caldera is similar in size and also consists of multiple collapse zones.

Shield volcanoes with summit calderas are by no means restricted to the Hawaiian Islands. They are commonly found where large outflows of fluid basalt occur, and prominent examples are found in Iceland and the Galápagos Islands. Summit shield calderas on what appear to be basaltic volcanoes are also found on Mars. The most impressive example is Olympus Mons, probably the largest volcano in the solar system. It is more than 600 kilometers across and 23 kilometers high, and its summit contains an 80-kilometer-wide caldera complex. Scientists speculate that Mars has hot spots but does not have independently moving plates. They believe Olympus Mons may have been volcanically active for 1.5 billion years as basaltic magma was fed upward from its mantle source.

Calderas are also associated with stratovolcanoes. Stratovolcanoes, with slopes of every grade, most closely resemble the stereotype of the volcano. Lava and pyroclastic material accumulate around a central vent to produce mountains rising as much as 5 kilometers above their bases. The rapid erosion of material from these lofty peaks, sometimes as disastrous mud-flows, produces aprons of sediment around the volcanoes' flanks. They are the most abundant type of large volcano on the earth's surface and the characteristic volcanic landform found on the island arcs and continental margins fringing the Pacific Ocean. Although andesite, a dark- to medium-colored volcanic rock with an intermediate silica content, is the most common type of rock erupted, stratovolcanic eruptions produce a wide range of magma types.

Stratovolcanoes show prolonged periods of dormancy broken by eruptive phases which range from mild degassing to catastrophic eruptions that greatly alter or destroy the volcano's shape. Large eruptions from stratovolcanoes are commonly associated with the emplacement of rocks called ignimbrites. During an ignimbrite eruption, parts of the cone may be blasted away, or the volcano may founder into an immense caldera.

Historic caldera-forming eruptions have been impressive. The eruption of Krakatau (more popularly known as Krakatoa) in 1883 took place on a deserted volcanic island, yet a giant wave produced by the volcano's collapse killed more than thirty thousand persons on neighboring shores. The great Tambora eruption of 1815 caused the deaths of more than ninety thousand, either directly by eruption or by the ensuing famine. Prehistoric eruptions must have been even more spectacular. The Bronze Age eruption of San-torini in the Mediterranean Sea, for example, has been linked to the decline of the Minoan civilization and, thus, may have changed the course of Western history.

Crater Lake, Oregon, has contributed much to the understanding of calderas and serves as a good example of caldera formation on andesitic stratovolcanoes. Crater Lake is a circular caldera approximately 10 kilometers in diameter. The average depth of the lake is about 600 meters, and the surrounding cliffs rise from 150 to 600 meters. Wizard Island, a small cinder cone, rises 225 meters above the level of the lake. The eruption that formed Crater Lake occurred approximately 6,845 years ago, following thousands of years of intermittent activity which built a large stratovolcano that geologists call Mount Mazama. It is estimated that the cone was approximately 3,500 meters high and was capped by glacial ice. Detailed field studies around Crater Lake have shown that the initial eruption was from a single vent, which fed ash and pumice into an eruption column that reached into the stratosphere and drifted with the prevailing wind. As the eruption intensified, so much material was emplaced into the cloud that, despite its heat, the cloud's density exceeded that of the surrounding air; it gravitationally collapsed to feed ground-hugging clouds of incandescent ash and pumice. These ash flows had great mobility and moved at hurricane speed to deposit ignimbrite around Mount Mazama. When about 30 cubic kilometers of magma had been expelled, the roof of the magma chamber collapsed to form a caldera. Venting, however, continued to eject another 20 cubic kilometers of magma from multiple vents located along the ring-fracture system bounding the caldera. The caldera continued to subside as venting progressed. Much of the ash fell back into the depression and mixed with rock that was sliding from the oversteepened walls to pile upon the caldera floor. A small cinder cone, Wizard Island, subsequently formed on the caldera floor, and Oregon's abundant rainfall produced the caldera lake.

Stratovolcanoes grade with increasing silica content to volcanoes composed mostly of rhyolite, a silica-rich volcanic rock that is usually light in color. Although eruption frequency of the more silicic volcanoes tends to decrease, their eruption volume and caldera size increase. Rhyolitic volcanoes are dominated by ignimbrite eruptions, and they tend to look very unlike volcanoes as most people picture them. A rhyolitic volcanic field consists of a rhyolitic ignimbrite plateau, punctuated here and there by large calderas. In most of these structures, the caldera floor has resurged or been uplifted and arched to form what is known as a resurgent dome. Resurgence, combined with the effects of sedimentation, continued volcanism, and erosion may even make the caldera difficult to detect. Hundreds of such calderas are known or await discovery, hidden among the rhyolitic ignimbrites of western North America. Many of these ancient calderas are associated with that region's important ore deposits. Most calderas of this type tend to be circular, but the largest examples are elongated; their irregular shape may be caused by their piecemeal collapse into larger, more complex magma chambers or may reflect stresses in the crust. The largest calderas are more properly called volcano-tectonic depressions. The Lake Toba caldera, a volcano-tectonic depression on the island of Sumatra, is 100 kilometers long by 35 kilometers wide, the largest caldera yet recognized. Dormant periods between minor eruptions at rhyolitic volcanoes may be measured in thousands of years. Repeated large eruptions may be separated by a million years and can form compound structures such as the Yellowstone caldera, a good example of a recently active, rhyolitic volcano.

Yellowstone National Park, famous for its hot springs and geysers, takes its name from volcanic rocks altered to bright colors by hot water and steam. The park's volcanic rocks belong to an igneous province composed mostly of basalt and rhyolite that extends southwest into Idaho's Snake River plain. Like Hawaii's, these volcanic rocks appear to be related to a hot spot. The Yellowstone hot spot, however, underlies the thicker and more silica-rich crust of the North American continent, and it is this continental crust that is believed to be the source for the enormous amounts of rhyolite.

Geologists of the U.S. Geological Survey have shown that the Yellowstone area has been the site of three large caldera-forming eruptions and numerous smaller eruptions during the past 2 million years. The ash emplaced within several tens of kilometers of the calderas was hot enough to anneal, or weld, into hard, rhyolite-capped plateaus. Remnants of the more widely dispersed ash have been found as far away as Texas. The last caldera-forming eruption occurred 600,000 years ago, when 1,000 cubic kilometers of magma was expelled, and a caldera 45 kilometers wide and 75 kilometers long was formed over the partly drained magma chamber. Within a few thousand years, magma elevated the caldera floor and arched it into the two resurgent domes contained in this complex structure. Over the past 600,000 years, much of the caldera has been filled with lava and sediment. Part of it is now covered by Yellowstone Lake, but the area remains thermally and seismically active.

Methods of Study

Geologists first became aware of the collapse origin of calderas in the late nineteenth century, through observations made at the summit calderas of Hawaii's basaltic shield volcanoes and especially as a result of the catastrophic 1883 eruption of Krakatau in Indonesia. By the first part of the twentieth century, the collapse origin of Crater Lake was recognized, and by the 1930's and 1940's, large calderas, both recent and ancient, were being described. A 1940's study of Crater Lake conclusively linked the formation of large calderas with the eruption of voluminous amounts of pyroclastic material. Subsequent work by the U.S. Geological Survey geologists in and around New Mexico's 1.1-million-year-old Valles caldera has provided a conceptual framework for the mapping and interpretation of voluminous pyroclastic deposits. It also led to the development of the resurgent caldera model, which describes the sequence of events leading up to and following caldera formation.

Since that time, the resurgent caldera model has been successfully applied to many of the world's large calderas. Most of these rhyolitic volcanoes experienced minor volcanic activity before the major rhyolitic eruption. The climatic eruption results in voluminous amount of welded rhyolite ash being emplaced as an ignimbrite apron around the caldera. The caldera subsides along nearly vertical boundary faults, with its area and depth roughly equal in volume to the amount of magma ejected. Approximately half of the erupted material falls back onto the subsiding caldera floor, where it may become interlayered with debris avalanching from the steep caldera walls. In most large calderas, renewed magmatic pressure uplifts and arches the floor to form the central dome that characterizes a resurgent caldera. Viscous, degassed magma may ooze from fractures in the caldera floor to form thick lava flows or domes. A lake may occupy a portion of the caldera floor, and a thick sequence of lake and stream sediments may be deposited between the resurgent dome and the caldera wall. Hydrothermal activity in the form of hot springs and geysers are produced by contact between the local groundwater and the cooling rock and magma. Hydrothermal activity related to the cooling magma or heated by some later igneous intrusion can continue to form ore deposits in and around the caldera for millions of years.

Work continues on many aspects of calderas and caldera-forming volcanism. The variations exhibited by some calderas reveal a range of caldera types related to the resurgent caldera theme. Events premonitory to a major pyroclastic eruption and caldera formation are of particular interest. Precursory volcanism, associated seismicity, ground deformation, and the mechanism by which the magma chamber's roof fails are all poorly understood yet essential to the appraisal of volcanic hazards. Additional detailed studies of ancient calderas and the monitoring of volcanoes with caldera-producing potential will help geologists to predict these eruptions. Experimental data and the modeling of calderas, especially with computers, will also contribute to the understanding of caldera formation.

Geologists have long struggled to understand the processes that take place in magma chambers, and large caldera-forming eruptions have given them a new tool. Major pyroclastic eruptions quickly tap large amounts of material from the top of magma chambers and deposit them as ignimbrite sheets. Geologists, by sampling these rock layers from their bases upward, are sampling material derived from progressively deeper levels in magma chambers. The chemical and physical variations noted in these zoned ignimbrite sheets give clues to the processes that occurred in their source magma chambers.

The enormous amounts of heat associated with large rhyolitic magma chambers cause widespread groundwater thermal effects, the surface manifestations of which are seen at places such as Yellowstone Park. Calderas are, therefore, important local sources of geothermal energy. In addition, the heated water or steam has the potential to dissolve some of the elements scattered throughout the rock and to concentrate them into ore deposits, especially as veins along caldera-related fractures and faults. Calderas are known to be important hosts for many kinds of ores, and drilling programs into the hydrothermal systems underlying recently formed calderas are leading to a better understanding of their resource potential.

Context

Calderas, except for those on basaltic shield volcanoes, form by the collapse of magma chamber roofs during enormous pyroclastic eruptions. Recently formed large calderas have large underlying magma chambers that can be important local sources of geothermal energy. One such geothermal plant is already operating in New Mexico's Valles caldera. The hot water set in convective motion by subcaldera magma chambers can produce important ore deposits of many metals. They are especially important sources for gold and silver, as in the precious metal deposits of Colorado's San Juan Mountains.

It is as volcanic hazards, however, that caldera-forming eruptions can have the most dramatic impact on human affairs. The largest documented volcanic eruption took place in 1815 on the island of Sumbawa, Indonesia. After hundreds, perhaps thousands, of years of dormancy, Tambora ejected about 175 cubic kilometers of ash. As many as ninety thousand people may have died from the direct effects of the eruption and from the starvation and disease that followed. What had been a cone 4,300 meters high was reduced by almost 1.5 kilometers, forming a caldera 6 by 7 kilometers in area and more than 1 kilometer deep. The ash injected into the stratosphere produced global atmospheric effects by reflecting some of the sun's energy back into space. Crops failed in New England because of frosts that continued through August, and similar crop failures were noticed in Canada and Europe.

The largest prehistoric eruptions studied produced up to 3,000 cubic kilometers of volcanic ash and formed calderas with a maximum dimension of up to 100 kilometers. The destructive potential of a major rhyolitic eruption cannot be overemphasized. Except for the rare impacts of asteroid-sized meteorites, these caldera-forming eruptions represent the most catastrophic geologic events known. Major rhyolitic centers such as Yellowstone, Wyoming, and Long Valley, California, are viewed as still capable of producing a large-volume ignimbrite. It should also be emphasized, however, that the frequency with which this type of volcano erupts makes it extremely unlikely that such an event will happen in the next few hundred, or even thousand, years. In addition, such an event would probably be preceded by measurable changes in surface elevations, swarms of earthquakes, increased thermal activity, and minor premonitory volcanic activity.

Bibliography

Berardelli, Phil. "Volcanic Catastrophe-in-Waiting Is Redefining 'Big Bang Theory'." *Insight on the News* 13, no. 30 (August 18, 1997): 40-42. Focuses on the possible threats posed by giant calderas and their potential global effect.
Bullard, Fred M. *Volcanoes of the Earth.* Austin: University of Texas Press, 1976. A widely available, college-level volcanology text that is generally accessible to readers with little or no background in the subject. Chapter 6 describes cones, craters, and calderas; it has a particularly good discussion of the term "caldera" and its confusion with the term "crater."

Cas, R. A. F., and J. V. Wright. *Volcanic Successions, Modern and Ancient: A Geological Approach to Processes, Products, and Successions.* London: Allen & Unwin, 1987. One of the most comprehensive of the college-level volcanology textbooks. Covers all aspects of pyroclastic volcanism and is available in university and large public libraries.

Decker, Robert W., Thomas L. Wright, and Peter H. Stauffer, eds. *Volcanism in Hawaii.* Denver, Colo.: U.S. Geological Survey, 1988. Published to commemorate the seventy-fifth anniversary of the Hawaiian Volcano Observatory, the two volumes that constitute this work are a treasury of information derived from years of research on all aspects of Hawaiian volcanism, and comprise the most comprehensive collection of scientific articles available on the topic, including the islands' calderas. Although written for professionals, many of the sixty-five reports are not above the level of the interested nonspecialist.

Green, Jack, and Nicholas M. Short. *Volcanic Landforms and Surface Features: A Photographic Atlas and Glossary.* New York: Springer-Verlag, 1971. This photographic atlas and glossary is a widely available, nontechnical treatment of all volcanic phenomena. The introductory text and glossary are somewhat dated, but the 198 photographs and captions illustrate well a wide variety of volcanic features, including calderas.

Schubert, Gerald, ed. *Journal of Geophysical Research* 89, no. B10 (1984). This special volume was published by the American Geophysical Union to commemorate the one hundredth anniversary of the eruption of Krakatoa. Of general interest are the introduction, a paper on extraterrestrial calderas, and several regional studies. The last paper in the volume is of special interest, because it is the most complete description available of large silicic calderas.

Simkin, Tom, L. Siebert, L. McClelland, D. Bridge, C. Newhall, and J. H. Latter. *Volcanoes of the World: A Regional Directory, Gazetteer, and Chronology of Volcanism During the Last 10,000 Years.* Stroudsburg, Pa.: Hutchinson Ross, 1981. An excellent, widely available general reference on geologically recent volcanic activity, this work is a compilation of information on all volcanoes known to be active during the past ten thousand years. Although not specifically about calderas, it contains the name, location, and known eruptive history of 1,353 volcanoes, including those with calderas.

Smith, Robert B., and Robert L. Christiansen. "Yellowstone Park as a Window on the Earth's Interior." In *Volcanoes and the Earth's Interior.* San Francisco: W. H. Freeman, 1982. This paper is one of a collection of articles on volcanology originally published in the journal *Scientific American.* The book is widely available and includes sections on volcanoes and plate tectonics, volcanic eruptions, and volcanoes as sources of information about the earth's interior.

Tilling, Robert I., Christina Heliker, and Thomas L. Wright. *Eruptions of Hawaiian Volcanoes: Past, Present, and Future.* Denver, Colo.: U.S. Geological Survey, 1987. One of a series of general-interest publications prepared by the U.S. Geological Survey to provide information on the earth sciences. Well illustrated and factual, it is probably the best source on Hawaiian volcanism available to those with no prior geological knowledge.

Eric R. Swanson

Cross-References

Plate Tectonics, 505; Pyroclastic Rocks, 512; Shield Volcanoes, 577; Volcanoes: Types of Eruption, 685.

Catastrophism

Historically, catastrophism was the doctrine that a series of sudden and violent events caused widespread or even global effects, producing the differences in fossil forms and other features found in successive geological strata. More recently, a new school of catastrophism has arisen, prompted by growing evidence that much of the geological record, including mass extinctions of living organisms, has been greatly influenced by rare events of large magnitude, such as widespread flooding, volcanic activity, and asteroid impacts.

Field of study: Geochronology and paleontology

Principal terms

ASTEROID: one of the numerous small rocky bodies bigger than about ten meters in size orbiting the Sun, mostly between Mars and Jupiter; some, however, cross the earth's orbit

CRATER: an abrupt circular depression formed by extrusion of volcanic material or by the impact of an asteroid or meteorite

CRUST: the outermost solid layer or shell surrounding the earth

FOSSIL: naturally preserved remains or evidence of past life, such as bones, shells, casts, and impressions

GEOLOGICAL COLUMN: the order of rock layers formed during the course of the earth's history

METEORITE: a fragment of an asteroid (less than ten meters in size) that survives passage through the atmosphere and strikes the surface of the earth

STRATIGRAPHY: the study of rock layers to determine the sequence of layers and the information this provides on the geological history of a region

UNIFORMITARIANISM: the doctrine that geological events are caused by natural and gradual processes operating over long periods of time

VOLCANISM: the processes by which magma is transferred from the earth's interior to produce lava flows on the surface and the ejection of gases and ash into the atmosphere

Summary

Although the term "catastrophism" is usually associated with the work of Georges Cuvier near the beginning of the nineteenth century, most theories of earth history before that time involved various ideas of catastrophism, emphasizing sudden and violent events rather than gradual processes. Based on the biblical account of creation in six days and genealogies of the descendants of Adam, most writers assumed that the earth was only about six thousand years old. Early theories of the earth's surface features were based on the biblical account of Noah's flood. In early eighteenth century England, Thomas Burnet and John Woodward used the idea of a universal flood to explain geological phenomena such as the formation of mountains and valleys, irregularities in strata, and the existence and location of fossils. These ideas stimulated the collection of fossils as evidence of biblical veracity.

In Italy, where volcanoes were active, the Venetian priest Anton Moro suggested in 1740 that Noah's flood was a more localized event and that rock strata formed from a series of violent volcanic eruptions that entombed plants and animals, forming fossils in the rocks. These catastrophic ideas were sometimes viewed as complementary; by the late eighteenth century, however, they led to a controversy between the Neptunists, who stressed the role of water and floods, and the Vulcanists, who emphasized fire and heat.

In 1749, Georges Buffon, keeper of the King's Gardens in Paris, suggested a speculative natural history of the earth with a vastly expanded time scale. Instead of a recent six-day creation, he proposed seven epochs of development over a span of about seventy-five thousand years. Using a calculation devised by Sir Isaac Newton for estimating the cooling of comets, Buffon experimented with the cooling of a red-hot globe of iron; he extrapolated his findings for a mass the size of the earth, arriving at an estimate that it would have taken 74,832 years for the earth to cool to its present temperature. By the time he finished his *Épochs de la Nature* (1779), Buffon had divided his history of nature into seven "epochs" as metaphors of the seven "days" of creation.

Although he did not refer to catastrophism, Buffon's epochs included catastrophic events of both fire and water. In the first epoch, the earth formed out of matter ejected from a collision of a comet with the Sun. As the earth solidified in the second epoch, its crust wrinkled to form the mountain ranges. In the third epoch, vapors condensed as the earth cooled, covering the earth with a flood in which fishes flourished and sediments formed, enclosing fossils and organic deposits such as coal. The fourth epoch began after further cooling produced subterranean openings, causing a rush of waters, earthquakes, and volcanoes that produced dry lands. Land animals and plants appeared in the fifth epoch, and the continents moved apart in the sixth after migrations of animals had separated various species. Finally, wrote Buffon, humans appeared in the seventh epoch.

By the end of the eighteenth century, the Neptunists and Vulcanists became more sharply divided. The British geologist Sir William Hamilton developed in more detail the implications of Vulcanism from the action of

French naturalist Georges Cuvier, an early advocate of the theory of catastrophism. *(Library of Congress)*

volcanoes. He identified basalt and other rocks found near volcanoes as products of lava flows. He argued that volcanic action played a constructive role in uplifting new land from the sea, shaping the landscape, and providing a safety valve for excess pressure below the crust.

A purely Neptunist school was established by the German mineralogist and geologist Abraham Werner. He accepted the idea of geological succession in sedimentary deposits but did not develop its historical implications, since he classified rocks by mineral content rather than by fossils. Secularizing earlier theories based on the biblical flood, Werner held that rock strata formed from a universal primeval ocean, which produced four types of rocks by sequential processes. Primitive rocks, such as granite, crystallized out of the primeval ocean and contained no fossils. Transitional rocks, such as micas and slate, contained only a few fossils. Sedimentary rocks such as coal and limestone were next and were rich in fossils. Derivative rocks such as sand and clay formed from the other three by processes of weathering. Werner believed that volcanoes resulted from the burning of underground coal and were not an important geological force.

Although Werner's theory about the origin of sedimentary rocks was largely upheld, most other rocks were eventually shown to have an igneous origin from a molten state. This idea was developed by the Plutonist school of geology, which stressed the geological activity of the internal heat of the earth, in addition to the volcanic eruptions of the Vulcanists. This view was developed by the Scottish amateur geologist James Hutton in his *Theory of the Earth* (1795). Hutton believed that the geological forces seen in the present operated in the same way and at the same rate in the earth's past and that this principle should be the basis of geological explanation: The present is the key to the past.

Hutton's "uniformitarian principle" contrasted with Werner's idea of a primeval ocean, which involved catastrophic events confined to the past and unobservable in the present. Hutton carefully observed the slow and steady erosion of the land as rivers carried silt into the sea. He examined the weathered beds of gravel, sand, and mud brought down by the rivers, as well as the crystalline granites of the Scottish mountains. He concluded that sedimentary rocks formed from beds of mud and sand compressed by overlying seas and heat pressure from below, while crystalline rocks came from molten material inside the earth brought to the surface by volcanic action.

Developing the idea that the interior of the earth is molten, Hutton suggested that molten rock pushes into cracks beneath the earth's crust, tilting up sedimentary strata and solidifying to form granites. Thus mountains were built with a crystalline core and sedimentary surface. This principle of injection was apparent in some granite intrusions into crevices in sedimentary rocks above, indicating that the granite was younger. The existence of granites of differing ages was contrary to Werner's assumptions. In some cases, Hutton found horizontal sedimentary strata covering tilted strata near the base of mountains, suggesting long periods of time since the strata tilted. The age of the earth appeared to be so long that catastrophic events did not seem to be necessary to account for its surface features.

Although these uniformitarian ideas found some early support, it was not enough to overcome religious objections to a theory that required such an ancient earth, delaying its eventual acceptance. In France, Georges Cuvier opposed Hutton's idea of slow geological processes with his theory of "catastrophism" in the introduction to his *Researches on Fossil Bones* (1812). Since there was no apparent continuity between successive strata and their fossils, he believed that a series of catastrophic floods must have occurred— rather than continuous forces—with each flood wiping out many species and eroding the earth. These catastrophes also tilted strata left by earlier floods, ending with Noah's flood some six thousand years ago. Cuvier's catastrophism applied Neptunism to the vast time scale of Hutton. His influence delayed the acceptance of both biological and geological ideas of evolution in France for several decades.

In England and France, the study of strata was made easier by many well-exposed horizontal layers rich in fossils. These were systematically studied in France by Cuvier and in England by William Smith, who discovered in 1793 that each stratum had its own characteristic form of fossils. Their work revealed that strata near the surface were younger than those farther down, and a history of life forms could be worked out from their fossils. Further stratigraphic studies by Adam Sedgwick and Roderick Murchison identified the Cambrian series of strata with the oldest fossil-bearing rocks, the Silurian series with the earliest land plants, and the Devonian series dominated by fish fossils.

The discoveries of this "heroic age of geology" (1790-1830) were summarized by the Scottish geologist Charles Lyell. Reviving Hutton's uniformitarian ideas, Lyell published three volumes entitled *The Principles of Geology: Being an Attempt to Explain the Former Changes of the Earth's Surface by Reference to Causes Now in Operation* (1830-1833). Assuming indefinitely long periods of time, he insisted that geological forces had always been the same as they are now. Yet few of Lyell's contemporaries accepted his ideas before Charles Darwin developed them in his theory of organic evolution. The combined influence of Lyell and Darwin caused many scientists to shift away from Cuvier's catastrophism.

One of Cuvier's later associates, the Swiss-American naturalist Louis Agassiz, helped to modify the extreme uniformitarianism of Lyell. From the distribution of boulders and the grooves scratched on solid rock in the Swiss Alps, he showed in 1837 that Alpine glaciers had once stretched from the Alps across the plains to the west and up the sides of the Jura Mountains. In 1847, he accepted a position at Harvard University, and in North America, he found evidence that glaciers had also overrun that continent's northern half. Agassiz's Ice Age theory gradually won acceptance over more catastrophic flood theories, and evidence for several long ages of advancing and retreating ice over millions of years was eventually found.

Methods of Study

By the end of the nineteenth century, uniformitarian logic had become the primary method of studying the history of the earth's surface. Early in the twentieth century, the development of radioactive dating techniques confirmed the enormous age of the earth, revealing some 4.6 billion years during which the same slow processes of erosion, eruption, sedimentation, and ground movement visible today could account for everything from the Grand Canyon to marine fossils in the Alps. Yet this very method has identified discontinuities and anomalies that reveal the importance of catastrophic events in earth history. It now appears that the planet and its life forms have been shaped by more than gradual processes still operating today. Evidence has accumulated that many sudden and violent events in the past had widespread consequences, including torrential flooding, massive volcanic activity, and huge asteroid impacts causing global disasters. A "new catastrophism" uses uniformitarian methods to show that past catastrophes may be the key to understanding the present.

An early attempt to revive catastrophism was made by J. Harlen Bretz in 1923 to explain certain landscape features. He suggested that some of the world's largest floods poured down the Columbia River Gorge from melting glaciers into the Pacific Ocean, scouring much of the Columbia Plateau down to bedrock and creating the Channeled Scabland of eastern Washington. His ideas were finally vindicated from aerial and satellite photographs. It was then shown that glacial Lake Missoula in Montana produced as many as seventy floods from about fifteen thousand to twelve thousand years ago, matching legends of several native American tribes in the Pacific Northwest. At that time, a glacial ice dam impounded a body of water some 250 miles long. Repeated emptying of Lake Missoula occurred when the ice dam floated and broke, releasing as much as ten cubic miles per hour for at least forty hours. This deluge, some ten times the combined flow of all the world's rivers, removed soil as deep as one thousand feet, inundating three thousand square miles as deep as 350 feet.

More obvious catastrophic events are associated with volcanic activity. The most devastating volcanic eruption in recorded history was that of the Tambora volcano on Sumbawa Island, Indonesia, in 1815. The explosion killed twelve thousand people, and another eighty-two thousand died of starvation and disease. Tambora ejected so much volcanic ash into the stratosphere that Europe and North America experienced a year without a summer. Snow blanketed New England in June, and frosts blighted crops throughout the growing season.

Much larger volcanoes changed the landscape in prehistoric times. Volcanic activity in the Yellowstone region began about two million years ago as the continental crust moved westward. Subterranean melting of the crust produced a large underground reservoir of magma that resulted in three major explosions over a period of a million years. The first produced one of the largest eruptions to occur on earth, ejecting more than 620 cubic miles of magma. After the roof of the magma reservoir collapsed, it subsided several thousand feet, producing a caldera (giant crater) that covered an area of a thousand square miles. The volcanic ash canopy from the last eruption annihilated life over much of the western United States, and its fallout is preserved in strata from California to Kansas.

Recent discoveries have revealed that asteroid or comet impacts are the probable source of even more widespread annihilation of life and may be associated with extensive volcanic activity. In 1958, the Estonian astrophysicist Ernst Öpik suggested that a sufficiently large asteroid collision could penetrate the continental crust, triggering the formation of huge areas of lava floods such as the Deccan Traps in western India and the Columbia River Plateau in the Pacific Northwest. In 1973, the American chemist Harold Urey proposed that a comet collision could cause sufficient heating of the biosphere to explain global extinctions. More recently, University of Montana geologists have shown that an asteroid impact 17 million years ago could account for the immense lava flows of the Columbia River Plateau, spreading as far as three hundred miles from their source to form the largest volcanic landform in North America. As the North American continental plate moved westward, the resulting hot spot shifted to form the Yellowstone volcanic region.

In 1980, a team of physicists and geologists headed by Luis Alvarez and his son Walter discovered that the thin global sediment that separates the end of the Cretaceous era (age of dinosaurs) from the Tertiary era (age of mammals) contained anomalous quantities of the element iridium, rare on the earth but common in meteorites. They suggested that this K-T boundary layer, dated at 65 million years ago, was evidence of an asteroid collision that ejected enough matter into the atmosphere to produce a "cosmic winter," killing the dinosaurs and many other species. Their estimate of at least a ten-kilometer asteroid was confirmed in 1990, when a two-hundred-square-kilometer crater was identified in the Yucatán region of Mexico, dating from 65 million years ago. Such a collision could have produced shock waves that came to a focus on the opposite side of the earth, explaining the 65 million-year-old eruptions that formed the earth's largest lava fields in the Deccan Traps. Satellite surveys have revealed at least a hundred large but weathered craters on the earth known as "astroblemes."

Context

The "new catastrophism" now seems to be fairly well established, giving rise to new concerns about threats to modern civilization from asteroid collisions, as well as volcanoes, earthquakes, and floods. Estimates indicate that asteroids about fifty meters in size and comets about one hundred meters in size penetrate the earth's atmosphere about once per century. Theoretical models show that such intruders, traveling at more than fifteen kilometers per second, tend to explode about ten kilometers above the surface of the earth as a result of shock waves generated by the atmosphere. Such an explosion seems to account for the Tunguska event that occurred over Siberia in 1908, flattening trees over an area more than one hundred kilometers in diameter in the biggest impact catastrophe of the twentieth century. Similar events have probably occurred in recent centuries, but few if any have entered the historical record.

Asteroids more than about one hundred meters in size can strike the earth before the shock wave propagates far enough to cause them to explode above the surface. Lunar studies show that such asteroids hit the earth about once every one thousand to two thousand years and can produce craters from one kilometer to two kilometers in diameter. Meteor Crater in the Arizona desert is of approximately this size, but most such craters can no longer be detected as a result of weathering effects. The greatest damage from such collisions would occur if such an object hit the ocean. Such an impact could produce a tidal wave that would rise to as much as two hundred meters in height, killing millions of inhabitants in coastal cities.

Asteroids of between one and two kilometers in size are expected to hit the earth about once every one hundred thousand years; such an impact could fill the stratosphere with enough light-reflecting dust to lower the surface temperature by several degrees for a few months to years. The resulting collapse of agricultural production could kill perhaps half of the world's human population. If temperatures fell far enough, mass annihilation of carbon-dioxide-consuming ocean plankton would cause a rapid increase of carbon dioxide in the atmosphere, causing enough global warming to melt glaciers and the polar icecaps and flooding coastal areas.

The greatest damage to the biosphere could be caused by asteroids of ten or more kilometers in size, such as the one believed to have killed the dinosaurs 65 million years ago. These are estimated to strike the earth about once every 100 million years. There is growing evidence in the fossil record that mass extinctions of species have occurred every 26 million to 31 million years, leading to theories that such large asteroids or comets are periodically disturbed by some regular astronomical event. One theory suggests that the sun is accompanied by a companion "nemesis" star, too dim to be seen but with a highly eccentric orbit that periodically brings it near enough to the solar system to send comets on Earth-crossing orbits. In spite of these possible catastrophes, analysis shows that the risk of dying from asteroid impact in an average life span is roughly one chance in ten thousand—compared to one chance in a hundred of being killed in an automobile accident, and one chance in thirty thousand of dying in a flood.

Bibliography

Briggs, John. "Biotic Replacements—Extinction or Clade Interaction?" *Bioscience* 48, no. 5 (May, 1998): 389-396. Discusses the shift away from catastrophism and toward uniformitarianism in the mid-nineteenth century in the context of biotic replacements.

Chapman, Clark, and David Morrison. *Cosmic Catastrophes.* New York: Plenum Press, 1989. A good review of classical catastrophism and uniformitarianism. Includes an authoritative discussion of the "new catastrophism" associated with asteroid collisions. The book is well illustrated and contains a good glossary.

Clube, Victor, and Bill Napier. *The Cosmic Winter.* Oxford, England: Basil Blackwell, 1990. A good history of meteorite collisions and their effects, including a chapter on assessing the risk from meteorite and asteroid collisions.

Harris, Stephen L. *Agents of Chaos.* Missoula, Mont.: Mountain Press, 1990. A good geological description of catastrophic events that have shaped the earth's crust, including earthquakes, volcanoes, floods, and asteroids. Many interesting illustrations and a good glossary are included.

Huggett, Richard. *Catastrophism: Systems of Earth History.* London: Edward Arnold, 1990. A careful geological assessment of the rise and fall of classical catastrophism and its modern neocatastrophist revival. Includes an extensive bibliography.

Lewis, John S. *Rain of Iron and Ice.* New York: Addison-Wesley, 1996. A survey of impact cratering in the solar system and the implications for bombardment of the earth by comets and asteroids. Includes a dozen photographs.

Raup, David M. *The Nemesis Affair: A Story of the Death of Dinosaurs and the Ways of Science.* New York: W. W. Norton, 1986. An excellent history of catastrophism and its revival in the light of asteroid-collision evidence, including a discussion of possible sources of periodic extinctions and the nature of scientific controversies.

Steel, Duncan. *Rogue Asteroids and Doomsday Comets.* New York: John Wiley & Sons, 1995. An interesting discussion of asteroid and comet impacts, past and future, including speculations about their historical influence and precautions that could be taken in the future. Contains a good glossary and bibliography.

Joseph L. Spradley

Cross-References

Caves and Caverns

Caves, large natural holes in the ground, are part of the earth's plumbing system. Groundwater passes through most caves at some point in time, creating many unusual features.

Field of study: Geomorphology

Principal terms

CALCITE: a common, rock-forming mineral that is soluble in carbonic and dilute hydrochloric acids

GROUNDWATER: water beneath the earth's surface

KARST: a landscape formed by the dissolution of rocks; characterized by sinking and rising streams, caves, and underground rivers

LAVA: molten rock extruded from a volcano

LIMESTONE: sedimentary rock, usually formed on the ocean floor and composed of calcite

PHREATIC: a zone in the ground below the level of complete water saturation

SPELEOLOGIST: a scientist who explores and studies caves

SPELEOLOGY: the exploration and scientific study of caves

SPELEOTHEM: a mineral deposit formed within a cave

SUPERSATURATED SOLUTION: a solution that contains more of the dissolved material (solute) than the water or other liquid (solvent) can hold in equilibrium

VADOSE: a zone in the ground above the level of complete water saturation

Summary

Caves, or caverns, are natural cavities in rock large enough for a person to enter. Most caves develop by the process of groundwater dissolving limestone, a common rock deposited on ocean floors. Gypsum, dolomite, and marble (metamorphosed limestone) are other rocks that dissolve readily to form caves. Rain and snow pick up a trace of carbon dioxide as they travel through the atmosphere. Where the ground has a thick layer of decaying vegetation, more carbon dioxide combines with the water, and a dilute solution of carbonic acid forms. (Carbonic acid, carbon dioxide dissolved in water, is also present in soda pop.) The water soaks into the ground and finds its way into cracks in the soluble rock (limestone, dolomite, gypsum, or marble). The acid dissolves the rock in a process similar to that of water dissolving table salt. Groundwater removes what was previously solid rock, and a hole, or cave, remains.

The longest cave in the world is the Mammoth-Flint Ridge Cave System, a solutional cave near Bowling Green, Kentucky. More than 560 kilometers of cave passage have been surveyed and mapped.

In mountainous areas such as the Alps in Europe or the Sierra Madre Oriental in Mexico, groundwater moves hundreds or even thousands of meters downward through cracks in the rocks. The resulting passageways are mostly vertical, with deep shafts. The deepest explored cave in the world is the 1,602-meter-deep réseau Jean Bernard in the French Pyrenees. Scientists have shown that water passes through a 2,525- meter-deep cave in southern Mexico, but explorers have not yet successfully followed much of its course.

Streams flow into or out of the entrances to many actively growing caves. Scientists refer to entrances where water flows into caves as "insurgences." Entrances that have streams or rivers flowing out are called "resurgences." Insurgences and resurgences often mark the boundary between soluble and insoluble rocks. Where water reaches insoluble rock, it flows onto the surface and the cave ends. Similarly, streams flowing over insoluble rocks commonly sink into caves upon reaching a limestone terrain.

In some places, acid-charged water comes from deep in the earth and not from rainwater. Pockets of carbon dioxide and hydrogen sulfide in the earth's crust can combine with deep flowing water to form carbonic acid and sulfuric acid, respectively. Caves form when these deep waters rise through cracks in soluble rocks. Water charged with hydrogen sulfide rose up through limestone and dissolved the spectacular Carlsbad Cavern in New Mexico.

Lava flows on the flank of a volcano can create another type of cave, commonly called a "lava tube." As lava flows down the slope of a volcano, the surface cools and solidifies while liquid lava continues to flow under the crust. As the flow cools, self-constructed pipelines under the crust continue to pass fast-moving, hot lava down the slope. Tubes drain when no more lava passes through, and a cave remains. The deepest known cave in the United States, which is 1,099 meters deep, is the 59.33-kilometer-long Kazumura Cave, a lava tube on the island of Hawaii.

A few caves are in insoluble rocks such as granite, sandstone, and volcanic tuff. These features are generally small and have varied histories leading to their development. In many cases, groundwater has carried individual grains of sand, one at a time, from the base of a cliff where a spring emerges. With time, the resulting hole at the cliff base is deep enough to be called a cave. Other caves have formed under blocks of rocks that fell or slid down adjacent hillsides. Pounding waves excavate caves in cliffs along some ocean shorelines. These caves are usually referred to as "sea caves."

Ice caves sometimes form under glaciers near their toes. Meltwater flowing under a glacier during summer may enlarge a passageway large enough to form a cave. Ice caves, however, are usually short-lived and are constantly changing size and shapes. Solutional caves are often divided into three categories: phreatic caves, vadose caves, and dry caves. Most phreatic caves are still actively forming. Their passages are below the water table and completely filled with water. Vadose caves are above the water table, but water passes through them as rivers and streams. Some caves form under vadose conditions. Other caves in the vadose zone were saturated when they formed. They now provide convenient paths for water to flow through unsaturated zone. Dry caves are no longer actively enlarging. The water that formed them has withdrawn. The air in a dry cave is usually humid, but the cave does not act as a conduit for water.

Water levels in caves commonly respond quickly to rain. A river may pass through an otherwise dry passage during the spring snow melt. A vadose passage with a small stream can become completely filled with water within a few minutes after a heavy rain commences. Tops of phreatic zones in caves have been observed to rise more than 50 meters within a few hours after the start of a surface deluge.

Once the void forms, nature commonly starts to fill caves. While most of the limestone (or dolomite or gypsum) dissolves, some impurities in the

Half Mile Hall in South Dakota's Wind Cave National Park. *(Wind Cave National Park)*

rock always remain as sediments on the cave floor. In addition, sand, mud, and gravel brought into the cave from outside by streams add to these sediments. Over hundreds, thousands, or even millions of years, caves can become completely choked by sediments.

If too much rock is dissolved and the rock is not strong enough to support the void, the ceiling collapses. Failure can occur one small rock at a time or in massive blocks. If the cave is still actively forming, groundwater may eventually dissolve the debris, and the passage will continue to grow upward. However, if the debris is not removed, the passage can completely fill with rubble, ending the existence of the cave.

Attractive deposits of minerals, called "speleothems," form from super-saturated water in a cave. Supersaturated waters contain more dissolved minerals, usually calcite, than they can maintain in solution. Supersaturation can occur when water evaporates and the dissolved minerals stay behind in the remaining liquid water. More commonly, supersaturation happens when cave water releases dissolved carbon dioxide. The less carbon dioxide dissolved in the water, the less acidic the water, and the less mineral the water can hold in solution. Cave water loses carbon dioxide—in the same manner that carbon dioxide bubbles escape from soda pop—when the surrounding pressure on the water drops (like opening a soda can) or when the tempera-ture of the water rises. In both cases, dissolved minerals solidify—a process called "precipitation"—as speleothems.

The most common speleothems are soda straws, stalactites, stalagmites, and flowstone. Soda straws look like their namesakes and hang from the ceilings of caves. Water is fed from a hole at the top of the soda straw, flows down through the hollow speleothem, and hangs on the end before falling. Calcite precipitates around the edges of the water as it slowly drips from the soda straw's end. Stalactites are typically cone-shaped deposits that hang from the ceiling. Originally soda straws, they grow as water deposits calcite around the outside of the speleothem. Stalagmites grow when drops of water from the ceiling hit the ground, lose carbon dioxide when they splash (like shaking a soda can), and precipitate calcite. They look almost like upside-down stalactites, but their ends usually are more rounded. Stalactites and stalagmites that grow together result in a "column." Flowstone, a sheet of calcite coating a sloping wall or floor of a cave, forms under flowing water.

Gypsum, composed of calcium sulfate, is commonly deposited within limestone, dolomite, and gypsum caves. Gypsum precipitates in a similar manner as calcite, but the process involves dissolved hydrogen sulfide instead of carbon dioxide. Gypsum speleothem shapes differ from those of calcite speleothems. One type of speleothem, a gypsum flower, looks like clear or white rock flowers growing out of cave walls. They form at the base of the "petals" and extrude earlier-formed deposits away from the wall in a manner similar to the squeezing of toothpaste out of a tube.

Ice forms many of the same speleothems as calcite. Ice stalactites (icicles), stalagmites, and flowstone are displayed in cold caves, particularly in winter and spring. Some caves in the Austrian Alps have moving glaciers and massive ice columns.

The temperature of most caves is the mean (average) annual temperature of the local area above ground. Temperatures typically fluctuate slightly near entrances and usually are a constant temperature a short way from the entrance. Thus, caves usually seem cool in summer and warm in winter. The moisture in the ground makes most caves very humid. Like temperature, humidity remains nearly constant year round away from entrances.

Caves try to adjust to changes in local barometric conditions. When the outside atmosphere changes from higher barometric pressure to a lower pressure, strong winds blow out the entrances of large, air-filled caves as they also try to lower their atmospheric pressure. When an area changes from lower pressure to higher pressure, large caves will suck air in from the outside. Just as on the surface, the temperature and air pressure of caves increase at greater depths. The change can be substantial in caves of more than 1,000 meters in depth. "Blowing caves" have entrances at different elevations. A chimney effect causes cold air to drop through the cave and blow out the lower entrance throughout the winter. The effect reverses as air blows out the upper entrance during the summer.

Methods of Study

Studying caves is one of the few scientific endeavors that still requires original geographic exploration. Although the continents' surfaces are almost thoroughly explored, most of the world's caves have never been entered. Exploration in caves can be hazardous. Since caves are the earth's natural storm sewers, flooding is common, and many inexperienced cave explorers have drowned. Falling is the second most common cause of serious accidents. Visits to undeveloped caves should always be in the company of an experienced caver. Cave diving is extremely hazardous, and only experienced scuba divers with extensive specialized training should enter a water-filled passage or cave.

Geologic research starts with exploring, surveying, and drawing detailed maps. Well-trained amateur explorers largely do this work. Using compasses, inclinometers (instruments that measure slope), and tape measures or similar devices, explorers measure the width, height, length, and depth of cave passages while drawing a detailed sketch of the floor, walls, and ceiling. Experienced explorers also record the geology of the passage, standing and flowing water, and speleothems.

The collected data must be adjusted using trigonometry to scale the true horizontal and vertical distances on the final map. Adjusted or "reduced" data are plotted to scale. Then, a cartographer uses the notes and sketches made in the cave to draw a plan and vertical cross sections. These maps give scientists important information about the unique qualities of the cave and provide a base for recording further observations.

Understanding water flow in caves is important to people living on the surface. Groundwater in karst areas behaves differently from groundwater elsewhere. Because water passing through a cave is, in effect, passing through a pipe rather than through sand or other soil, there is no opportunity for the ground to filter out pollutants. Contamination dumped in a stream that enters a cave may come out in a spring tens of kilometers away.

When speleologists cannot follow the water in a cave for part of its underground journey, they use non-toxic dyes or other markers to determine where the water goes. Before pouring dye into a disappearing stream, scientists put traps in all springs and wells that may be fed by the stream. The traps are monitored during the weeks and months following the injection of dye to detect it. In karst regions, one cave can be the source of several springs many kilometers apart.

Speleothems are sources of important information about the climate in the past. Molecules that make up calcite contain oxygen. Most oxygen atoms have eight neutrons, but some oxygen atoms have ten neutrons in their nuclei. Both types are "isotopes" of oxygen. These isotopes do not decay, but the ratio of the two oxygen isotopes to each other indicates the temperature of the water that precipitated the speleothem. Since the temperature of the cave water is generally the mean surface temperature, speleologists can learn about the past temperatures of an area.

Many speleothems contain trace amounts of a uranium isotope that decays to lead. Using radioactive dating techniques, speleologists can calculate the age of a speleothem. If the oxygen isotope ratio is also known, the past average temperature of the area at a specific time can be determined. Radioactive dating of speleothems near coastlines also gives information on sea-level rises and falls. Stalactites and stalagmites only grow in air-filled environments. Some Caribbean "blue holes," caves under the ocean floor, have stalactites and stalagmites. The ages of these speleothems prove to scientists that the ocean in the area was lower thousands of years ago.

Caves are particularly delicate environments. Once damaged, most caves will never return to their natural state. Scientists are very careful to avoid damaging a cave. Even touching many speleothems will harm or destroy them. Speleologists collect speleothems only when they have a specific purpose and permission from the landowner. They then collect the smallest piece necessary for the job. Samples for radioactive dating and oxygen isotope analysis are usually small cores drilled into the speleothem. Amateurs should never touch, break, or remove a speleothem. One of the greatest values of caves is their pristine beauty.

Context

Caves are sites of great beauty and adventure. They are the only continental areas left where individuals can truly be the first to explore and map. Underground streams sculpt beautiful, smooth walls. Bizarre and spectacular speleothems sparkle against the brown and gray walls of rock. A visit to a cave is an escape from the civilized world.

The constant temperature and humidity and protection from rain, snow, and wind in caves made them valuable resources to early humans. Earlier people used caves as art galleries, temples, shelters, refrigerators, sources of minerals, burial sites, and fortresses. The protected environment of caves preserves their paintings, shrines, pottery, baskets, and skeletal remains. Many of the world's most important archaeological sites are in caves. Animals also seek the shelter of caves. Much of the knowledge of many extinct mammals is based on the remains left in caves. Cave explorers have found bones, feces, and even mummies of extinct animals. Knowledgeable speleologists do not touch or remove archaeological (ancient human) or paleontological (ancient animal) material. Such items may have been preserved for hundreds or even thousands of years, and removing them from the cave environment will usually destroy them. Instead, experienced explorers work with specialists from a museum or college to document their finds.

Today, caves are used as sites of recreation, natural science classrooms, and preserves for endangered animals such as bats. In caves, scientists collect information about past climates and sea levels and examine the interior of the earth firsthand.

Caves can also affect human lives on the surface. When the roof of a cave collapses, a sinkhole develops. Homes and businesses are occasionally lost over a period of a few hours when a cave collapses underneath them. While caves are generally stable and roof collapses are rare, the process is often greatly accelerated in urban areas when a town pumps groundwater out faster than it is replenished. As a previously filled cave drains, water no longer partially supports the ceiling. This is the time when a cave is most vulnerable to roof collapse.

Water traveling through caves moves through the ground much faster than water moving through insoluble rocks. Polluted water entering a cave can travel through a natural pipeline without experiencing the filtration and cleansing that occurs when water passes through sand. The contaminated water may cross under surface drainage divides and re-emerge at a spring tens of kilometers away. Learning the course of water flowing through caves can help prevent a town from drinking contaminated water after a toxic spill into a distant source stream. Researchers in a karst area with a contaminated water supply may learn the source of the contamination by tracing the path of the underground water through caves.

Many endangered species, particularly bats, depend on the unique environment of caves to live. If the atmosphere of their cave home is altered by closing or enlarging of entrances, these animals may die or be forced to try to find another home. Earth scientists working with biologists determine the conditions necessary for the welfare of the animals. They strive to ensure that cave habitats are not adversely affected by the gating of an entrance, by quarrying near the cave, or by developing it as a public attraction.

Bibliography

Courbon, Paul, et al. *Atlas: Great Caves of the World.* St. Louis, Mo.: Cave Books, 1989. Describes and provides maps of the deepest and longest caves in the world. Includes all countries and all types of caves. An essential reference book for understanding the world's greatest caves.

Davies, W. E., and I. M. Morgan. *Geology of Caves.* U.S. Geological Survey, 1991. A brief, inexpensive brochure published by the U.S. government. Explains how most caves form and discusses the common speleothems within them.

Erickson, Jon. *Craters, Caverns, and Canyons: Delving Beneath the Earth's Surface.* Chicago: Facts on File, 1993. Covers structural geology and geomorphology, including caves, at a high-school level. A basic explanation of caves.

Exley, Sheck. *Caverns Measureless to Man.* St. Louis, Mo.: Cave Books, 1994. Sheck Exley was the greatest scuba diver ever to explore caves. This book documents his explorations and the water-filled caves he explored. Although the book focuses on his explorations, a good feeling for how caves develop and their significance can be achieved by reading about his adventures.

Furness, Adrian. "Caverns Measureless to Man." *Focus: Technology, Life, and Outrageous Science,* July, 1996, pp. 92-97. An illustrated examination of the world's largest caves.

Hill, Carol A., and Paolo Forti. *Cave Minerals of the World.* Huntsville, Ala.: National Speleological Society, 1986. The definitive book on speleothems. Describes them and explains how they form; filled with beautiful pictures.

Jagnow, David H., and Rebecca Rohwer Jagnow. *Stories from Stones.* Carlsbad, N.M.: Carlsbad Caverns-Guadalupe Mountains Association, 1992. Describes the geology of the Guadalupe Mountains in New Mexico and gives an excellent description of how the spectacular caves in the area evolved. Carlsbad Cavern and other area caves formed in an unusual manner, and the Jagnows are experts on their origins.

Middleton, John, and Tony Waltham. *The Underground Atlas: A Gazetteer of the World's Cave Regions.* New York: St. Martin's Press, 1986. Describes the major caves and karst areas of nearly every country in the world. The potential for finding caves in countries without presently known caves is also discussed.

Moore, George W., and G. Nicholas Sullivan. *Speleology: The Study of Caves.* 2d ed. St. Louis: Cave Books, 1978. Small, easy-to-read book clearly explains the fundamentals of cave geology and biology. Excellent, brief discussions on cave atmospheres, speleothems, evolution of blind cave animals, interactions of microorganisms with the cave walls, and human uses of caves.

Rea, G. Thomas, ed. *Caving Basics: A Comprehensive Guide for Beginning*

Cavers. 3d ed. Huntsville, Ala.: National Speleological Society, 1992. A comprehensive book on the geology, biology, archaeology, and exploration of caves written by members of the world's largest organization dedicated to caves. Each chapter was written by a leading expert on the subject. Useful bibliographies accompany each chapter.

Louise D. Hose

Cross-References

Karst Topography, 348; Water-Rock Interactions, 692; Weathering and Erosion, 701.

Coastal Processes and Beaches

The shoreline is the meeting place for the interaction of land, water, and atmosphere, and rapid changes are the rule rather than the exception.

Field of study: Sedimentology

Principal terms

BEACH: an accumulation of loose material, such as sand or gravel, that is deposited by waves and currents

LONGSHORE CURRENT: a slow-moving current between a beach and the breakers, moving parallel to the beach; the current direction is determined by the wave refraction pattern

LONGSHORE DRIFT: the movement of sediment parallel to the beach by a longshore current

OSCILLATORY WAVE: a wind-generated wave in which each water particle describes a circular motion; such waves develop far from shore, where the water is deep

TSUNAMI: a low, rapidly moving wave created by a disturbance on the ocean floor, such as a submarine landslide or earthquake

WAVE BASE: the depth to which water particles of an oscillatory wave have an orbital motion; generally the wave base is equal to one-half the distance of the length of the wave

WAVE REFRACTION: the process by which the angle of a wave moving into shallow water is changed; the bending which results is also termed wave refraction

Summary

The processes that create, erode, and modify beaches are many. Marine processes, such as waves, wave refraction, currents, and tides, work concurrently to modify, create, or erode a beach. This suggests that a beach is a very sensitive landform, and indeed it is. Generally a beach is a deposit made by waves and related processes. Beaches are often regarded as sandy deposits created by wave action; however, beaches may be composed of broken fragments of lava, sea shells, coral reef fragments, or even gravel. A beach is composed of whatever sediment is available. Beaches have remarkably resilient characteristics. They are landforms made of loose sediment and are constantly exposed to wave and current action. On occasion, the coastal processes may be very intensive, such as during a hurricane or tropical storm. Yet, in spite of the intensity of wave processes, these rather thin and narrow landforms, although perhaps displaced, restore themselves within a matter of days. Coastal scientists are inclined to believe that the occurrence and maintenance of beaches are related to their flexibility and rapid readjustment to the varying intensity of persistent processes.

Sediment deposited on beaches is derived from the continents. Rivers are one source of beach sediment and the coast is another. As rivers erode the land, the sediment they carry ultimately finds its way to the shore. Because the sediment may be transported several tens or hundreds of kilometers, it is refined and broken down even further to finer-sized particles. Once the river reaches the sea, the sediment is distributed by longshore currents along the shoreline as a beach. Beaches also occur where the shoreline is composed of cliffs, such as along the Pacific coast of North America. Here, waves erode the sea cliffs and the sediment is deposited locally as a beach. Under these conditions, the beach deposit is most often

gravelly because the sediment is transported only a short distance and has not had an opportunity to break up into finer-sized sediment such as sand.

The most obvious process and energy source working on beaches is waves. Waves approaching a beach are generally created by winds in storm areas at sea. As wind velocities increase, a wave form develops and radiates out from the storm. An oscillatory motion of the water occurs as the wave form moves across the water surface. It is important to note that the water movement within a wave is not the same as the movement on a wave form. In a wave created in deep water which is approaching a beach, the water particles move in a circular orbit, and very little forward move-

A beach and offshore rock formations along the rugged Oregon coast. *(Herb Noseworthy/ Archive Photos)*

ment of the water occurs. The water at the surface moves from the top of the orbit (the wave crest) to the base of the orbit (the wave trough) and then back up. Thus the water particles form an oscillatory wave motion; this motion continues down into the water. Although the size of the orbits in the water column decreases, motion occurs to a depth referred to as the wave base. At this point, the depth of the wave base is less than the depth of the water. As wave crests and troughs move into shallow water, the water depth decreases to a point where it is equal to the wave base. From this point, the orbital motion is confined because of the shallowness of the water and takes on an elliptical path. The ellipse becomes more confined as waves enter shallower water and eventually becomes a horizontal line. At this point, there is a net forward movement of water in the form of a breaker.

As a wave enters shallow water, many adjustments occur, such as a change in the orbital path of water particles described above and a change in the velocity of the wave form. Since the wave base is "feeling the bottom" in shallower water, friction occurs, slowing the wave down. As seen from an airplane, waves entering shallow water do so at an angle, not parallel to the shoreline. Therefore, one part of the wave enters shallow water and slows down relative to the rest of the wave. Thus, a part of the wave crest is feeling bottom sooner than the rest of the wave. Since the wave crests and troughs have different velocities, the wave refracts or bends. In so doing, the wave crests and troughs try to parallel the shallow bottom topography, which they have encountered.

The wave refraction is seldom completed, and the breaking wave surges obliquely up the slope of the beach and then returns perpendicular to the shoreline. The result is a current that basically moves water in one direction parallel to the beach in a zig-zag pattern (a longshore, or littoral, current). It is a slow-moving current which is located between the breaking wave out in the sea and the beach. Because longshore-current movement operates in shallow water and along the beach, it is capable of transporting sediment along the shoreline. "Longshore drift" refers to the movement of sediment along beaches. In a sense, the longshore current is like a river moving sand and other material parallel to the shore. Beaches are always in a state of flux. Although they appear to be somewhat permanent, they are constantly being moved in the direction of the longshore current. Along any shoreline, several thousand cubic meters of sediment are constantly in motion as longshore drift. Along the beaches of the eastern United States, about 200,000 cubic meters are transported annually within a longshore drift system.

A different type of ocean wave is one which is generated on the ocean floor rather than by the wind. Such waves are properly termed tsunamis. Although they popularly have been coined "tidal waves," they are completely unrelated to tides or the movement of planets. Some type of submarine displacement, such as the creation of a volcano, a landslide on the sea floor, or an earthquake beneath the ocean bottom, causes a displacement of water, which triggers waves. The waves are low, subdued forms traveling thousands of kilometers over the ocean surface at extremely high velocities, often in excess of 800 kilometers per hour. As a tsunami approaches shallow water or a confined bay, its height increases. There is no method for direct measurement of heights of tsunami waves; however, in 1946, a lighthouse at Scotch Cap, Alaska, located on a headland 31 meters above the Pacific Ocean, was destroyed by waves caused by a landslide-generated tsunami.

The changing character of a beach is very dynamic because the properties of waves are variable. During storm conditions or when more powerful waves strike a beach during the winter season, for example, the beach commonly is eroded, has a steeper slope, and is composed of a residue of coarser sediment, such as gravel, that is more difficult to remove. During fair-weather or summer conditions, however, beaches are redeposited and built up.

Methods of Study

The study of beaches and coastal processes is not particularly easy because of constant wave motion and changes in the beach shape. Scientists have, however, developed field methods as well as laboratory techniques to study these phenomena. To study wave motion and related current action, several techniques have been devised to include tracers, current meters, and pictures taken from satellites and aircraft. Two types of tracers are commonly used to determine the direction and velocity of sand movement: radioactive isotopes and fluorescent coatings to produce luminophors. (Luminophors are sediment particles coated with selected organic or inorganic substances which glow under certain light conditions.) In the former type, a radioactive substance such as gold, chromium, or iridium is placed on the surface of the grains of sediment. Alternatively, grains of glass may be coated with a radioactive element. The radioactive sediment can then be detected relatively easily and quickly with a Geiger counter. Both techniques can trace the direction and abundance of sediment along the seashore. Bright-colored dyes or current meters can also be used to document the direction of longshore currents. Surveying instruments are used to determine the high and low topography of beaches and adjacent sand bars. By measuring beach topography before and after a storm, for example, scientists can record changes. In this way, they can document the volume and dimensions of beach erosion or deposition which has taken place during a storm. Measurement of wave height and other wave characteristics can be done with varying degrees of sophistication. Holding a graduated pole in the water and visually observing wave crest and troughs is simplest and cheapest. To achieve more refined measurements, scientists place pressure transducers or ultrasonic devices on the shallow sea floor to record pressure differences or fluctuations of the sea surface. These more precise instruments also record wave data on a graph for later study and analysis.

All the above methods are detailed field techniques. By comparing aerial photographs, detailed maps, satellite pictures, and in some cases government studies in a coastal sector, changes over long periods of time may be detected. Finally, because wave motion cannot be controlled on a shoreline, wave tank studies are used to derive wave theories. Normally, an elongated glass-lined tank with water 1 to 3 feet in depth is used. A wave machine creates waves at one end of the wave tank and sediment is introduced at the other end. The heights and other characteristics of the waves can be varied, as can the type of sediment, to form the beach. In this controlled way, the various beach and wave relationships can be studied.

Context

More than 70 percent of the population of the United States lives in a shoreline setting, and thus an understanding of how waves interact with beaches is important. Shoreline property is highly prized and hence valuable because of the demand for it. On many beaches, the great investments made in hotels and condominiums suggest that beaches are the most sought-after environment on earth; shoreline frontage is sold by the foot or meter, not by the acre or hectare. Currently, however, such demand and investment are threatened with rising sea levels and continued coastal erosion.

Beaches represent the line of defense against wave erosion. Waves generated in the open sea slow down in shallow water, and the beach deposit absorbs the impact of waves. Beaches are therefore constantly changing and are one of the most ephemeral environments of earth. Thus, a sound knowledge of longshore currents and beach development is necessary prior to nearshore or marine construction. Sea walls and groins, for example, interfere with waves and longshore currents and may cause considerable erosion in selected areas: Sea walls are often constructed perpendicular to a shoreline to slow longshore currents so that a beach can be deposited. Downcurrent, beyond the area of beach deposition, erosion will take place. Similarly, rivers—a major source of the sediment that creates beaches—are sometimes dammed, thus depriving beaches of sediment and resulting in their erosion. Unless planners, developers, and builders understand these processes, major damage can result: Failure to understand coastal processes has on occasion caused a riparian property owner to sue a neighbor who caused beach erosion to take place.

Beaches are in a sense climatic barometers that record changes of sea level. The warmer atmosphere has led to a rising sea level as a result of glacial melting and warmer water temperatures. The impact of rising ocean levels was first noted on eroding beaches along the eastern United States many years ago. That was followed by the discovery of changes in the ozone content of the atmosphere. Beaches do indeed detect changes in the local sand supply, as well as global changes in the atmosphere.

Bibliography

Bascom, Willard. "Beaches." *Scientific American* 203 (August, 1960).
_____. "Ocean Waves." *Scientific American* 201 (August, 1959).
  Although older, both of these articles are still current. Well illustrated
  with photos and diagrams. Technical concepts are explained in language a nonscience layperson can understand. Included is a discussion
  of the tsunami or tidal wave, wave properties, and wave refraction. The
  more recent article is a continuation of the wave article; it presents
  aerial photographs of the impact of seawalls and related structures on
  the beaches.
_____. *Waves and Beaches.* Garden City, N.Y.: Anchor Books, 1964.
  A good introductory pocket book for nonscientists by an expert.
  Numerous examples and illustrations have made this softcover book
  popular. Nonmathematical and nonscience laypersons should have no
  difficulty with this book.
Bird, Eric. *Coasts: An Introduction to Coastal Geomorphology.* 3d ed. New York:
  Basil Blackwell, 1984. Includes chapters on waves and on beaches.
  Although most examples are Australian, the book covers fundamental
  concepts. A good introduction for anyone who has had a high-school-
  level earth science course.
"How the Coast Advances." *Oceans,* April, 1987, pp. 8-18.
Kaufman, Wallace, and Orrin Pilkey. *The Beaches Are Moving: The Drowning
  of America's Shoreline.* Garden City, N.Y.: Anchor Books, 1979. This
  thought-provoking book, written in a nontechnical style, deals with the
  processes working in the coastal zone, such as winds, waves, and tides.
  The impact of rising sea levels and the modification and urbanization
  of the coast are highlighted. A narrative text suitable for all ages.
Komar, Paul D. *Beach Process and Sedimentation.* Englewood Cliffs, N.J.:
  Prentice-Hall, 1976. Extensive treatment of waves, longshore currents,
  and sand transport on beaches. Equations and mathematical relationships are presented and elaborated upon. College-level material. This
  book is for those interested in the specifics of coastal processes.
Leatherman, Stephen P. *Barrier Island Handbook.* Amherst: University of
  Massachusetts Press, 1979. Based on actual field studies along the East
  Coast of the United States. Numerous photographs, diagrams, and
  tables. Most suitable for coastal managers and government employees;
  however, very readable and suitable for nonscientists as well as the
  general scientist. Emphasizes the dynamic nature of beaches, recreation and construction impacts, and nearshore processes.
Leonard, Jonathan Norton. *Atlantic Beaches.* New York: Time-Life Books,
  1972. A regional travel description of the shoreline from Cape Cod,
  Massachusetts, southward to Cape Hatteras and the Outer Banks,
  North Carolina. Information is presented in a nonscience narrative
  form. Color photography is excellent. Useful in planning trips along

the East Coast of the United States, as the emphasis is on the scenery of the seascape.

Pethick, John. *An Introduction to Coastal Geomorphology*. Baltimore, Md.: Edward Arnold, 1984. A thorough survey of coastal processes, this 260-page book is divided into three sections: wave energy on the coast and its characteristics, the relationship between currents and the movement of beach material along the shore, and the landforms, such as beaches, mud flats, and estuaries. Most suitable for anyone needing an equation or a technical explanation of selected processes operating in a coastal zone.

Thorsen, G. W. "Overview of Earthquake-Induced Water Waves in Washington and Oregon." *Washington Geologic Newsletter* 16 (October, 1988). A nine-page introduction to the impact of a tsunami on the coasts of Washington and Oregon. An earthquake occurred in March, 1964, in Alaska, and the waves traveled southward along the Pacific coast of North America. The tsunami is discussed in nontechnical language. Damage and economic losses are estimated. Maps and tidal gauge records presented. A good review of wave activity, intended for interested laypersons and teachers.

Walker, H. J. "Coastal Morphology." *Soil Science* 119 (January, 1975). A nontechnical overview of coastal landforms. Discussion includes beaches, deltas, and lagoons. The worldview is taken, and maps illustrating processes are included. This fifteen-page article is useful for a nonscientist interested in the causes and distribution of coastal features from a geographical perspective.

C. Nicholas Raphael

Cross-References

Continental Crust

Continental crust underlies the continents, their margins, and isolated regions of the oceans. Continental crust is distinguished from its counterpart oceanic crust by its physical properties, chemical composition, topography, and age. The creation and eventual modification of continental crust is a direct function of plate tectonics.

Field of study: Tectonics

Principal terms

ASTHENOSPHERE: a layer of the earth's mantle at the base of the lithosphere

CRUST: the outermost shell of the lithosphere

LITHOSPHERE: the outer, rigid shell of the earth, overlying the asthenosphere

MANTLE: the region of the earth's interior between the crust and the outer core

MOHOROVIČIĆ (Moho) discontinuity: the seismic discontinuity, or physical interface, between the earth's crust and mantle

OROGENESIS: the process of mountain-range formation

PLUTON: a deep-seated igneous intrusion

TECTONICS: the study of the assembling, deformation, and structure of the earth's crust

Summary

The earth's crust exists in two distinct forms: continental crust, or sial, and oceanic crust, or sima. Oceanic crust is characterized by the dense, basic, igneous rock basalt, while continental crust is an assemblage of sedimentary, metamorphic, and less dense, silicon-rich igneous (granitic) rocks. Oceanic crust makes up the floors of the earth's ocean basins. Continental crust underlies the continents and their margins and also small, isolated regions within the oceans. The total area of all existing continental crust is 150×106 square kilometers. In total, continental crust covers about 43 percent of the earth's surface and makes up about 0.3 percent of its mass.

Continental crust is distinguished from its counterpart, oceanic crust, and from underlying mantle by its physical properties and chemical composition. In addition, the continental crust and oceanic crust contrast in topography. The earth's major topographic features range from the highest mountain on the continental crust (Everest: 8,848 meters) to the deepest ocean trench (Marianas: 10,912 meters). The difference in average elevation between the two crustal forms is quite pronounced. Continental crust varies in thickness from 10 kilometers along the Atlantic margin to more than 90 kilometers beneath the Himalaya mountain system. On average, continental crust is about 35 kilometers thick. Seismic studies of the Mohorovičić (Moho) discontinuity indicate that oceanic crust is on average 5-8 kilometers thick. Continental crust averages a height of 0.9 kilometer above mean sea level, while oceanic crust averages a depth of 3.8 kilometers below that datum. This difference in levels is attributed to the fact that despite being thin, oceanic crust comprises the majority of the earth's crust and has a density (3.0-3.1 grams per cubic centimeter) greater than that of continental crust. While continental crust is thicker than oceanic crust, it is less dense (2.7-2.8 grams per cubic centimeter) and comprises less crustal surface area. For the most part, continental crust lies near sea level or above it. It is thickest where it underlies places of great elevation, such as mountain ranges. It is thinnest where it lies below sea level, such as along continental shelves. There are exceptions to this pattern of thickening and thinning. The relatively flat basins of the oceans are transversed by 2-kilometer-high ridge systems, and areas of continents where intraplate volcanism is active often display thinning where the crust is stretched by rising hot mantle material. Rising hot material also makes for more buoyancy and raises the surface elevation yet maintains a thin crust. The Basin and Range province of the western United States is a good example; the crust beneath the mountains is of relatively normal thickness, yet the elevation is high.

Differences in the vertical structure and rock composition between oceanic and continental crust are pronounced. The structure of oceanic crust has a prominent layered effect that seismic waves can readily detect. The layers are attributed to petrologic differences between basalt, gabbro, and peridotites that comprise the layers. Continental crust has a more complex layered structure, and the contacts between layers are not well defined. Continental crust is separated into upper and lower zones (see figure). The upper zone is usually highly variable in composition, with the top few kilometers of material being any combination of unmetamorphosed volcanic or sedimentary rocks, to medium-grade metamorphics such as quartzites and greenschists. Below this immediate layer, the upper zone of continental crust is typically regarded as either granodiorite or quartz diorite. This assumption is based on seismic wave travel times. The upper zone of the crust is separated from the lower zone by a change in seismic velocity similar to that which separates the asthenosphere from the crust itself. This intracrustal boundary is called the Conrad discontinuity. The composition of the lower continental crust is less well known because of the relatively few places where outcrops are available for study. Observations made on rocks in the most deeply eroded regions of Precambrian shields lead researchers to believe that the lower zone is composed of granulite. Granulite is a rock of intermediate-to-basic composition, containing mainly pyroxene and calcium feldspars. The velocities of seismic waves through granulite compare favorably with seismic velocities observed passing through rocks of the lower zone. Such circumstantial evidence favors granulite as the composition of the lower continental crust.

When taken on average, the overall chemical composition of continental crust corresponds to that of an intermediate igneous rock with a composition between andesite (tonalite) and dacite (granodiorite). Because igneous rocks of this type are added to continental crust from the mantle at convergent (destructive) plate boundaries, igneous activity of this kind is thought

Comparison of Zones of Oceanic and Continental Crust

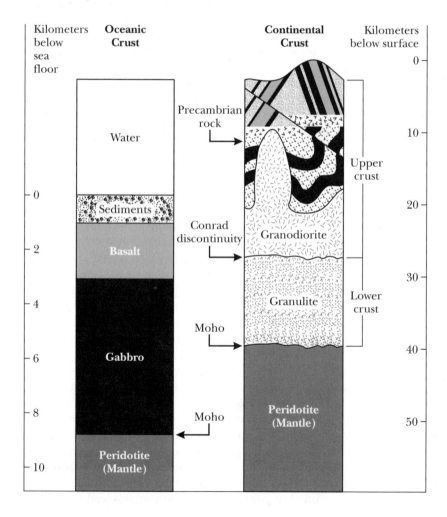

to be responsible for the majority of growth to the continental crust. The accretion rate for new continental crust forming at destructive margins is estimated at 0.5 cubic kilometer per year. Based on calculation of existing continental crust surface areas, ages, and thicknesses, an accretion rate of 0.5 cubic kilometer per year can account for only about half of the existing continental crust. It is concluded that while the formation of continental crust has happened throughout geological time, the accretion rates for new continental crust must have occurred at higher rates at different times during the past.

Rocks of the continental crust formed throughout nearly the complete 4.6-billion-year history of the earth. These rocks can be grouped into three main components: orogenic belts, Precambrian shields, and continental platforms. Orogenic belts (orogens) are long, broad, linear-to-arcuate (curved) areas of deformed rocks. The deformation occurs to the crust during uplift and usually includes faulting and sometimes the formation of plutons and volcanoes. The result of the deformation is the creation of a mountain system. The deformations affect thick sections through the crust and leave permanent scars that can be recognized long after the uplifted mountains are eroded away. Precambrian shields consist of deformed crystalline igneous and high-grade metamorphic rocks more than 570 million years old. These shields are the eroded roots of ancient orogens. Continental platforms are regions of relatively underformed, younger sedimentary or volcanic rocks overlying Precambrian basement. These platforms, while nestled within the continental interior and isolated from internal strain, still typically warp into broad regional structures, usually basins or domes. Shields and platforms can form a stable nucleus to continental masses. These stable regions are called cratons. Examples of a shield and sedimentary platforms forming stable craton regions are the Canadian Shield, the Michigan Basin, and the Ozark Uplift.

The earth's crust is a solid, rigid layer of mobile plates that comprise the uppermost part of the lithosphere. There are seven major plates, several minor plates, and numerous microplates. These plates appear to float on the plastic upper mantle of the earth, called the asthenosphere. The vertical boundary between the asthenosphere and the crust is called the Mohorovičić discontinuity, or the Moho. The Moho is a zone less than 1 kilometer thick in some places but several kilometers thick in others, where the velocity of seismic waves changes from about 7 kilometers per second in the crust to about 8 kilometers per second in the mantle. This change in seismic velocity is caused largely by a change in composition between the crust and mantle. Rocks of the mantle are rich in iron and magnesium but poor in silicon, making them denser than the silicon-enriched overlying crust.

The movement of crustal plates upon the denser asthenosphere is believed to be caused by complex convection currents deep within the mantle. The upper zone of the earth's crust, in which plate movement takes place, is called the tectonosphere. As the plates move about the tectonosphere, they interact with one another. The plates tear apart (rift), collide, slide under (subduct), or slide against each other (transform fault). The active edges of the plates are called plate boundaries. The interaction of crustal plates at plate boundaries, in addition to the cyclic phenomena of sedimentation, metamorphism, and igneous activity, makes the crust the most complex region of the earth. These activities process and reprocess crustal material and lead to the diversity of physical and chemical properties observed in crustal rocks. The rocks of the crust indicate that these processes have taken place throughout geological time and further suggest that the crust has grown in bulk at the expense of the upper mantle.

The oceanic and continental crusts interact along their margins. The margin may be passive, in that stresses are no longer deforming it, or the margin may be active, in which case it is a zone of seismic and tectonic activity. Along passive margins, the transition between continental and oceanic crust is gradual; a good example is the Atlantic shelf along the eastern coast of North America. The best example of an active margin between continental and oceanic crustal plates is the Pacific basin. Around the margin of the Pacific basin, relatively dense oceanic crust is being actively subducted beneath the lighter continental crust. On the continental side, mountain ranges rise (the Andes) and island arcs are formed (the Aleutians), both dominated by active volcanism. Active margins also exist where two continental plates collide. When continents come together, there is little subduction because both plates are of low density. While igneous activity is less prominent than along convergent plate boundaries, the degree of deformation along the margins can be extreme; there is often considerable uplift involved. The contacting continental plates can either slide past each other along a transform fault (such as the San Andreas fault) or act like two cars in a head-on collision. As the plates collide, the crust shortens and the intervening sea floor is uplifted, folded, faulted, and overthrust. The most dramatic example of such an interaction of continental masses is the Himalaya mountain system.

Regions such as the Himalaya of Asia and the Alps of Europe are known as suture zones. They mark the boundaries where two plates of continental crust have collided. At suture zones, oceanic crust is subducted until the ocean basin separating the two continents disappears, and a violent collision takes place. Moving only centimeters per year, the two continents ram into each other. The deformation to the plates during such a collision can be quite dramatic. The two continental plate margins that collide already have thick, mountainous continental margins along their active subduction zones. As the collision takes place, mountain range meets mountain range, and a new, higher, and more complex set of mountains is created. The already thicker-than-average crusts beneath the two colliding continental edges combine to form an even thicker crust to support the newly uplifted mountains. This process is complicated by secondary magmatic activity that adds buoyancy and uplift. At such suture zones, the continental crust is at its thickest, and mountain peaks reach their most spectacular heights.

Continental crust thus forms convergent (subducting) boundaries with oceanic crust and can also collide or slide alongside other continental plates at transform boundaries. Continental crust can form one other tectonic boundary within its plate margin: It can split and form a spreading zone (rift) similar to the spreading ridges of oceanic crust. Plate interaction at a continental margin may influence the crust hundreds of kilometers inland. If forces within the plate work to stretch the crust, thinning it markedly, crustal faults may develop along the thinning zone. The crustal fault blocks that form will begin to subside as the crust continues to be stretched. Because the upper mantle is also being stretched, material from the lower mantle rises to take its place. This material is hotter and raises the temperature of the surrounding rock. The result is the formation of a magma zone beneath the thin crust of the rift zones. If the magma reaches the surface, volcanic activity similar to that seen at ocean ridge systems develops. Basaltic lava flows to the surface and begins to force the sides of the rift apart. Sometimes the divergence ends as a result of a shift in the overall dynamics of the plate. If that happens, the rift may leave a scar only a few tens of kilometers wide. Some examples are the Midcontinental rift system of North America, the Rhine Valley of Europe, and the East African rift valley. If the rift continues to expand, a new ocean basin/plate is formed. The Atlanic basin is an example of continued rifting of a continental plate to form an active oceanic plate. In some instances, rifting of two continental bodies occurs near the margin of an older continental margin, and fragments are rifted away from the main continental body. When that happens, small plateaus of continental crust (microplates) become partially submerged in the ocean or become surrounded by oceanic crust. One example is the Lord Howe Rise of the South Pacific. The highest part of the Lord Howe microplate surfaces above the ocean as New Zealand.

Methods of Study

Earth scientists have used studies of seismic velocity waves to define the boundaries and limits of continental crust and, through exhaustive field investigations and geophysical analysis, have made reliable estimates of the crust's composition. The processes responsible for the formation and dynamic nature of continental crust, however, have remained elusive. To explain their observations, earth scientists have come to rely on their present understanding of plate tectonics and the related processes of volcanism and orogenesis.

The Andesite Model is a tectonics-based explanation for the formation and growth of continental crust. The model can be stated as follows: The growth of continental crust results from the emplacement or extrusion of largely mantle-derived magmas formed at destructive plate margins. The process begins at the ocean ridges, where melted mantle peridotite rises to the surface as basaltic lava, forming new oceanic crust. The oceanic crust moves away from the ridge by way of sea-floor spreading. The spreading is caused by the constant extrusion of more basaltic lava at the ridge. Eventually, the oceanic plate encounters a continental plate and, because of the oceanic crust's greater density, descends below the continental plate. The oceanic crust descends at an angle of 30-60 degrees, forming deep trenches along the continental margin. The descending plate eventually reaches a seismically active region of the mantle known as the Benioff zone. At the Benioff zone, the subducting plate melts, producing a chemically complex, destructive margin magma (andesitic). This lighter, less dense andesitic melt rises through the mantle and into the overriding continental crust. The rising melt creates large plutons within the crust or breaks through to the surface to form andesitic volcanoes. Many large andesitic stratovolcanoes surround the Pacific basin and form the Ring of Fire. Around the Ring of Fire, andesitic lava is erupted and added to the surface of the continents. The volcanoes of the Andes, Cascades, Indonesian Arc, Japan, and Alaska, having such familiar names as Krakatoa, Rainier, and Fujiyama, are the birthplaces of new continental crust.

Context

The study of continental crust and its related processes is important to earth scientists because the development of continental crust appears to be a terrestrial phenomenon, that is, one not observed on other planets in the solar system. Furthermore, the continental crust of the earth provided a platform on which the later stages of the evolution of animal and plant life occurred. Without it, life would have been restricted to ocean basins and isolated volcanic islands, and evolution would have taken a drastically different course.

The memory of early earth history can only be found in the continental crust. Since oceanic crust records an age no older than 200 million years, the continental crust is scientists' only link with the 4.1-billion-year-old geological record of the earth. Studies of the continental crust also allow scientists to venture educated speculations as to the beginnings of the solar system some 4.6 billion years ago.

Although the mass of continental crust is small compared to the overall mass of the earth, it contains substantial amounts of all minerals and elements that are necessary for life to continue on earth. Additionally, continued investigations into the dynamic processes that form continental crust aid scientists in understanding many of the geological hazards that plague humans. Earthquakes and volcanoes, two of the earth's most destructive forces, are directly related to the processes that form and shape continental crust. By establishing a more complete understanding of the nature and functions of continental crust, scientists can better prepare and warn citizens of impending geological hazards.

Bibliography

Brown, G. C., and A. E. Mussett. *The Inaccessible Earth.* Winchester, Mass.: Allen & Unwin, 1981. An excellent source of general information about continental crust and crustal processes. For the undergraduate student.

Foster, R. J. *Geology.* 2d ed. Westerville, Ohio: Charles E. Merrill, 1971. Excellent general reference on continental crust and crustal processes. Written specifically for the undergraduate nonscience major.

Meissner, R. *The Continental Crust: A Geophysical Approach.* San Diego, Calif.: Academic Press, 1986. Highly technical approach to the study of continental crust. Graduate student reading level.

Sylvester, Paul J., and Ian H. Campbell. "Niobium/Uranium Evidence for Early Formation of the Continental Crust." *Science* 275, no. 5299 (January 24, 1997): 521-524.

Taylor, S. R., and S. M. McLennan. *The Continental Crust: Its Composition and Evolution.* Oxford, England: Blackwell Scientific, 1985. Highly technical, more advanced text about the continental crust. For the graduate-level geology student.

Windley, B. F. *The Evolving Continents.* New York: John Wiley & Sons, 1977. A good reference on plate tectonics, crustal processes, and crustal evolution. Written for the college-level reader.

Randall L. Milstein

Cross-References

Continental Drift

Continental drift is the modern paradigm that describes and accounts for the distribution of present-day continents and associated geological formations and phenomena, including mountain ranges, mineral deposits, volcanoes, and earthquakes.

Field of study: Tectonics

Principal terms

CONTINENTAL CRUST: the outermost part of the lithosphere, consisting of granite and granodiorite

CONTINENTAL DRIFT: the horizontal displacement or rotation of continents relative to one another

CONVECTION (CELL): a mechanism of heat transfer in a flowing material in which hot material from the bottom rises because of its lesser density while cool surface material sinks

EARTHQUAKE: the violent motion of the ground caused by the passage of a seismic wave radiating from a fault along which sudden movement has occurred

FAULT: a fracture in the earth's crust along which there has been relative displacement

GONDWANALAND: a hypothetical supercontinent made up of approximately the present continents of the Southern Hemisphere

LAURASIA: a hypothetical supercontinent made up of approximately the present continents of the Northern Hemisphere

LITHOSPHERE: the outer layer of the earth, situated above the asthenosphere and containing the crust, continents, and tectonic plates

OCEANIC CRUST: the outer part of the lithosphere, consisting mostly of basalt

PALEOMAGNETISM: the science of reconstruction of the earth's former magnetic fields and the former positions of the continents from the magnetization in rocks

PANGAEA: a hypothetical supercontinent, made up of all presently known continents, which began to break up in the Mesozoic era

PLATE: a large segment of the lithosphere that is internally ridged and moves independently over the interior, meeting in convergence zones and separating at divergence zones

PLATE TECTONICS: the study of plate formation, movement, and interactions

TECTONICS: the study of the movements and deformation of the earth's crust on a large scale

Summary

Continental drift is the guiding model for the mechanisms driving the geologic forces near the surface of the earth. This theory is the simplest explanation for the behavior of the earth's crust and the distribution of continents and their associated topographic features. The theory is useful not only in decoding the history of the earth but also in predicting future observations.

The idea that continents may have occupied different geographies in the past was developed by Alfred L. Wegener as early as 1910. Wegener observed that the coastlines of the Americas corresponded with those of Europe and Africa in a jigsaw-puzzle fashion. He was encouraged to learn that similar fossils had been discovered on both sides of the Atlantic Ocean, and he proposed that the splitting up of a supercontinent and the drift of its pieces could explain this data. He continued to refine his ideas and published a book, *The Origins of Continents and Oceans,* in 1915. By his account, about 200 million years ago, at the end of the Permian period, there existed a single supercontinent that Wegener called "Pangaea." This supercontinent, he theorized, broke apart, and the various pieces drifted; for example, North America and South America moved westward from Europe and Africa, creating the Atlantic Ocean.

Wegener had an American rival, Frank B. Taylor, who in 1910 published his own theory of mobile continents. Interestingly, Taylor's starting point was not the physical similarity of the Atlantic coastlines but the pattern of mountain belts in Eurasia and Europe. Yet Taylor's hypothesis, like Wegener's, soon faded from scientific memory.

There was, however, much physical evidence to suggest that continental drift was a feasible theory. First, the physical fit was a good one; in addition, the discovery of similar fossils, mineral deposits, glacier deposits, and mountain ranges seem to show a correlation across the oceans. Wegener described these correlations: "It is just as if we were to refit the torn pieces of a newspaper by matching their edges and then check whether the lines of print run smoothly across." Efforts to confirm the hypothesis were interrupted by World War I and, soon thereafter, the Depression and World War II. The theory of continental drift thus retained its marginal status until the 1950's.

The major stumbling block to its acceptance was the need to describe a plausible mechanism for driving the continents. Pushing continents around requires a tremendous amount of energy, and no model proposed was acceptable to the geophysics community. The only plausible suggestion came from British geologist Arthur Holmes, who tentatively proposed that thermal convection within the mantle could split the continents and drive them across the surface. Holmes was one of the most respected geologists of his time, and the scientific community did pay attention to his idea. Yet he could offer no evidence to support it, and continental drift thus remained merely an interesting possibility in the eyes of most scientists.

Wegener had a more difficult time gathering an audience. To most of the geologic community, he seemed to be an outsider attempting to restructure the science. For example, in the publication of the 1928 American Association of Petroleum Geologists symposium, R. T. Chamberlain quotes a remark made by a colleague: ". . . if we are to believe Wegener's hypothesis we must forget everything which has been learned in the last 70 years and start all over again." (This, however, is exactly what happened in the 1950's, as the strength of the hypothesis eventually became evident.) The hypothesis thus lived on the fringes of the scientific community and was supported by a minority of geologists, most of whom worked in the Southern Hemisphere.

Worldwide interest in the origins and evolution of the planet's features culminated in the observation of the International Geophysical Year from July, 1957, to December, 1958. The result of this effort was that in almost every area of research, and especially in geology, scientists found the earth and particularly its oceans to be very different from what they had imagined. One of the most interesting features studied was the Mid-Atlantic Ridge, an investigation that would lead to the understanding of plate tectonics.

The existence of a submarine ridge in the Atlantic had been recognized in the 1850's by Matthew Maury, director of the U.S. Navy's Department of

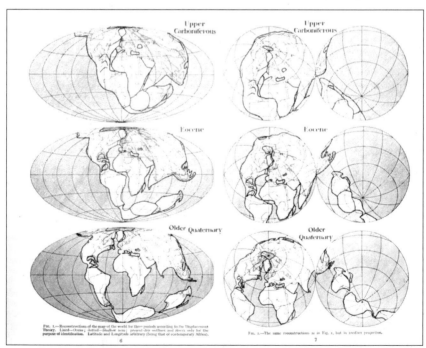

Drawings by Alfred L. Wegener illustrating his conception of the theory of continental drift. *(Library of Congress)*

Charts and Instruments. The British expedition aboard HMS *Challenger* (1872-1876) also recorded a submarine mountain. The next advance came in the 1920's with a German expedition led by Nobel laureate Fritz Haber. The expedition utilized an echo sounder to map the ocean floor. In 1933, German oceanographers Theodor Stocks and Georg Wust produced a detailed map of the ridge, and they noted a valley that seemed to be bisecting it. In 1935, geophysicist Nicholas H. Heck found a strong correlation between earthquakes and the Mid-Atlantic Ridge.

Oceanic exploration resumed after World War II as a predominantly American venture. The data collected pointed to an array of seemingly unrelated phenomena. In 1950, Maurice Ewing of the Lamont-Geological Observatory discovered that no continental crust existed beneath the ocean basins. In 1952, Roger Revelle, the director of the Scripps Institute of Oceanography, and his student A. E. Maxwell measured the heat flow from the earth's interior and discovered that it was hotter over the oceanic ridges. Additional data from Jean P. Rothe, director of the International Bureau of Seismology, revealed a continuous belt of earthquake centers associated with this submarine mountain range, which extends from Iceland through the mid-Atlantic, around South Africa, and into the Indian Ocean to the Red Sea. In 1956, Maurice Ewing and Bruce C. Heezen mapped a large area of this submarine mountain range and confirmed the existence of a rift valley bisecting the mountain crest. A peculiar faulting style was discovered in association with the range in 1959 by Victor Vacquier. The mountain range was offset by a large transverse fault that ran for hundreds of miles but did not extend into the continents. In 1961, Ewing and Mark Landisman discovered that this ridge system extended throughout the world's oceans, was seismically and volcanically active, and was mostly devoid of sediment cover.

The paleomagnetic researches of University of Manchester scientist Patrick M. S. Backett and his student Keith Runcom proved central to understanding the relationships among these phenomena. Their studies of fossil magnetism suggested that the position and polarity of the earth's magnetic field had once been very different than its present orientation. This data could only make sense if one assumed that the continents had shifted relative to the poles and to one another.

By the end of the 1950's, it was clear that then-current geologic theories had failed to predict or explain these seemingly unrelated phenomena, and the new data required a new theory. In 1960, Harry Hammond Hess proposed a simple model to explain the data. He suggested that sea-floor spreading powered by convection currents within the mantle might be the cause of the motion of the continents. Hess's theory, though simple, was radical; it bore out Chamberlain's earlier insight that previous geologic models would have to be discarded and that the geologic community would have to reinterpret and test all of its data in the light of the new model. This did not come easily to the science community; eventually, however, the theory of continental drift did emerge as the dominant paradigm for the earth sciences.

Methods of Study

The lengthy process leading to the acceptance of the theory of continental drift is not unusual in the history of science. Often, a hypothesis has to wait upon the development of technology or upon accidents of timing for its observations to be tested.

The observation that there was a relationship between the coastlines of the Americas and those of Europe and Africa can be traced to Francis Bacon, who in 1620 noted the similarities of shape. The idea was further enhanced by a French monk, François Placet, who in 1666 suggested that the earth's land masses had split as a result of the biblical flood, thus separating Europe and Africa from the Americas. The idea was repeated by a German theologian, Theodor Lilenthal, in the 1700's. In 1800, Alexander von Humbolt suggested that the oceans had eroded the land to further divide the continents. The drift theory was then supported by Evan Hopkins, who in 1844 proposed the existence of a "magnetic fluid" that circulated to drive the continents. In 1857, American geologist Richard Owen published *Key to the Geology of the Globe*, in which he proposed that the earth was originally a tetrahedron that had expanded in a great cataclysm, breaking the crust and expelling the moon from the Mediterranean.

These ideas and hypotheses were interesting but largely unsupported by physical evidence; Wegener's and Taylor's hypotheses proposed in the early twentieth century were thus fundamentally different from those of the past. Still, it was the development of new technological tools that illuminated the phenomena and eventually supported the drifting-continent theory. Much of this new technology was developed as a result of World Wars I and II and the technological race of the Cold War.

Wegener collected data from paleontology to show a correlation between continents and to illustrate that the continents had been in different latitudes in the geologic past. Further, he suggested an experiment to confirm the theory; the experiment failed not because the idea was inappropriate but because the experimental error resulting from his crude equipment was greater than the phenomena he was trying to measure. His experiment, conducted in 1922 and again in 1927 and 1936, involved the measurement of the time it took radio signals to travel across the Atlantic. The measurements failed to reveal a widening of the Atlantic through progressively longer travel times. Upon the advent of satellite and laser technology, however, widening was detected.

Eventually, wartime technology such as sonar was applied to scientific applications. Sonar is the underwater version of radar; an energy pulse is sent out, and its reflection from the sea floor is recorded. This data can be translated via computers (another new technology) to create either a profile or contour map. Literally thousands of soundings were made over thousands of kilometers of ocean in an effort to construct a map of the ocean floor in the greatest possible detail. These data began to produce a map that revealed rather remarkable features, including a continuous 64,000-kilometer mountain range that had a valley running along its crest. Princeton's Harry Hess realized the significance of the valley on top of the mountain range: It was a tensional, or "pull-apart," feature. The same forces that formed the mountain chain were also pulling it apart.

Other technologies were also contributing to the investigation. After World War II, a worldwide network of seismographs was deployed, not so much for recording earthquakes as to listen for atomic explosions. These new and sensitive instruments mostly recorded earthquakes at plate boundaries and revealed their outlines. (Interestingly, no nuclear powers camouflaged their atomic blasts as earthquakes by detonation at a plate boundary). The earthquake pattern was a fingerprint of plate activity and evidence of a dynamic crust.

Other compelling data were those of monitoring internal heat flow. The temperature of the earth increases with depth. At the core, the temperature is above 4,000 degrees Celsius. This heat flowing from the interior can be measured; the hottest crustal areas were found to be above the junction of plates that are spreading centers.

More traditional geologic sampling of the subsurface was conducted by retrieving core samples from below the ocean depths. These physical samples were analyzed according to the type of sediment and the age of fossils present. The findings were surprising: The oceans are very young compared to the continents, and the sediments on the midoceanic ridges is thin to nonexistent, while the sediments next to the continents are kilometers thick. Therefore, not only are the oceans young, but they are also youngest in the middle and oldest next to the continents.

The straw that broke the back of opposition to the theory of continental drift came with the study of fossil magnetism. Sedimentary and igneous rocks offer a record of the orientation of the earth's magnetic field through time, as iron particles within them are incorporated into their structures as they form. The decoding of these fossil magnetic fields suggested that the magnetic poles have reversed themselves and that the continents have wandered through the latitudes. If the poles reversed and the continents wandered, then a mirror image of polar reversal correlating with submarine topography should be present on both sides of a spreading center such as the Mid-Atlantic Ridge. This was exactly what was observed. The model was no longer merely an explanation of previously observed phenomena; it had also been shown to be capable of predicting future observations. Earth scientists thus came to perceive the idea of drifting continents and plate tectonics as the unifying model of geologic phenomena.

Context

The development of the continental drift theory is a story of how science works in a period of paradigm revolution. This period began with Alfred Wegener's work in 1912 and ended with Harry Hess's discoveries in 1960.

The model Hess developed gave a new explanation of virtually all geologic phenomena at or near the surface of the earth. Within its field, the theory has had an impact comparable to that of Charles Darwin's theory of evolution in the field of biology. In essence, the theory of continental drift accounts for the global distribution of the continents, the birth and death of oceans, the distribution of earthquakes, volcanoes, and mountain ranges, and leads explorers to mineral and fossil-fuel deposits.

Bibliography

Engle, A. E., H. L. James, and B. F. Leonard, eds. *Petrologic Studies: A Volume in Honor of A. F. Buddington*. New York: Geological Society of America, 1962. The primary source for Hess's theory of continental drift.

James, Harold Lloyd. *Harry Hammond Hess*. National Academy of Sciences Biographical Memoirs, vol. 43. New York: Columbia University Press, 1973. This biographical sketch yields insight into Hess and his synthesis of the oceanic data to form a new paradigm.

LeGrand, H. E. *Drifting Continents and Shifting Theories*. New York: Cambridge University Press, 1988. Traces the development of the continental drift paradigm and the work of people who contributed to the data collection and debate.

Scientific American. *Continents*. San Francisco: W. H. Freeman, 1973. An excellent resource for landmark papers from 1952 to 1970. Illustrates the progression of the geologic "revolution."

Silver, Paul G., et al. "Coupling of South American and African Plate Motion and Plate Deformation." *Science* 279, 5347 (January 2, 1998): 60-64.

Sullivan, W. *Continents in Motion: The New Earth Debate*. New York: McGraw Hill, 1974. Well written and illustrated with drawings and photographs. Develops like a detective story rather than a textbook.

Young, P. *Drifting Continents, Shifting Seas*. New York: Impact Books, 1976. An introductory book to the story of continental drift. The emphasis is on the researchers involved and the chronological development of an idea.

Richard C. Jones

Cross-References

Continental Crust, 78; Earth's Crust, 179; Earth's Mantle, 195; Ocean Basins, 456; Plate Tectonics, 505; Rock Magnetism, 542; Subduction and Orogeny, 615; Supercontinent Cycle, 623; Transform Faults, 638.

Continental Glaciers

Continental glaciers once covered much of northern North America and Europe, but now only Greenland and Antarctica have such huge masses of permanent ice and snow. Because continental glaciers are so large, they affect the climate of large regions outside their boundaries by cooling of air and water temperatures. Continental glaciers of past ice ages have produced a wide variety of erosional and depositional features in northern latitudes.

Field of study: Glacial geology

Principal terms

ABLATION: the result of processes, primarily melting (evaporation is also involved), that waste ice and snow from a glacier

COLD POLAR GLACIER: a glacier that is below the pressure melting temperature of ice throughout

EQUILIBRIUM LINE: the line or zone that divides a glacier into the upper zone of accumulation and the lower zone of wastage

ISOSTATIC ADJUSTMENT: the adjustment of the crust of the earth to maintain equilibrium by subsiding when loaded and uplifting when unloaded

MASS BALANCE: the annual equivalency between the total amount of incoming accumulated snow and outgoing meltwater and evaporation

PLEISTOCENE ICE AGE: the time from about 2 million years ago to about ten thousand years ago, during which large continental glaciers covered much of northern North America, Europe, and other parts of the world

PRESSURE MELTING TEMPERATURE: the temperature at which ice will melt under a specified pressure; under pressure, water can exist even at temperatures below freezing

RESPONSE TIME: the time it takes a glacier to respond to changes in its mass balance, which it will generally do by advance or retreat of its terminus (tip)

WARM TEMPERATE GLACIER: a glacier that is at the pressure melting temperature throughout

Summary

Continental glaciers are ice sheets of huge extent. These continent-sized masses of ice overwhelm nearly all the land surface at their margins. Modern continental ice sheets occur only in Greenland and Antarctica and comprise nearly 96 percent of all glacier ice on the earth. From about 2 million to 10 thousand years ago, during the Pleistocene glacial period of the world's geological history, continental glaciers spread over much of northern North America and Europe.

Continental ice sheets tend to create their own weather—weather that is naturally favorable for glaciers. A huge ice sheet can even have worldwide climatic effects. The greatest part of the world's ice in Greenland and the Antarctic occurs in high latitudes characterized by very low winter temperatures, low summer temperatures, small annual precipitation, and minimal ablation. It does not snow much, but it is so cold that what falls tends to remain for a very long time. As a result of all these factors, such ice masses are relatively inactive and stable. Greater movement activity is noticeable in areas of less extreme cold and moderate precipitation.

Large ice sheets are complex and consist of several domes from which the ice flows radially to the ice margin or to broad interdome saddles, where the ice flow diverges downslope. The location of ice domes and ice saddles determines the flow path of ice on an ice sheet, but such features can change location over time as an ice sheet grows or shrinks in size. Continental glaciers are rarely more than 3,000 meters thick. Ice does not have the strength to support the weight of an appreciably thicker accumulation. If more ice is added by increased precipitation, the glacier simply flows out from the domal centers of accumulation more swiftly. Also, as the pressure at the base increases sufficiently, basal melting occurs that further decreases ice thickness.

Although snow accumulates over much of a continental ice sheet and gradually transforms into granular firn (snow that has survived a melt) and finally to glacial ice, it is from the highest interior domal areas that the main ice streams flow. The ice surface is built up at the interior and slopes outward on all sides. The glacier moves down and out in all directions. At the edge of the sea, if the area is mountainous, such ice sheets will break up into narrow tongues resembling valley glaciers that wind through the mountains to the sea. Otherwise, the ice may end in giant ice ramparts that calve, or break off, icebergs into the ocean, or as floating ice shelves over large continental embayments.

Ice shelves occur at several places along the margins of the Greenland and Antarctic ice sheets, as well as locally in the Canadian Arctic islands. They are nourished by ice streams flowing off the land, as well as by direct snowfall on their surface, and perhaps by freezing of the sea ice on their undersides as well. The largest ice shelves extend hundreds of kilometers seaward from the coastline and can reach a thickness of at least 1,000 meters. The Ross Ice Shelf in Antarctica, for example, is about as large as the state of Texas.

The continental ice sheet of Greenland covers an area of about 1.8 million cubic kilometers, or about 80 percent of the total land area there.

The ice sheets of Greenland represent one of earth's two remaining continental glaciers. *(Archive Photos)*

The volume of the ice is about 2.8 million cubic kilometers. In cross section, the ice has the shape of an extremely wide lens, convex on both the smooth upper surface and the rough lower boundary with the ground. The center of the ice sheet is more than 3,200 meters thick. The greatest area, 18 percent, occurs between 2,440 and 2,740 meters and 6.5 percent lies between 3,050 and 3,390 meters. The equilibrium line, or boundary between the upper accumulation area and the lower wastage area, is at about 1,400 meters, and 83 percent of the total area lies in the accumulation zone. Measurements of ice velocity in Greenland show that the main ice cap advances at approximately 10 to 30 centimeters per day, but the outlet glaciers near the coast can move as fast as one meter per hour. In some places, the ice can actually be seen to move.

The continental glaciers that once covered much of North America, Europe, and elsewhere during the Pleistocene ice age rivaled Antarctica in size and also exerted a widespread effect when they were at their maximum extent. The Laurentide ice sheet of North America at its largest covered an area similar in size to the present Antarctic ice sheet, but the Scandinavian ice sheet of Europe covered only about half this area during the maximum of glaciation.

The Laurentide and Scandinavian ice sheets did not extend to the seas in their southern limits as do the ice caps of Greenland and Antarctica. Instead, these continental glacial systems covered a large part of the northern continents and caused a number of significant peripheral changes of the regional physical setting outside their limits. The weight of the ice depressed the ground surface isostatically, just as it does in Greenland and Antarctica, so that the land sloped toward the glacier. Consequently, glacial lakes formed in the depressions along the ice margins, or arms of the ocean invaded the depressions. The preglacial drainage systems were greatly modified, as the streams that flowed toward the ice margins were impounded to form lakes. Later, as the ice dams melted away and the land again rose isostatically after the weight of the ice was removed, many such lakes and arms of the sea drained away, leaving behind extensive lake and marine clays and silts.

Glaciers produce many different erosional and depositional features as a result of their interaction with the ground beneath or at the front of the ice. During the Pleistocene ice age, the great thick continental ice sheets moved over the flat low-lying areas of the northlands, removing the existing soils and eroding up to several meters into the bedrock. As a result, many thousands of square kilometers of northern North America and Europe have little or no soil cover and the effects of the former glaciation are seen everywhere in the polished and grooved fresh bedrock. Continental glaciers are so large that they can produce widespread or massive abrasion and streamlined forms, as well as erode huge lake basins, all of which are exposed after the ice melts away. Large parts of the north-central United States and much of central and eastern Canada have such landforms plentifully displayed.

In many areas that lie near the outer edges of former continental ice sheets, the land surface has been molded into smooth, nearly parallel ridges that range up to many kilometers in length. These forms resemble the streamlined bodies of supersonic aircraft and offer minimum resistance to glacier ice flowing over and around them. The most common of these forms is the drumlin, which is a smooth, streamlined hill or ridge consisting of glacially deposited sediment that is elongated parallel with the direction of ice flow. Some drumlins are composed of contemporaneously deposited and smoothed sediment; others are of older sediment that was eroded long after its initial deposition. Bedrock can also be eroded in this fashion. In some places with steeply rising mountains that were overridden by continental ice, the up-ice side of such mountains will be streamlined and smoothed by the ice abrasion, whereas the down-ice side will be plucked and quarried into a rough and jagged cliff. Such asymmetrical mountains are called flyggberg ("flying mountains"), or stoss-and-iee, topography. Smaller such forms a few meters in height are referred to as roche moutonnée, or "wig-shaped" rocks, after their fancied resemblance to the curls of the smooth and powdered periwigs of the eighteenth century.

Continental glaciers erode by freezing on of blocks to the base of the ice and by abrasion with these blocks against the bedrock further along in the ice stream. The processes produce large quantities of sedimentary debris. Sediments deposited by continental glaciers can be more than 300 meters thick, so that they blanket most of the preglacial topography upon which they rest, thus modifying, disrupting, and obliterating previously established drainage systems.

Most of the rock debris that is transported by glaciers is deposited near the terminus, where melting dominates. The material accumulates as a moraine ridge marking the former front edge of the glacier. As a glacier retreats from an area by backwasting, it may deposit a series of recessional moraines in loops or ridges one behind the other. The sediment of such recessional and terminal moraines is made up of a jumbled mixture of all sorts of rock materials, ranging from clay to boulders, that are collectively referred to as till.

As continental glaciers develop large quantities of meltwater from wastage of the ice, streams of meltwater begin to flow in tunnels within and beneath stagnant ice and carry a large load of sediment in their ice-walled beds. When the ice melts away, such bedloads can be deposited beneath the ice to form a long sinuous ridge called an esker.

The plentiful meltwater at the terminus of a continental ice sheet also will flow through the terminal moraine landforms and erode away much of the till. This debris will be transported and reworked by the meltwater before being sorted into different sizes and deposited by rivers beyond the ice margins. The resulting layers or strata of sorted sediment can be spread out in broad outwash plains pockmarked with kettle holes where blocks of ice have later melted away. Both the unsorted, unstratified, or unlayered till and the sorted and stratified outwash materials are collectively called drift, the name being a heritage of the time when such materials were thought to have drifted to their present locations during Noah's flood.

Methods of Study

Continental glaciers are studied by scientists who measure climate controls on the glaciers, and glacier controls on climate. Continuous weather records are maintained in remote outposts across Greenland and Antarctica. As the ice and snow build up over the years, their layers contain sensitive records of precipitation, temperature, and atmospheric gases of past years. Snow pits and ice cores drilled from deep within continental glaciers provide such mass balance records for tens of thousands of years back into the past.

The distribution of all glaciers is a complex function of the distribution and amount of accumulating snow in winter and of the nature and amount of incoming energy and its utilization at the glacier surface during the summer ablation season. This period is when snow and ice are lost through melting, evaporation, wind erosion, and other forms of wastage. Thus, an ice mass exists where the average amount of snow-accumulation mass equals or exceeds that mass depleted in the summer. In most cases, the snow accumulation areas are higher elevation sites and the wastage areas are lower. Flow of the glacier ice transports ice from the accumulation to the wastage sites. The overall, or net, mass balance of a glacier is thus positive if gaining mass, negative if losing more mass than gaining, and exactly balanced if the annual increase in glacier mass is exactly balanced by the annual loss. Mass balance studies are an integral part of glaciological studies and allow analysis of response time to climate change.

Small glaciers can respond to climate change that affects mass balance fairly quickly, but there is a lag between the change in mass balance and the glacier response in retreating or advancing. Because of this lag, glaciers can respond out of phase with a present climate, and they do so as a function of size; the bigger the glacier, the longer the response time to a climate event that happened long ago. Response times to mass balance changes of continental glaciers are thought to be on the order of thousands of years.

Temperature of ice in continental glaciers is another important measurement and has been much collected from boreholes in the past few decades. Temperature in continental glaciers depends upon ice thickness and pressure, atmospheric climate controls, and bedrock heat flow. Highly variable ice temperatures can have a profound effect upon its interaction with the substrate upon which it rests. Basal meltwater allows a glacier to slip and slide over the bedrock beneath, whereas if the ice is frozen down to the bedrock, all flow motion must be entirely up within the ice itself. The presence or absence of meltwater at the base of the ice, therefore, can control depth of erosion and quantity of deposition.

At high latitudes and altitudes, where the mean annual air temperature is below freezing, ice temperature drops below the pressure melting point, which is the temperature at which ice can melt at a particular pressure. Ice is an unusual substance that, when enough pressure is applied to it, will melt as when an ice skater glides over the ice on the thin film of lubricating water produced by the skate blades. Polar, or cold, glaciers tend to have all ice temperatures below the pressure melting point, and so little or no seasonal melting can occur. Even so, the very cold ice in parts of Antarctica is thick enough to cause very high pressure at the glacier bed so that the ice reaches the pressure melting point and forms basal meltwater in spite of a temperature well below zero degrees Celsius.

Temperate or warm glaciers have ice at the pressure melting point everywhere, so that meltwater and ice can exist together in equilibrium throughout such glaciers. During the Pleistocene ice age, some of the continental glaciers may have been cold-based and frozen to their beds in the north and warm-based with plentiful meltwater in the south.

Large, flat-topped or tabular icebergs produced by marginal breakup of floating ice shelves can form huge ice islands. Some of these drifting in the Arctic Ocean were used as remote scientific stations by polar scientists in the 1960's. In early October, 1987, an iceberg measuring approximately 6,500 square kilometers, or twice the size of Rhode Island, broke free of the Ross Ice Shelf. Such colossal icebergs can pose a serious threat to shipping and are monitored closely by satellites and other observation techniques. It has been suggested that at some time in the future such huge bergs be wrapped in plastic and towed to arid areas such as Saudia Arabia as a source of fresh water.

Context

Continental glaciers are significant in two chief ways: The ice of Greenland and Antarctica is both a valuable record of past climate and a control of much high-latitude weather. The Antarctic continent contains by far the largest amount of snow and ice in the world. Its ice sheet measures about 12.6 million square kilometers, an area about seven times that of Greenland's ice sheet, or one and a half times the area of the United States. At present, 85 percent of the world's ice is in the Antarctic and 11 percent in Greenland. The overwhelming size of the Antarctic ice sheet compared with any other ice mass alters the whole climate and life of the Southern Hemisphere.

Ice thicknesses in Antarctica exceed 3,000 meters in much of the central area. The ground beneath the ice is mountainous over much of the continent. In some places, the rock floor beneath the ice is more than 2,500 meters below sea level because it has been pushed down by the great weight of the ice overhead. On many maps, the Antarctic ice sheet seems to be a single vast glacier, but in actuality it consists of two large ice sheets that meet along the lofty Transantarctic Mountains. The East Antarctic Ice Sheet is the largest of the two and is the ice cap upon which the South Pole is located. The smaller West Antarctic Ice Sheet overlies numerous islands. The combined estimated volume of about 24 million cubic kilometers of the two ice sheets would be sufficient to raise world sea level by nearly 60 meters if the ice were to waste away entirely. Because the bedrock base of Antarctica would then be unloaded from the heavy ice, however, it would rise in crustal, or isostatic, adjustment, increase the nearby ocean depth in linked compensation, and thus reduce the sea level rise to only about 40 meters, which could still make significant changes in the world environment.

The Antarctic ice sheet seems largely frozen to its bed in the peripheral parts but is at the pressure melting point in the interior. There is a possibility that the ice sheet could reach the pressure melting point at its base over a wide area if the thickness of the ice sheet were to increase. The ice sheet might then suddenly surge forward on basal meltwater and extend rapidly from its existing margins at about 60° south latitude to as far north as 50° south latitude. This change in position would cause a sudden worldwide rise of sea level of about 33 meters. The cooling resulting from such an increase of Southern Hemisphere ice could lead to a renewed northern ice advance that could lead to a new ice age. Such ideas about continental glaciers help scientists guide research programs about potentially hazardous situations that humankind might have to face in the future. Few expect such hazards to occur any time soon, but such forewarnings about possible problems enable better overall contingency planning.

Bibliography

Andrews, J. T. *Glacial Systems: An Approach to Glaciers and Their Environments.* North Scituate, Mass.: Duxbury Press, 1975. This work contains a wealth of information on glacial classification, formation, mechanics, mass balance, and landforms. Andrews has spent many years in the Arctic working on glaciers and their effects.

Embleton, C., and C. A. M. King. *Glacial Geomorphology.* New York: John Wiley & Sons, 1975. The chief English language source on glaciers past and present, this work has great detail on all aspects of glaciers and their effects. The language is reasonably nontechnical, but the work is quite thorough. Dominant coverage is on landforms produced by past glaciation.

Eyles, N., ed. *Glacial Geology.* Oxford, England: Pergamon Press, 1983. This book of edited papers is almost entirely devoted to the effects of glacier erosion and deposition on the land. Engineering applications are stressed and include foundation engineering, road construction, dam and reservoir construction, and groundwater studies in glaciated areas.

Flint, R. F. *Glacial and Quaternary Geology.* New York: John Wiley & Sons, 1971. This famous work is by a master in the field. It is a thoroughly scholarly, dependable, and detailed compilation and discussion of data, features, and processes related to glaciation and, especially, the worldwide history of past glaciations.

Paterson, W. S. B. *The Physics of Glaciers.* Elmsford, N.Y.: Pergamon Press, 1981. This work is one of the best available references dealing with the basic mechanics of glacier formation, nourishment, structures, flow, and behavior. Despite the considerable amount of rigorous mathematical material, a discerning nonprofessional reader can glean a world of useful reliable information on glaciers from the intervening parts of the book.

Schultz, Gwen. *Glaciers and the Ice Age.* New York: Holt, Rinehart and Winston, 1963. Though a short book, it is written for the layperson. It provides a good review of basic aspects of glaciers; it also emphasizes their impact on human interests and activities.

Sharp, R. P. *Living Ice: Understanding Glaciers and Glaciation.* New York: Cambridge University Press, 1988. A most readable and detailed book on glaciers, it is very well illustrated. It also includes a comprehensive glossary with more than three hundred entries.

Sowers, Todd, and Michael Bender. "Climate Records Covering the Last Deglaciation." *Science* 269, no. 5221 (July 14, 1995): 210-215.

Weiner, Jonathan. "Glacier Bubbles Are Telling Us What Was in Ice Age Air." *Smithsonian* 20 (May, 1989): 78-87. This article details the use of gas bubbles trapped in glacial ice to find out what the atmosphere was like long in the past. This information is used to understand ice-air interactions in order to better find out what is happening to the present climate of the earth and what the future holds in store. A very readable and popular account.

Williams, R. S. *Glaciers: Clues to Future Climate.* Reston, Va.: U.S. Geological Survey, 1983. This pamphlet provides a brief and popular summary on glaciers, causes of glaciation, glacial history, and climatic relationships.

John F. Shroder, Jr.

Cross-References

Alpine Glaciers, 8; Glacial Landforms, 281.

Continental Growth

Continents are believed to have increased in size during earth history by accretion of additional crustal material along their margins. This process has played a significant role in the formation of valuable mineral deposits such as gold, silver, copper, gas, and oil.

Field of study: Tectonics

Principal terms

FAULT: a fracture in rock strata with relative displacement of the two sides
FOLD: an upward or downward bend in layered rock strata
GEOSYNCLINE: an elongate subsiding trough in which great thicknesses of sedimentary and volcanic rocks accumulate
GRANITE: a light-colored crustal rock produced by the underground cooling of molten rock
LATERAL ACCRETION: the process by which crustal material is welded to a shield by horizontal compression
MANTLE: the earth's 2,900-kilometer-thick intermediate layer, which is found beneath the crust
PLATE TECTONICS: a theory that describes the earth's outer shell as consisting of individual moving plates
SHIELD: a continental block of the earth's crust that has been stable over a long period of time
SUBDUCTION: the process by which one crustal plate slides beneath another as a result of horizontal compression

Summary

Geologists believe that the continents have increased in size during geologic time by the accretion of additional material along their margins. This additional material usually consists of younger rocks deposited in a deeply subsiding belt (known as a geosyncline), which is then welded to the continent by compressive forces. In some cases, the additional material may represent portions of a preexisting continent—or even an entire continent itself—that has been "drifted in" by the mechanism known as plate tectonics.

The idea of geosynclines dates back to the work of two nineteenth century American geologists. In the 1850's, James Hall pointed out that the crumpled strata of mountain ranges along the continental margins were thicker than the equivalent strata in the continental interiors, and in 1873 J. D. Dana suggested the term "geosyncline" (literally, "great earth down-fold") for elongated belts of thick sedimentary rocks deposited along the continental margins. Further geologic fieldwork, primarily in the Alps and in the Appalachians, showed that the sedimentary rocks of the geosynclines had been deformed by compressive forces emanating from the ocean basins. By the 1950's, the generally accepted picture of continental growth was that of a stable continental interior, called the shield or craton, surrounded by increasingly younger belts of deformed rock. Each belt was believed to represent a geosynclinal sequence that was deposited in a bordering trough and then welded to the shield by lateral accretion. Yet, no satisfactory mechanism for the source of the compressive forces from the ocean could be discovered.

In North America, the central shield is called the Canadian Shield. Its exposed portion occupies the eastern two-thirds of Canada, the U.S. margins of Lake Superior, and most of Greenland. The Canadian Shield is the largest exposure of Precambrian rocks in the world, consisting predominantly of igneous and metamorphic rocks. There is also a buried portion of the Canadian Shield extending westward to the Rocky Mountains and southward to the Appalachian Mountains, the Arbuckle Mountains in Oklahoma, and into Mexico. This buried portion of the shield has a thin cover of largely Paleozoic sedimentary rocks deposited in shallow transgressing seas. These Paleozoic rocks are still flat-lying except where they have been gently warped into broad domes and basins. Along the margins of the Canadian Shield, four belts of deformed sedimentary rocks represent the former geosynclines. Geologists named the deformed belt on the eastern side of the shield the Appalachian geosyncline, the belt on the south side the Ouachita geosyncline, the belt on the west side the Cordilleran geosyncline, and the belt on the north side the Franklin geosyncline. Because of compressive forces emanating from the ocean basins, overturned folds and thrust faults are present in all four geosynclinal belts.

Similar patterns of continental growth are found elsewhere in the world. Each continent has at least one shield. These include shields in South America, Africa, northern Europe, Siberia, eastern Asia, India, Australia, and Antarctica. The remnants of geosynclinal belts are located adjacent to these shields. The most famous of these geosynclinal belts is the Tethyan geosyncline, found along the southern margin of Europe and Asia. The present-day Alps and Himalayas have risen out of this geosynclinal belt.

During the 1960's, the concept of plate tectonics gradually emerged as the result of the work of oceanographers trying to explain the origin of the planet's major sea-floor features. These features include midocean ridges rivaling the largest mountain ranges on earth and volcanic island chains with associated deep oceanic trenches that rim the Pacific. The plate tectonics theory has revolutionized not only the field of oceanography but also the

103

field of geology. According to the plate tectonics theory, the surface of the earth is covered by a series of rigid slabs or plates that are capable of moving slowly over the earth's interior. Geologists recognize six or seven major plates, each one usually containing a continent. The plates are presumed to behave as separate units, and where plates jostle each other, intense geologic activity occurs along their boundaries.

Three different types of activity are believed to take place at plate boundaries. Divergence occurs where two plates are moving apart from each other. The result of the divergence of two continental plates is believed to be the formation of a new ocean basin, a process referred to by geologists as sea-floor spreading. When two plates are moving toward each other, the result is convergence. In this case, three possibilities arise, depending on the nature of the plate boundary. If two continental plates converge, the intervening oceanic sediments are believed to be compressed into a new mountain range, such as the Himalaya. If a continental plate runs into an oceanic plate, the oceanic plate is believed to slide beneath the overriding continental plate, producing a deep oceanic trench. Geologists refer to this process as subduction. Finally, one oceanic plate may override another oceanic plate, producing a trench and adjoining volcanic island arc. The third type of movement found along plate boundaries occurs when two plates slide

The Alps were produced by compressive forces caused by the northward drift of Africa. *(Robert McClenaghan)*

past each other horizontally, just as trains pass each other in opposite directions on adjacent tracks. Such movement may proceed continuously or in a series of abrupt jerks, depending upon the amount of friction encountered along the plate boundary. The abrupt jerks result in the type of earth movement known as earthquakes, and these are particularly associated with the famous San Andreas fault in California.

The underlying causes for plate movements are not well understood. Geologists speculate that convection cells of rising and sinking material in the mantle (the earth's intermediate layer) may carry the plates slowly along. Nor is the mechanism by which rigid plates are able to slide across the earth's interior well known either. Presumably there is a "plastic" layer in the upper mantle that provides the necessary lubrication for the crustal plates to move.

The significance of plate tectonics for the concept of continental growth has been recognized by scientists. Because most of the jostling takes place at plate boundaries, that is where they find the downwarped geosynclinal belts, earthquakes, volcanic activity, and recently formed mountains. On the other hand, the stable plate interiors are places where little jostling takes place and thus where the quiescent continental shields are located.

An example of a present-day geosynclinal belt that is being squeezed between two converging shields is the Tethyan geosyncline. This geosyncline is believed to have originated between the Eurasian and African shields during the Mesozoic era. It must have resembled a broad tropical seaway extending from the Caribbean eastward through the Mediterranean and the Himalaya to Indonesia on the borders of the Pacific. Thick marine sediments accumulated on the floor of this seaway, and they are preserved today as richly fossiliferous limestone sedimentary rocks. During Cenozoic time, geologists believe, the convergence of several continental plates initiated the destruction of this seaway. The Indian subcontinent, for example, is believed to have drifted northward until it collided with Asia, producing the Himalaya, the highest mountain chain on earth. A second collision occurred as the Arabian plate (a minor subplate) drifted north to collide with Asia Minor, forming the Zagros and other mountains. Finally, Africa is believed to have drifted northward, resulting in the compressive forces that have produced the Alps. The result of these collisions is the near obliteration of the old Tethyan seaway. The only relics of it that survive are the Caspian Sea, the Black Sea, and the Mediterranean Sea. In addition, the sedimentary rocks that accumulated in the geosyncline have been folded and thrust northward against the Eurasian continental platform. Scientists have a clear picture, therefore, of continental growth taking place as a result of crustal plates moving toward each other, with the intervening sediments being welded to the shields by lateral accretion.

As indicated earlier, continental growth can also result when a portion of a preexisting continent, or even an entire continent, collides with another continent. An example of such a collision is believed to be provided by the formation of the Ural Mountains at the end of the Paleozoic era. Geologists now believe that these north-south trending mountains, which have been eroded down to their roots, were formed by the compression of sediments deposited in a seaway lying between Europe and Asia. In other words, the present-day continent of Eurasia, which is twice as large as any other continent, was once two separate continents.

In the cases of the destruction of the Tethyan geosyncline and the formation of the Ural Mountains, the role of plate tectonics seems clear because the plates that did the moving can be identified. Sometimes, however, relationships are not so apparent—for example, in the case of the deformation of North America's Appalachian geosyncline and the thrusting of its Paleozoic sediments against the shield. No continental plate lies along the east coast of North America that might account for the compression. As a result, students of plate tectonics have postulated an elaborate scenario that involves two stages. They assume, first, the closing of the Atlantic Ocean at the end of the Paleozoic era as a result of the collision of Europe with North America, and then, that Europe moved eastward again, resulting in the reopening of the Atlantic because of sea-floor spreading.

The deformation of the Cordilleran geosyncline along the west coast of North America offers no such problems. It can be explained by the collision of the North American plate with the Pacific Ocean plate. According to plate tectonics, the Pacific Ocean is a separate plate even though it lacks a continent. Thus, the deformation of the Cordilleran geosyncline has resulted from the sliding of the Pacific Ocean floor beneath the North American continent in the process known as subduction.

Methods of Study

Scientists have studied the subject of continental growth in many ways. Foremost among them has been field investigations in the rock strata found along the margins of the shields. Using the fossils contained within these rocks, as well as radioactive dating, geologists have pieced together a detailed history of geosynclinal accretion. By analyzing the geometry of the folds and faults, scientists have also been able to infer the direction from which the compressive forces came.

An example of such geologic field investigations is seen in the deciphering of the rocks of the Canadian Shield. These rocks constitute the largest outcrop of Precambrian strata exposed anywhere in the world today, and to early workers they appeared to be a hopeless tangle of similar-looking igneous and metamorphic rocks. After years of painstaking research, however, scientists have been able to identify an orderly sequence of mappable rock units within the Canadian Shield, so that at least four distinct cycles of deposition and mountain making during Precambrian time are recognized.

Another way in which scientists have approached the subject of continental growth is by examining the rock types that compose the shields and ocean floors. They have found that the continental rocks are largely granitic and are rich in silica, aluminum, and potassium. These rocks also have a slightly lower average density than do the rocks underlying the ocean basins, and they stand higher, as if both were floating on interior layers of the earth. By contrast, the rocks of the ocean floors consist of slightly heavier basalt lava and related volcanic rocks.

To everyone's surprise, the granitic rocks of the continents have not proved to be the oldest rocks on earth. Radioactive dating indicates that slivers of sea-floor rocks incorporated in the granites claim this distinction. Thus, scientists have concluded that the shields do not represent parts of the earth's original crust but have been built up through time by a process of lateral accretion. Their granites may have come from the reworking of sea-floor rocks.

A third way in which scientists are investigating the subject of continental growth is through detailed study of the sea floor itself. This study began with the Deep Sea Drilling Project in 1968, using the *Glomar Challenger,* a drillship that was retired in 1984. Subsequently the program continued under the name Ocean Drilling Project, utilizing the *JOIDES Resolution,* a 143-meter drillship which is capable of recovering samples of rock and sediments from depths of 9,000 meters under water. The findings of these ships have been of great significance for the study of plate tectonics.

All evidence seems to point to a very young age for the ocean basins—less than 200 million years old, which is less than one-twentieth the earth's presumed age of 4.7-5.0 billion years. Furthermore, it appears that the Atlantic and Indian Oceans are widening, while the Pacific is shrinking, which, if true, means that the crustal plates will eventually collide, thus providing the mechanism for further continental growth.

Context

The same processes that have thrust former geosynclinal belts against the shields have also produced rich mineral deposits in the resulting mountain chains. These mineral deposits fall into three categories: metals, such as gold, silver, and copper; nonmetallic deposits, such as certain abrasives, gemstones, and the building stones granite, marble, and slate; and the important energy resources petroleum, natural gas, and anthracite coal.

A good example of a metal deposit found in the deformed rocks of a former geosyncline is California's famous Mother Lode. This zone of gold veins, which is more than 200 kilometers long but barely a kilometer wide, can be traced along the western slopes of the Sierra Nevada, a mountain range that has risen out of the former Cordilleran geosyncline. The gold discoveries—which attracted the "forty-niners" to California and led to the rapid growth of San Francisco and neighboring cities—were nuggets and flakes of gold derived from these veins and washed down into the sand and gravel deposits of rivers at the foot of the mountains. The early settlers realized the gold was from a source upstream, and they called this source the Mother Lode (literally, "parent vein"). Eventually, the settlers traced the streams up to their headwaters in the Sierra Nevada and discovered the Mother Lode itself.

The fabulously rich oil deposits of the Middle East are another example of an economic resource related to continental growth. The Tethyan seaway, which stretched from the Caribbean to the Pacific during Mesozoictime, was the site of extensive deposits of thick limestone sedimentary rocks. These limestones are now oil-bearing and have been caught in the closing vise between the northward-moving Arabian plate and the portion of the Eurasian continent known as Asia Minor. Because of this compression, the limestones have been shaped into a series of gently undulating folds. Migrating oil has been trapped in the crests of the upfolds (technically known as anticlines), where it is obtained by drilling wells down into the anticlinal structures.

In 1988, more than 40 percent of the world's oil imports came from the Middle Eastern oil fields, a situation that has enabled the Arab nations to wield a political and economic influence far out of proportion to their geographic size or number of inhabitants. A dramatic example of this influence was provided in the 1970's, when these nations paralyzed the free world's economic system with an oil embargo. Even though the major consumers of Middle Eastern oil were Western Europe and Japan, the dislocation in world oil supplies had severe consequences in the United States as well. Americans were asked to turn down their thermostats, the nationwide speed limit was reduced to 55 miles per hour, and automobile companies were told to improve the gas mileage of their cars. As the ripple effects of the oil shortage spread through the United States' economy, a recession was triggered that cost people their jobs and set off a major stock market decline.

Bibliography

Cloud, Preston. *Oasis in Space: Earth History from the Beginning.* New York: W. W. Norton, 1988. A definitive, one-volume synthesis of earth history by a distinguished geologist and gifted writer. Illustrated with more than three hundred maps, photographs, and diagrams. Continental growth is a major theme, with heavy emphasis on the Precambrian era. Suitable for college-level readers and laypersons with some technical background.

Dott, Robert H., and Roger L. Batten. *Evolution of the Earth.* 4th ed. New York: McGraw-Hill, 1988. A well-written and well-illustrated text. Presents an up-to-date account of earth history from the viewpoint of plate tectonics and explains the analytical methods used for obtaining this knowledge. The emphasis is on world synthesis, with coverage of all the continents. The Tethyan seaway is also documented. Suitable for college-level readers.

Grocott, John, and Michael Brown. "Mechanisms of Continental Growth in Extensional Arcs: An Example from the Andean Plate Boundary." *Geology* 22, no. 5 (May, 1994): 391-395.

King, P. B. *The Evolution of North America.* Rev. ed. Princeton, N.J.: Princeton University Press, 1977. The leading work on the overall geology of the North American continent. Originally written in the 1950's, it reflects the ideas then prevalent but has been revised to incorporate the concepts of plate tectonics. Suitable for college-level readers and interested laypersons.

Scientific American. *The Dynamic Earth.* New York: W. H. Freeman, 1983. A collection of eight outstanding articles written by leading scientists who have been involved in the development of the plate tectonics theory as well as the unified view of the earth. Excellent color photographs, maps, and line drawings. Suitable for college-level readers and the interested layperson.

Sullivan, Walter. *Continents in Motion: The New Earth Debate.* New York: McGraw-Hill, 1974. As science editor of *The New York Times,* the author has the ability to make this introduction to plate tectonics understandable and exciting for the average reader. He uses the historical approach, which gives the reader the feeling of participating in the various discoveries. Well written although photographs are sparse.

Windley, Brian F. *The Evolving Continents.* New York: Wiley, 1977. A good source book for the serious student of continental growth. Data are provided for selected shields and fold belts throughout the world, with major emphasis on the Precambrian, Europe, the Alps, and areas outside North America. Photographs, however, are lacking. Suitable for college-level readers with some technical background.

Wyllie, Peter J. *The Way the Earth Works: An Introduction to the New Global Geology and Its Revolutionary Development.* New York: Wiley, 1976. A concise introduction to plate tectonics, suitable for high-school-level readers and the interested layperson. The supporting evidence for plate tectonics is explained in more detail than is customary in most textbooks. Many excellent diagrams have been especially prepared for this text in order to illustrate the basic concepts.

Donald W. Lovejoy

Cross-References

Continental Crust, 78; Oil and Gas: Distribution, 481; Plate Tectonics, 505; Subduction and Orogeny, 615.

Continents and Subcontinents

Continents are large land masses with elevations that are considerably higher than that of the surrounding crust. Subcontinents are smaller land masses that converged over time to form the large continents familiar today. Because of this, continents have a wide variety of terrains and landforms.

Field of study: Tectonics

Principal terms

CONTINENT: a large land area consisting of a variety of terrains

CRATON: a stable, relatively immobile area of the earth's crust that forms the nucleus of a continental land mass

CRUST: the thin layer of rock covering the surface of the earth; solid and cool, the crust makes up the continents and floor of the ocean and may be covered with thick layers of sediments

DENSITY: the weight per unit volume of a substance

MAFIC ROCKS: rocks that contain large amounts of magnesium and iron, found mainly in the oceanic crust and upper mantle

MANTLE: the region of the earth between the dense core and the thin crust; the mantle makes up most of the volume of the earth

OROGENIC BELT: an area where mountain-forming forces have been applied to the crust

PLATE TECTONICS: the process that causes the continents and large unbroken land areas within the oceanic crust, called "plates," to move slowly along with currents of rock in the upper mantle

SEDIMENTS: rocks and soil that have been eroded from their original position by forces of weather

SUBCONTINENT: an area of land that is less extensive in size and has a smaller variety of terrains than a continent

Summary

The word "continent" comes from the Latin *continere*, which means "to hold together." Continents are the large land masses composed of lighter rocks that ride on top of the more dense rocks in the mantle, somewhat like a cork in water. This results in areas on the earth's surface that are higher than sea level, producing dry land. The earth was not formed with these continents in place; a long, complicated process resulted in the formation of the land masses familiar today.

When the earth was first formed approximately 5 billion years ago, it was a molten ball composed of all the elements. The high temperature of the planet was the result of the heat released from several sources: the process that formed the planet; decay of radioactive elements; and intense meteoritic bombardment. This molten state allowed the different elements, and the compounds they form, to differentiate, or separate.

This differentiation process is similar to mixing different kinds of oil with water in a bottle. Shake up the bottle, and the different liquids will be mixed together. Let it sit, and they will begin to form layers. The water is densest and will form a layer on the bottom. Each of the different oils will then form a separate layer above the water, with the layer of the densest oil being on top of the water and the least-dense oil forming the uppermost layer.

Similarly, when the earth underwent differentiation, the densest material, mainly iron and nickel, sank toward the interior and formed the planetary core. Compounds and elements that were medium in density would lie on top of the denser core and form the layers of the planetary mantle. The least-dense compounds floated to the surface and eventually formed the crust, seawater, and atmosphere. These least-dense compounds consisted principally of the elements silicon, oxygen, aluminum, potassium, sodium, calcium, carbon, nitrogen, hydrogen, and helium, with lesser amounts of other elements.

The earth today has two kinds of crust: the heavier, thinner crust under the oceans and the thicker, lighter continental crust. The oceanic crust was created between 4.2 billion and 4.5 billion years ago, has an average density of 2.9 grams per cubic centimeter, and consists mainly of mafic rocks. Mafic rocks are made of minerals that consist mainly of magnesium and iron. The most common kind of mafic rock in the oceanic crust is basalt, a dark, hard stone. The dark maria on the face of the Moon are the result of basalt that was able to reach the Moon's surface after large meteor impacts. Beneath the crust is the upper mantle, which has a density of approximately 3.3 grams per cubic centimeter and consists of mafic rocks that contain an even larger percentage of magnesium and iron; hence, they are called "ultramafic." This layer formed at about the same time as the oceanic crust.

Mixed with these two layers were even lighter materials, mainly compounds of silicon, oxygen, and aluminum, but the high temperature of the planet would not allow these materials to start solidifying until about 4 billion years ago. When this occurred, the first continental rocks began to form, although they were being continuously broken up. The oceanic rocks were mainly basalt; the continental rocks were mainly granite with a density of about 2.7 grams per cubic centimeter. The cooling process was slowed by the formation of crystal structures within the oceanic crust and upper mantle that forced out certain rare-earth elements, including the radioactive elements. These elements had to go somewhere and are thus found concentrated in continental rocks. Continental granites contain about ten times as much uranium as the oceanic basalts and about a thousand times as much as the upper mantle rocks. Heat released by the decay of this concentration of radioactive elements helped keep the continental rocks molten longer than the oceanic rocks.

This period in the earth's history was also characterized by a large amount of volcanic activity; large chains of volcanoes formed archipelagoes of islands. As a result of plate tectonics, these islands would move around on the surface of the earth and would eventually be reabsorbed back into the earth's interior at subduction zones. Subduction zones occur where one plate in the earth's crust meets another. The denser, heavier plate will be forced under the other plate and into the upper mantle, where it will be melted and returned to the surface through volcanic activity. Given enough time, this process will completely recycle the oceanic crust; today there is none of the original oceanic crust remaining.

Sometime about 4 billion years ago, the intense meteoritic bombardment suddenly came to an end; the surface of the planet began to cool more

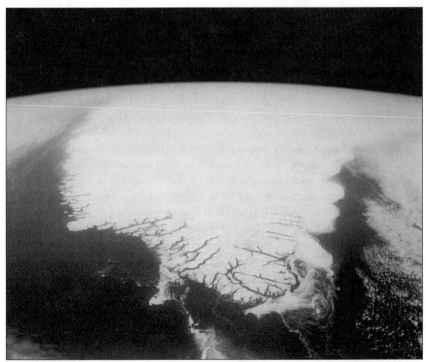

A satellite view of Greenland, a present–day example of a subcontinent. *(National Aeronautics and Space Administration)*

quickly, and more island chains were formed. However, the still-molten, lighter continental rocks would sometimes flow into large cracks, called "fissures," in the volcanic islands and provide them with additional buoyancy. When these islands, riding on the surface of a plate, reached the subduction zones, they would be too light to be subducted and would instead be scraped off by the other plate. As the other plate continued to move along, it would continue to scrape off more of the light islands; over time, a large amount of this lighter material would accumulate in front of the plate. Eventually, that plate would be subducted by another plate, and the light continental material it had collected would be added to any collected by the new plate. In this way, the size of continental crustal material grew until it was large enough to be a subcontinent.

A subcontinent is an area of land that is too large to be pushed simply by the movement of oceanic crustal plates. Instead, these large pieces of land ride on top of moving mantle rock deep beneath the surface. Yet while subcontinents are extensive, they are still not large enough to be considered continents. Modern examples of subcontinents include the island of Greenland and the Indian subcontinent. The importance of subcontinents is that they will eventually collide with one another. When they do, they can stick together, forming even larger areas of land and, eventually, continents. This process is called "accretion."

When subcontinents or continents collide, the event is something like an automobile crash in very slow motion. The two large bodies are moving and do not stop immediately; they continue to plow into each other, causing the rock to bend, fold, and lift, forming mountain ranges. Areas where this has occurred in the past are called "orogenic belts." The Himalayas are an example of the result of this process. The Indian subcontinent took millions of years to move from southeast Africa to its current position on the southern side of Asia. When it collided with Asia, the force was enough to raise a giant plateau, with the towering Himalayas on top. Large areas between the orogenic belts are called "cratons." The American Midwest between the Appalachian and Rocky Mountains is an example of a craton.

This movement of the land masses continues even after the formation of continents. The continents continue to ride on top of currents of rock in the mantle, slowly making their way across the earth's surface. Currently, continents are moving at a rate of about five to ten centimeters (two to four inches) per year, but this rate was faster in the past. Over millions of years, the continents have been able to move great distances, and the shape and distribution of continents in the past did not resemble the global features of today. At times, the continents have been together to form supercontinents that last for millions of years before breaking apart.

In addition to accretion, two other major forces at work on continents are volcanism and erosion. Volcanism is a result of plate tectonics, and most volcanism thus occurs along the edges of the land masses where subduction is occurring or where fault lines are found. Volcanism recycles material that has been subducted and adds to the mass of the continents.

Erosion, meanwhile, wears down landforms. Rain, ice, heat, wind, and flowing water all work to break apart the rocks and slowly wash the surface material away. Some of these sediments are washed to sea, while others collect in low-lying areas on the land. These sedimentary deposits can be several kilometers thick and will eventually turn into sedimentary rock.

Continents are complicated structures consisting of a patchwork of a large variety of landforms. The processes that formed the continents of today have been going on for billions of years and are continuing. The forces of plate tectonics are still slowly moving the continents, forming mountains, subduction zones, fault lines, and volcanoes, while the effects of weather work to wear down the land masses.

Methods of Study

Continents are vast in size and complexity. Likewise, the study of these land masses is also vast and complex. Much of the work in learning about continents is hampered by the fact that scientists can easily sample only the thinnest top layer of the crust; moreover, much of the evidence of past activity is destroyed through erosion. As a result, the study of the continents is a slow process involving many scientists using a large variety of techniques and instruments.

The most basic method of studying the continents involves studying the layers of rocks. Sometimes these layers can be seen from the surface, and other times scientists must use drills to remove core samples. Sometimes these layers lie flat, while other times they are at all angles. By studying these layers, it is possible to learn what they are made of, how they were made, and even when they were made. Eventually, it becomes possible to conclude that various layers are related and sometimes even constitute the same layer. For example, it is possible that a layer of rock found in North Dakota is identical to a layer of rock found in Nebraska, hundreds of miles away. In this way, geologists are able to build maps showing where these layers of rock can be found.

Maps like this then reveal much about the past of the land. If a type of rock found in the desert is made of material found only on the bottom of swamps, geologists can deduce that the desert was once a swamp and that the climate in the area was once different. It might even be possible to track the change from swamp to desert by examining the different layers, although sometimes the layers are destroyed through erosion. If the ages of the different layers are known, it is possible to build a storyline showing how the swamp changed to desert over a period of time.

Also, the angle of the layers tells about what happened to the land. If the layer is horizontal, then it has probably been undisturbed since it formed. If it is tilted or folded over on itself, then some kind of forces were applied to the rocks. Also, if a layer of rock is found on one continent and also found on another continent, then it can be concluded that the two continents were together when the layer was formed.

Rocks themselves also provide clues. What are the rocks made of? If they are sedimentary rocks, then the material in them existed in some other rocks before. Where were those rocks? If the rocks are basalts, it is possible to conclude that there was volcanic activity in the area at one time. The chemical structure can tell much about the temperatures and pressures to which rocks have been exposed over the years.

Today, instruments on spacecraft can make measurements over very large regions. This not only speeds up the process but provides new views and evidence not possible before.

All this information can be included into computer models in attempts to determine the forces at work in the formation of continents. In this way, clues can be found that were previously unsuspected. Once something is suggested by a computer model, scientists can investigate it to find scientific evidence to support or refute it. Through these and other processes, geologists gradually learn more and more about the earth and its history.

Context

Continents are obviously an important part of everyday life, as the vast majority of the world's population lives on them. Yet they are also important in less obvious ways.

Continental landforms such as mountain ranges strongly influence the weather, which affects nearly every aspect of daily life. The extent of these effects can be graphically demonstrated by comparing the weather over land masses to the weather over the oceans. People on continents may be having freezing weather, while people on islands at the same latitude may be enjoying tropical weather.

With ever-increasing populations, proper management of increasingly strained freshwater resources becomes more and more important. Knowledge gained through the study of continental landforms makes it possible to determine where underground water is and how fast it can be replenished when it is tapped. The shape of the surface of the land also helps to determine where water will flow. This not only helps people better to use the water that is available but also helps to minimize the damage caused by floods and droughts.

Continents also have a great effect on the oceans. The very existence of continents helps to determine the nature of oceanic currents, and most marine life lives in the shallow waters of the Continental Shelf. Increased understanding of the continents therefore increases the understanding of these and other phenomena.

How the land is used is also of great importance. The effects of human activities on the earth are vitally important. By improving the understanding of how continents are formed and how they fit within the global environment, scientists make it possible to predict the planetwide results of human actions with greater accuracy.

Likewise, the processes of the continents have a great impact on human life. By understanding the nature of earthquakes and volcanic eruptions—phenomena that result from plate tectonics—researchers can suggest ways to minimize the damage such events cause. Increased understanding also increases the ability to provide warning that eruptions or earthquakes are likely to occur.

Satellites in space make it possible to use natural resources more efficiently. Satellites can track the motion of sediments as they are washed away by storms, providing information that helps to manage valuable soil resources more efficiently. It is also possible to use scientific instruments from space to identify the nature of mineral deposits, simplifying the process of prospecting for necessary minerals.

Bibliography

Dott, Robert H., Jr., and Donald R. Prothero. *Evolution of the Earth.* 5th ed. New York: McGraw-Hill, 1994. Chapters 6 through 8 provide a good account of the early history of the planet and the formation of the crust. Chapters 10 and 11 discuss the formation of cratons and orogenic belts. A well-written, well-illustrated text suitable for college and advanced high-school readers.

Moores, Eldridge, ed. *Shaping the Earth: Tectonics of Continents and Oceans.* New York: W. H. Freeman, 1990. This collection of readings from *Scientific American* magazine presents a variety of well-written articles by experts in their respective fields. Individual articles are devoted to plate tectonics, mountain forming, and crustal formation, among other topics. Good illustrations and an index covering all articles is provided. Suitable for college and advanced high-school readers.

Stanley, Steven M. *Earth and Life Through Time.* New York: W. H. Freeman, 1986. Chapter 7 provides a good discussion of plate tectonics, while chapter 8 covers mountain building. The formation of the earth, the crust, and the continents is discussed in chapters 9 through 11. Well written and well illustrated; also has a good index and several appendices that provide additional details about specific topics. Suitable for college and advanced high-school students.

Taylor, S. Ross, and Scott M. McLennan. "The Evolution of Continental Crust." *Scientific American,* January, 1996, 274. A comprehensive discussion of the origin of the continental crust and the evolution of the continents. Suitable for high-school readers.

Weiner, Jonathan. *Planet Earth.* Toronto: Bantam Books, 1986. This companion volume to the Public Broadcasting Service television series *Planet Earth* covers many aspects of the earth. Well illustrated; suitable for high-school readers.

Christopher Keating

Cross-References

Archaeological Geology, 23; Continental Crust, 78; Continental Drift, 86; Continental Growth, 102; Earth's Composition, 162; Earth's Crust, 179; Earth's Differentiation, 188; Earth's Structure, 224; Elemental Distribution, 232; Igneous Rocks, 315; The Lithosphere, 375; Plate Tectonics, 505; Subduction and Orogeny, 615; Supercontinent Cycle, 623.

The Cretaceous-Tertiary Boundary

The Cretaceous-Tertiary boundary, 66 million years ago, is the junction between the Mesozoic and Cenozoic eras. This boundary coincides with a major extinction of marine and terrestrial organisms, the most conspicuous of which were the ammonoid cephalopods in the sea and the dinosaurs on the land. A 10-kilometer-diameter bolide that collided with Earth at this time has been invoked by some as the cause of these extinctions.

Field of study: Geochronology and paleontology

Principal terms

BOLIDE: a meteorite or comet that explodes upon striking the earth

CENOZOIC ERA: the youngest of the three Phanerozoic eras, from 66 million years ago to the present; it encompasses two geologic periods, the Tertiary (older) and the Quaternary

CRETACEOUS PERIOD: the third, last, and longest period of the Mesozoic era, 144 to 66 million years ago

ERA: a large division of geologic time, composed of more than one geologic period

EXTINCTION: the disappearance of a species or large group of animals or plants

FAMILY: a grouping of types of organisms above the level of a genus

MESOZOIC ERA: the middle of the three eras that constitute the Phanerozoic eon (the last 570 million years), which encompasses three geologic periods—the Triassic, the Jurassic, and the Cretaceous—and represents earth history between about 250 and 66 million years ago

STRATUM (pl. STRATA): a single bed or layer of sedimentary rock

TERTIARY PERIOD: the earlier and much longer of the two geologic periods encompassed by the Cenozoic era, from 66 to 1.6 million years ago

Summary

The Cretaceous-Tertiary boundary is a point in geological time located 66 million years before the present. It corresponds to the junction between the geological eras known as the Mesozoic, of which Cretaceous is the youngest subdivision, and the Cenozoic, of which Tertiary is the oldest subdivision. This boundary coincides with (and, using fossils, is recognizeds by) a major extinction of marine and terrestrial organisms. This extinction is not the most massive extinction in the history of life; the Paleozoic-Mesozoic extinction, 250 million years ago, holds that honor. The extinction at the end of the Cretaceous is, however, the most talked about extinction in earth history, because it was during this time that the dinosaurs disappeared.

When British paleontologist John Phillips coined the terms Mesozoic and Cenozoic in 1840, he already knew that they represented time intervals in earth history characterized by very different types of organisms. It was not until the beginning of the twentieth century, however, that paleontologists recognized the full significance of the boundary between the Mesozoic and Cenozoic eras. By 1900, about a century of scientific collecting and study of fossils demonstrated that many types of organisms had become extinct at or just before the Cretaceous-Tertiary boundary. This extinction thus ended what is popularly termed "the age of reptiles," setting the stage for the appearance and proliferation of the types of organisms that have inhabited earth for the last 66 million years, or what is popularly called "the age of mammals."

In examining the extinctions that took place in the seas at the end of the Cretaceous, scientists have learned that about 15 percent of the families (or approximately one hundred families) of shelled invertebrates became extinct. Particularly hard-hit groups were the ammonoid cephalopods, relatives of living squids and octopi, who suffered total extinction; clams and gastropods (snails), who endured significant losses; and the marine reptiles, the mosasaurs (giant marine lizards) and plesiosaurs (long-necked reptiles), who vanished altogether. Major changes also occurred in the marine plankton, and the foraminiferans (microscopic shelled protozoans) also suffered heavy losses. On land, the flying reptiles (pterosaurs) and the dinosaurs became extinct; many types of marsupial mammals disappeared; and a few types of flowering plants, especially broad-leafed forms and those living in low latitudes, died out.

After the extinction, the land surface was populated by many placental mammals, which rapidly diversified during the early Tertiary; by turtles, crocodiles, lizards and snakes, and reptiles little affected by the extinction; and by birds and flowering plants, groups not seriously impaired by the extinctions. In the sea, the most conspicuous Mesozoic denizens—ammonoids, mosasaurs, and plesiosaurs—were gone, as were some types of clams, especially the reef-building rudists and the platelike inoceramids.

However, many other clams survived, as did representatives of the other hard-hit invertebrate groups. The plankton and bony fishes recovered, and sharks remained unscathed by the extinctions.

The Cretaceous-Tertiary boundary is almost always identified by the extinctions that took place at that time. Thus, in the sequence of strata, certain fossil groups (for example, dinosaurs) are present in Cretaceous rocks but are absent in Tertiary rocks. Using the criterion of extinction, however, to identify the Cretaceous-Tertiary boundary, produces two significant problems.

The first of these problems stems from the inherent diachrony of extinction—in other words, the fact that an extinction almost never occurs simultaneously across the geographic range of an organism. For example, hippopotamuses, which have been undergoing extinction for thousands of years, disappeared from Europe and Asia a few thousand years ago. They are now restricted to small areas in Africa, where they will probably suffer extinction within the next few thousand years unless human intervention saves them. With the exception of a possible pervasive global catastrophe at the Cretaceous-Tertiary boundary, why should not the extinction of many Cretaceous organisms have taken place in the same diachronous fashion as the ongoing extinction of the hippopotamus? Indeed, some paleontologists believe that there is evidence that dinosaurs became extinct in South America after their extinction in North America. If this is correct, then what is identified as the Cretaceous-Tertiary boundary in North America is older than what is identified as the boundary in South America. This presents a serious problem for placing the Cretaceous-Tertiary boundary, which, ideally, should represent the same point in time everywhere.

The second problem faced when using extinctions to identify the Cretaceous-Tertiary boundary is the circularity of reasoning that can result; that is, if one identifies the Cretaceous-Tertiary boundary by the extinction of dinosaurs, one must be careful in saying that dinosaurs became extinct at the Cretaceous-Tertiary boundary. What if, as some believe, dinosaurs survived longer in some parts of the world than in others? To determine if this was the case, another criterion (usually another group of fossils) must be used to determine the age of the youngest dinosaur fossils.

Without question, the most intriguing aspect of the Cretaceous-Tertiary boundary is the question of what caused the extinctions. In order to answer this question, the timing of these extinctions must be determined. Did they occur simultaneously and suddenly? If so, then a major catastrophe of global proportions apparently was their cause. If, however, the extinctions were not simultaneous, and if some groups of organisms were already in decline prior to the Cretaceous-Tertiary boundary, then a single catastrophe alone cannot explain the extinctions.

In 1979, Nobel physics laureate Luis Alvarez, his geologist son Walter Alvarez, and two nuclear chemists, Frank Asaro and Helen Michel, proposed that a bolide (a comet or meteorite) the size of a mountain (10 kilometers in diameter) collided with Earth 66 million years ago and caused the extinction of the dinosaurs and other groups of organisms that died out at the end of the Cretaceous. They initially based this proposition on chemical analysis of a clay layer at Gubbio in northern Italy. This clay layer was deposited at the bottom of the sea 66 million years ago, and the chemical analysis revealed that it contains an unusually large concentration of the platinum-group metal iridium. Such a high concentration of iridium, reasoned Alvarez and his colleagues, could not be produced by known terrestrial mechanisms and thus must have settled in the dust produced by a huge bolide impact.

Geological studies at other localities worldwide where 66-million-year-old rocks are preserved have confirmed the Alvarez team's proposition of a bolide collision with Earth 66 million years ago. Their claim that the bolide impact is linked directly to the Cretaceous-Tertiary-boundary extinctions has not fared as well. Indeed, the fossil evidence indicates that many groups of organisms in the sea (for example, the ammonoids and inoceramid clams) and on the land (dinosaurs) were declining millions of years before the Cretaceous-Tertiary boundary. Furthermore, some groups of organisms (rudist clams are an example) became extinct a million or more years before the boundary. Also, there is some evidence, hotly debated, that a few types of dinosaurs may have survived into the earliest Tertiary. Nevertheless, the fossil evidence is not without its detractors, since many fossils remain to be discovered and the suddenness and synchrony or diachrony of some extinctions still is subject to debate.

A dispassionate reading of the existing fossil evidence does not support a single, mass extinction at the Cretaceous-Tertiary boundary. Instead, it suggests that, as a result of changing climates and sea levels, a period of extinction beginning three to five million years before the Cretaceous-Tertiary boundary was culminated by the final disappearance of several groups of or-

Nobel laureate Luis Alvarez, who collaborated with his son Walter to advance the theory that a meteor impact led to the mass extinctions at the end of the Cretaceous era. *(The Nobel Foundation)*

ganisms at (or perhaps just after) the end of the Cretaceous. Perhaps the bolide impact at the Cretaceous-Tertiary boundary is best interpreted as the last piece of bad luck encountered by a Mesozoic biota already doomed to extinction.

Methods of Study

Research on the Cretaceous-Tertiary boundary first must focus on locating the boundary in strata of a given region. To facilitate this, there are two places—Stevns Klint in Denmark and Gubbio in Italy—where by international agreement the position of the Cretaceous-Tertiary boundary is fixed in the strata. Identifying the boundary elsewhere on earth has thus been reduced to a problem of stratigraphic correlation, the method by which the equivalence in age or position of strata in disparate areas is determined. The goal of the fieldworker then has to be identifying criteria (usually fossils) by which correlation with the Cretaceous-Tertiary boundary in Denmark and/ or Italy can be demonstrated.

Since the Cretaceous-Tertiary-boundary rocks in Denmark and Italy were deposited at the bottom of the sea 66 million years ago, it is sometimes difficult to identify good criteria for stratigraphic correlation in 66-million-year-old rocks that were deposited on land. In these rocks, the youngest dinosaur fossils usually are believed to mark the Cretaceous-Tertiary boundary until other evidence demonstrates otherwise. This other evidence sometimes comes from fossil pollen grains, numerical ages, or other geophysical techniques, such as studying the magnetic properties of the rocks in order to determine their age.

Once the boundary has been placed with confidence, other aspects of studying the Cretaceous-Tertiary boundary are even more complex. They focus on the extinctions themselves and their potential causes. Data and techniques from many fields are brought to bear here, including paleontology (the study of fossils), sedimentology (the study of how sediment is transported and deposited), and geochemistry (the study of rock chemistry). At its simplest, in a given sequence of strata that encompasses the Cretaceous-Tertiary boundary, the goal of research is to collect and document the vertical ranges of all fossils, their relative abundances, and how their ranges and abundances correspond to environmental changes indicated by the sediments and rock chemistry.

One of the problems these studies face is the incompleteness of the fossil record. For example, when paleontologists think that they have found the youngest dinosaur fossil in a local sequence of strata, how can they be sure? Maybe younger dinosaurs lived in the area and their fossils were not preserved, or, if they were preserved, the fossils may not have yet been found. This caveat makes it difficult, especially in rocks deposited on land, where fossil occurrence often is very spotty, not only to be certain of the position of the Cretaceous-Tertiary boundary but also to be confident of the correspondence between fossil range, fossil abundance, and environmental

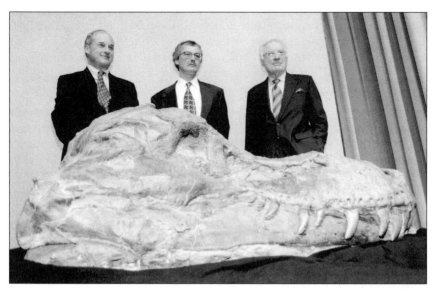

Officials of the Field Museum of Natural History pose with the fossilized head of a Tyrannosaurus Rex, one of the many dinosaur species that died out at the end of the Cretaceous era. *(Reuters/Jeff Christensen/Archive Photos)*

changes indicated by sediments and rock chemistry. The potential for new fossil discoveries always exists. This is only one reason that research on the Cretaceous-Tertiary boundary continues at a fast pace, and that the cause of the extinctions at and around this boundary remains a subject of heated debate.

Context

One of the most interesting aspects of the extinctions at the Cretaceous-Tertiary boundary is the disappearance of the dinosaurs. Dinosaurs included the largest land animals of all time and dominated earth's surface for 150 million years. Why such large and seemingly successful reptiles died out has captured the imagination of scientist and layperson alike for more than a century. More important, understanding extinctions in the past, such as those that took place at the Cretaceous-Tertiary boundary, may allow humankind to understand the causes and effects of massive extinctions. This understanding may, in turn, help us to avoid extinction in the future. Insight into these past extinctions would also provide some basis for understanding the potential effects of the ongoing extinction of species in the tropical regions of the globe.

Finally, there is seemingly incontrovertible evidence that a large bolide impacted the earth 66 million years ago. The effects of this impact have been likened to the "nuclear winter" that might result from a global thermo-nuclear war. Such a nuclear winter would be a period of intense cold when all incoming sunlight is blocked out by the smoke accumulated in the atmosphere from continent-wide forest fires. Analogous conditions may have existed on earth during the first ten to one hundred years that followed the bolide impact at the Cretaceous-Tertiary boundary. Studying the effects of this impact thus provides insight into a global disaster of horrific proportions and, if nothing else, is an inducement to the human species to avoid such a cataclysm.

Bibliography

Alvarez, Luis W. "Mass Extinctions Caused by Large Bolide Impacts." *Physics Today* 40 (July, 1987): 24-33. This very polemical article presents a strong argument for the bolide impact at the Cretaceous-Tertiary boundary having caused sudden and simultaneous mass extinctions. It also relates a very readable chronology of the Alvarez team's work on the iridium-rich clay layer and the other lines of evidence and arguments that ensued. This is the late Luis Alvarez's last written word on the subject. Well illustrated and referenced.

Archibald, J. D., and W. A. Clemens. "Late Cretaceous Extinctions." *American Scientist* 70 (July/August, 1982): 377-385. This article examines the Cretaceous-Tertiary-boundary extinctions on land in eastern Montana, where an outstanding fossil record of these extinctions is preserved. It argues that a complex pattern of extinctions is evident here, not a sudden and simultaneous extinction at the end of the Cretaceous, and thus represents a good contretemps to Alvarez's article. Well illustrated and referenced.

Bralower, Timothy J., and Charles K. Paull, et al. "The Cretaceous-Tertiary Boundary Cocktail: Chicxculub Impact Triggers Margin Collapse." *Geology* 26, no. 4 (April, 1998): 331-335.

Hsü, Kenneth J. *The Great Dying: A Cosmic Catastrophe Demolishes the Dinosaurs and Rocks the Theory of Evolution.* New York: Harcourt Brace Jovanovich, 1986. Hsü extensively reviews and accepts the ideas of the Alvarez team. He then argues against typical notions of Darwinian evolution to support the idea that major crises (extinctions) are the driving force of evolution. Some very debatable ideas are wrapped up in this well-written, novel-like book. Indexed but lacks illustrations and references.

Russell, Dale A. "The Mass Extinctions of the Late Mesozoic." *Scientific American* 246 (January, 1982): 58-65. In this well-illustrated article, Russell presents evidence for a sudden and simultaneous extinction of life on land (including dinosaurs) at the end of the Cretaceous. His evidence has subsequently been picked apart by other scientists, but this article articulates well the viewpoint of one of the few paleontologists to have readily accepted the Alvarez team's propositions. Lacks references.

Stanley, S. M. *Extinction.* New York: Scientific American Books, 1987. A very readable, extensive treatment of the subject of extinction. Chapter 7 reviews the extinctions at the end of the Cretaceous and elegantly reduces the welter of data and viewpoints to explain why the fossil record does not support a single, massive extinction at the Cretaceous-Tertiary boundary. Well illustrated and indexed.

Ward, Peter. "The Extinction of the Ammonites." *Scientific American* 249 (October, 1983): 136-141. A very readable and extensively illustrated article that presents the evidence that ammonoids were declining well before the end of the Cretaceous. Ward sees this decline as a losing battle against more mobile, shell-crushing predators. No references.

Spencer G. Lucas

Cross-References

Biostratigraphy, 37; The Fossil Record, 257; The Geologic Time Scale, 272; Stratigraphic Correlation, 593.

Deltas

Deltas are dynamic sedimentary environments which undergo rapid changes over very short periods. Found in lakes and shallow ocean waters, they are rich in fossil fuels and provide food and shelter for fish and wildlife.

Field of study: Sedimentology

Principal terms

CREVASSE: a break in the bank of a distributary channel causing a partial diversion of flow and sediment into an interdistributary bay

DELTA: a deposit of sediment, often triangular, formed at a river mouth where the wave action of the sea is low

DISTRIBUTARY CHANNEL: a river which is divided into several smaller channels, thus distributing its flow and sediment load

GEOARCHAEOLOGY: the technique of using ancient human habitation sites to determine the age of landforms and when changes occurred

INTERDISTRIBUTARY BAY: a shallow, triangular bay between two distributary channels; over time, the bay is filled with sediment and colonized with marsh plants or trees

NATURAL LEVEE: a low ridge deposited on the flanks of a river during a flood stage

PRODELTA: sedimentary layer composed of silt and clay deposited under water; it is the foundation on which a delta is deposited

SEDIMENT: fragmented rock material composed of gravel, sand, silt, or clay which is deposited by a river to form a delta

WAVE ENERGY: the capacity of a wave to erode and deposit; as wave energy increases, erosion increases

Summary

Deltas contain many valuable resources. Government agencies such as the U.S. Fish and Wildlife Service study the surface properties of deltas because of the enormous wetlands and abundant wildlife that occupy these landforms. Geologists study deltas because they are favored places for the accumulation of oil and gas resources. This low topographical feature serves society in many ways, and that is why it has been the object of intense study.

Deltas are deposits of sediments, such as sand or silt, which are carried by rivers and deposited at the shoreline of a lake, estuary, or sea. As the river meets the water body, its velocity is greatly decreased, causing the river sediment to be deposited. If the accumulated sediment is not removed by waves or currents, a delta will accumulate and continue to extend itself into the lake or ocean. The term "delta" is used to describe this depositional landform because it is often triangular. It is believed that the Greek historian Herodotus coined the term with the shape of the Greek letter delta in mind. Herodotus visited Egypt, where the Nile delta is located, and he correctly defined the shape of that delta; not all deltas, however, are triangular.

The Ganges delta, the Colorado River delta, and many other deltas have different shapes. The shape postulated by Herodotus is, in fact, somewhat unusual, but it is applicable to the Nile and the Mississippi river deltas. The Nile delta has a smooth but curved shoreline—it is an "arcuate" delta— whereas the Mississippi delta has spreading channels and resembles the digits of a bird's foot. Occasionally, current and wave action at the shoreline causes sediment to be distributed to the left and right of a river channel, forming smooth beaches on either side. Such a delta is shaped like a cone, the point of which projects toward the sea, and is called a "cuspate" delta. The Tiber River, which empties into the Mediterranean Sea, is a classic example of this type of delta. Rivers such as the Seine, in France, may deposit sediment in elongated estuaries, forming shoals and tidal flats.

Earth scientists have noted that the shapes of deltas are associated with several conditions, such as the character of river flow, the magnitude of wave energy and tides, and the geologic setting. The bird-foot delta of the Mississippi River has extended itself well into the Gulf of Mexico, because the river carries and deposits a high volume of sediment on a shallow sea floor or continental shelf. The wave and tidal forces are low, and the delta deposit is not redistributed along the shoreline or swept away. Conversely, a cuspate delta, such as that of the Tiber, is a product of strong waves moving over a steep continental shelf. The persistent high wave energy redistributes the sediment often, forming beaches and sand dunes along the delta shoreline. The Nile delta is an arcuate delta characterized by moderately high wave energy and a modest tide range. Occasional high wave conditions deposit beaches and sand dunes along the arc-shaped delta front at the Mediterranean Sea. Tides also play a direct role in the creation of deltas. Deltas in estuaries are formed because of a high tidal range coupled with low wave conditions. The Seine estuary, with its distinctive mud flats exposed at low tide, provides a good example.

Although deltas have different shapes which reflect differences in the intensity of river, wave, current, and tidal processes, certain landforms may be identified as characteristic of delta formation. Submarine features are deposited below sea level, and subaerial features form at or just above sea level. As a river empties into the sea, the finest sediments, usually very fine silt or clay, are deposited offshore on the sea floor. This submarine deposit forms the foundation on which the delta sits and is appropriately referred to as a "prodelta deposit." The deposit can often be detected on navigation maps as a relatively shallow, semicircular deposit under the water.

As deposition continues, the prodelta deposit is covered with the extending subaerial delta, which is composed of coarser sediments. Deltaic extension occurs along the distributary channels. During higher river flow, the distributary channels overflow, depositing natural levees along their sides. The digitate distributary pattern of the Mississippi delta illustrates this process very well. As the distributaries extend to deeper water, the shallow areas between the distributaries are better developed. These areas, known as interdistributary bays, are shallow landforms colonized by aquatic plant life. Over time, deposition occurs in the interdistributary bays through breaches in the natural levees. As the river mouth distributaries enter a flood stage, the lower regions of a natural levee are broken and fine suspended sediments introduced into the interdistributary bay area. Such overbank splays, or crevasse splays, are primarily responsible for the infilling of a delta. The crevassing is usually a very rapid but short-lived process, occurring during a high river stage and operating over a ten- to fifteen-year period. With the passage of time, the openwater interdistributary bays are silted and colonized. Eventually, however, the marshy bays may subside, because of the compaction of the sediment, creating water areas once again.

Although the geologic history of large deltas such as the Mississippi is complex, the succession and behavior of shifting deltas have been determined in some detail. Over the past twenty thousand years, the large continental glaciers that occupied much of the upper Midwestern United States began to melt. As the climate of the earth continued to warm, the melting ice was returned to the oceans and the seas rose some 100 meters, inundating valleys that had previously been cut by streams. The oceans reached their present approximate level about five thousand years ago. The Mississippi and similar valleys were flooded and became elongated bays. Over time, the shallow water bays were choked with sediments which formed broad floodplains extending down the valleys. Once a depositing river extended beyond the confines of its valley, a delta was deposited in deeper water. Because the river was no longer confined, it was free to shift over great distances. The Mississippi River delta is actually composed of seven distinct delta lobes extending over an approximate distance of 315 kilometers. The oldest delta, Salé Cypremort, was deposited some 4,600 years ago; the most modern delta was deposited within the past 550 years. Older Mississippi delta lobes, such as the Teche delta, have subsided since they were deposited, giving an opportunity for a more recent delta (in this case, the Lafourche) to be deposited on top and more seaward of the older feature. The different delta lobes making up the enormous deltaic plain have resulted from a shifting of the Mississippi River well upstream in its valley. This process is analogous to a hand movement occurring because of a shoulder movement.

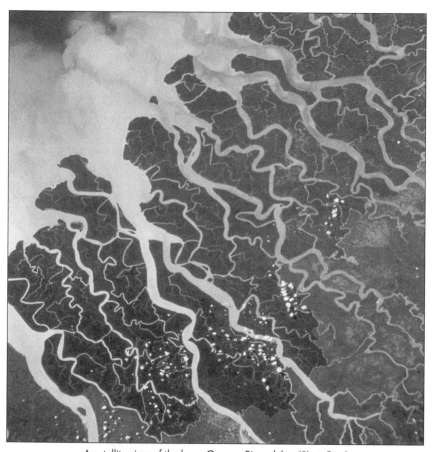

A satellite view of the huge Ganges River delta. *(PhotoDisc)*

Because of significant changes in the shoreline environments, many deltas in marine costal zones are eroding. The Mississippi River is at the edge of the continental shelf and cannot build out into deep water. Also, subsidence and a slight rise in sea level are causing the delta to erode. The Nile delta is eroding as well. With the construction of the High Aswan Dam upstream, there has been a decrease in the sediment supplied to the Nile delta. This lack of sediment, along with a slight rise in sea level, has led to erosion. Some earth scientists have suggested that the wave action in coastal Egypt is cutting back the Nile delta at a rate of 1.5 to 3 meters per year in some areas.

Methods of Study

In some ways, deltas are difficult to study, because most of the features are very flat, marshy, or under water. Since deltas change rapidly over only a few decades, however, maps are an important tool with which to determine changes. Navigation maps and maps that illustrate the topography of coastal areas around the world have been made for generations. By comparing the size and location of a delta on old maps and new maps, changes can be analyzed. Also, aerial photographs and pictures taken from satellites aid in identifying the erosion and deposition of delta landforms.

Often, older delta lobes were settled by ancient peoples. Through the science of geoarchaeology, it can be determined when changes occurred. As deltas, such as the Mississippi, shift from side to side over time, the human population follows the deltas from place to place. By examining the location or archaeological sites over the past fifteen hundred years, scientists have determined the minimum age of the several deltas forming the Mississippi deltaic plain. The Indian pottery found there reveals that the delta framework was deposited very recently. Cultural remains indicate other changes—in salinity, subsidence, and delta deterioration.

By boring holes into the soft sediments of a delta, geologists can decipher its subsurface aspects. Because oil and natural gas are often associated with deltas, oil companies have bored holes in many delta landforms. Information derived from this method of study reveals the composition of the thick delta sediments and the rate of delta accumulation. In fact, boreholes in some deltas have indicated that older deltas once existed and are now buried beneath younger deltas.

Finally, deltas can be created in the laboratory. In nature, deltas are often very large and complex. To make the study of deltas less difficult, scientists use tanks filled with water and sediments. Experiments can be performed which, for example, control the amount of sediment used to build deltas. Relationships between sediments and current velocities may be studied to gather information on such properties as the rate of delta growth. By controlling the phenomena that cause deltas to form, geologists can gain an overview of the behavior and processes of delta development.

Context

Deltas, with their marshes and bogs, are not aesthetically pleasing; however, depositional landforms have been useful to prehistoric and historic populations in many ways. Deltas, along with estuaries, are perhaps the most biologically productive areas on the earth's surface. Most deltas are colonized with wetland swamps or marshes, which are breeding areas for wildlife. In the United States, for example, duck hunting is a popular sport which is associated with wetlands. In marine deltas, where there is tidal influence, freshwater and saltwater mix. The river brings oxygen and nutritive substances into the delta, and the result is an enormous production of sea life. High biological productivity attracted humans to this land feature. Deltas have often been centers of civilization; the deltas of the Nile, in Egypt, and the Tigris-Euphrates, at the head of the Persian Gulf, have supported important societies. Soils in delta regions are nourished through seasonal flooding, and water tables are high, guaranteeing adequate water with which to irrigate crops, even in the dry season. Food and crop production from tropical deltas is significant because most tropical soils are not very productive. Deltas such as those of the Mekong and the Ganges are outstanding examples.

Deltas are transition zones between the land and the sea, between river and marine processes. Their rivers are also links between ocean and continent. Cities such as New Orleans, Venice, Amsterdam, and Rotterdam owe their prosperity to their delta geography. Such cities, known as *entrepôts*, thrive on marine traffic entering a country or on overland traffic exiting the country.

Since deltas are areas of vast accumulations of sediments, they generate building material for future mountains. The young mountains of the world, such as the Alps and the Himalayas, parallel coastal areas and are composed of sedimentary rocks. Marine fossils frequently found in such rocks reveal that they not only are composed of sediment but also were once deposited under water, later to be thrusted to great heights.

Bibliography

Bird, Eric. *Coasts: An Introduction to Coastal Geomorphology.* 3d ed. New York: Basil Blackwell, 1984. An introductory text on coastal zones and processes. Most examples are taken from Australia. The chapter on deltas is well illustrated with maps and diagrams. The text is nontechnical and comprehensible to readers with little scientific background.

Coleman, J. M. *Deltas: Process of Deposition and Models for Exploration.* Boston: International Human Resources Development Corporation, 1982. A detailed review of deltaic processes, including an overview of the Mississippi River delta and discussions of other deltas and their variability. Numerous maps and diagrams illustrate specific points. For readers with some background in the subject this text can be a useful supplement.

Davis, R. A., Jr. *Coastal Sedimentary Environments.* 2d rev. ed. New York: Springer-Verlag, 1985. A book on deposition in coastal areas. Topics include deltas, beaches, marshes, and estuaries. The treatment of deltas is generally narrative, and equations are sparingly used. A comparative presentation of deltas is instructive and not difficult to understand; some background in physical geology is useful but not necessary.

LaBlanc, R. J., ed. *Modern Deltas.* Tulsa, Okla.: American Association of Petroleum Geologists, 1976. A college-level text describing some of the world's deltas. The treatment is generally nonmathematical and descriptive, but it will be most useful to those who have had a course in geology. Well illustrated with diagrams, maps, and pictures.

Morgan, J. P. *Deltaic Sedimentation Modern and Ancient.* Tulsa, Okla.: American Association of Petroleum Geologists, 1970. An old but not outdated series of chapters by different authors on deltas around the world. Thorough in its treatment but not overly technical, the book emphasizes sedimentary differences and the biological and physical character of deltas. Well illustrated with maps and aerial photographs. Some knowledge of geology or earth science would be useful.

_____, ed. "Deltas: A Resume." *Journal of Geologic Education 18* (1970): 107-117. An excellent introductory article on deltas. The emphasis is on the Mississippi River delta, which the author studied for many years. Different processes and their influence on delta development are covered. An excellent and not overly technical paper. Includes maps and tables.

Peterson, J. F. "Using Miniature Landforms in Teaching Geomorphology." *Journal of Geography* 85 (November/December, 1986): 256-258. This paper discusses small-scale landforms and their advantages in the classroom. Deltas and related features, such as alluvial fans, are highlighted. The nontechnical text is supplemented with photographs. A good article for those interested in reconstructing a delta on a small scale. Suitable for high school students.

Raphael, C. N., and E. Jaworski. "The St. Clair River Delta: A Unique Lake Delta." The Geographical Bulletin 21 (April, 1982): 7-28. A delta in the Great Lakes is described and compared with the Mississippi River delta. The paper suggests that although many deltas look the same, different processes are at work in them. Good aerial photographs, maps, and cross sections. A nontechnical treatment of delta processes, forms, and vegetation.

Thornbury, W. D. *Principles of Geomorphology.* 2d ed. New York: John Wiley & Sons, 1969. A well-written introductory textbook on landforms. The various landforms of deltas and related river features are discussed in detail. The book emphasizes form rather than process and is nontechnical in its presentation. Suitable for those who want a rapid, non-mathematical introduction to deltas and related features, such as floodplains and coasts.

C. Nicholas Raphael

Cross-References

Alluvial Systems, 1; Archaeological Geology, 23; Biostratigraphy, 37; Drainage Basins, 141; Floodplains, 241; River Flow, 526; River Valleys, 534; Weathering and Erosion, 701.

Discontinuities

Discontinuities are boundaries within the earth that divide the crust from the mantle, the mantle from the core, and the outer core from the inner core. The term is also used to describe the less dramatic boundaries within layers.

Field of study: Structural geology

Principal terms

CRUST: the top layer of the earth, composed largely of the igneous rock granite; it ranges from 3 to 42 miles in thickness

DISCONTINUITY: a boundary between two adjacent earth layers, such as the Mohorovičić Discontinuity between the crust and the mantle

EARTHQUAKE: a tremor caused by the release of energy when one section of the earth rapidly slips past another; earthquakes occur along faults or cracks in the earth's crust

EARTHQUAKE WAVES: vibrations that emanate from an earthquake; earthquake waves can be measured with a seismograph

INNER CORE: the innermost layer of the earth; the inner core is a solid ball with a radius of about 900 miles

MANTLE: the largest layer of the earth, about 1,800 miles in thickness; the mantle is within 3 miles of the earth's surface at some locations

OUTER CORE: the outer portion of the core, about 1,300 miles in thickness; it is believed to be composed of molten iron

SEISMOGRAPH: a device that measures earthquake waves

Summary

Discontinuities are underground boundaries between layers of the earth. The closest discontinuity to the earth's surface is the Mohorovičić Discontinuity, which divides the earth's crust from the mantle underneath. Other discontinuities divide the mantle from the outer core and the outer core from the inner core. Minor discontinuities are found within these layers.

The interior of the earth has been the object of much speculation and interest for thousands of years. Because direct observation of the earth's interior is usually impossible, however, inferences about its structure and characteristics must be made from phenomena seen or felt at or near the earth's surface. Several phenomena do give indications of the subsurface earth: caves that are often cool and damp, cool water emanating from springs and artesian wells, hot water spewing upward from geysers, and volcanoes from which extremely hot lava erupts. These phenomena give a mixed and incomplete picture of the earth beneath the surface.

The structure and composition of the interior of the earth can, however, be inferred from the study of earthquake waves. Seismographs can detect three types of vibrations: surface waves (the ones that can cause damage when there is an earthquake), P (primary) waves, and S (secondary) waves, which are also generated by every quake. P waves are compressional (pushing) waves, in which earth or rock particles move forward in the direction of wave movement; S waves are shear waves, in which the particle motion is sideways or perpendicular to the direction of wave movement. The more efficient P waves travel twice as fast as S waves and thus are always detected first by a seismograph. Seismographs record these waves on charts, called "seismograms," attached to moving drums. By noting the arrival times of the various waves, seismologists can determine the distance to an earthquake and can see the effects on these waves caused by the type of rock through which the waves have moved.

Seismic waves travel through rock layers at specific speeds, which are different for each type of mineral or rock. For example, waves travel through basalt at 5 miles a second and through peridotite at 8 miles a second. Seismogram study has shown that the earth's interior is not homogeneous, but rather is composed of several major layers and many sublayers.

In 1906, Richard Oldham discovered that S waves are never detected on the opposite side of the earth from any earthquake. As he already knew that S waves cannot travel through liquid substances, Oldham postulated that the center of the earth must be composed of a molten core and that the materials above this core are not molten. The depth of the boundary between this core and the material above it was discovered eight years later by Beno Gutenberg. Now called the "Gutenberg Discontinuity," it is located about 1,800 miles beneath the earth's surface.

When Oldham made his discovery of a central core, Andrija Mohorovičić was the director of the Royal Regional Center for Meteorology and Geodynamics at Zagreb, one of the leading seismological observatories in Europe. In 1909, his meticulous study of a Croatian earthquake showed that some of the P waves from that quake had traveled faster than others. He already knew that other waves speed up or slow down when they move from one medium into another (as when light moves from air into water) and that this change in speed can result both in reflection, a bouncing back of waves, and in refraction, or a change in wave direction through the new medium. He deduced that the faster-moving P waves had traversed down through the earth, through a discontinuity to a material of a different density, and then had come back up to the surface. Deep in the earth was a material that allowed for faster transmission of P waves. Above this discontinuity, seismic waves travel at about 4.2 miles a second; below the boundary, they travel at about 4.9 miles a second.

When Mohorovičić's results were replicated by other seismologists, it was concluded that the discontinuity was a global phenomenon. Data from these studies showed that there were two very distinct layers of the earth: an upper, less-dense layer now called the "crust," and a denser layer below called the "mantle." Thus Andrija Mohorovičić had discovered what is now called the Mohorovičić Discontinuity, the boundary between the earth's crust and mantle (it is often called the "Moho").

The crust of the earth is made up of continents and ocean basins that are very different from one another. Continental crust is made primarily of granite. Covering this granite over much of the earth's continents may be found layers of younger sedimentary rock such as sandstone, limestone, and shale. Ocean basins, on the other hand, are composed of the dark, heavy rock basalt.

Mohorovičić believed the discontinuity between the crust and mantle to be about 30 to 35 miles below the surface of the earth. Subsequent studies have shown that it is usually at a depth of about 21 miles. However, the Moho has an irregular shape that is roughly a mirror-image of the surface of the earth. Under the continents, the Moho is much deeper; under the oceans, the crust is very thin, and the Moho is as close as from 3 to 5 miles from the surface. The greatest depth of the Moho is probably beneath the Tibetan Plateau, where it reaches a depth of 42 miles.

The continents are higher because they are composed of granite, which is a lower-density rock than basalt or the materials of the mantle. Even though the mantle is composed of solid rock, under long-term stresses, the rock moves slowly like a liquid. Thus, just as ice floats in water, the continents actually are floating upon the heavier mantle rock. The Moho is the boundary between continental granitic rocks and the denser peridotite rock of the mantle.

The mantle extends from the bottom of the crust to a depth of 1,800 miles. It appears to be made of the rocks somewhat similar chemically to those in the earth's crust but more "basic,"—that is, having more of the heavy iron and magnesium minerals such as olivine, and less lightweight aluminum. The mantle also appears to be composed of layers with discontinuities about 220 and 400 miles beneath the earth's surface. Although mantle rock is solid, it can, under certain conditions, behave somewhat like a liquid. Under long-term pressures, the molecules of this solid rock can move like liquids, but under sharp, short-term stresses, mantle rock fractures like a brittle solid.

Heat within the earth is created through the decay of radioactive isotopes. Although this generated heat is very small when compared to the heat received from the Sun, it is well insulated and is enough to create volcanoes and the convection currents of the mantle. Mantle rock is extremely hot; because of the pressure on it from the crust above, however, it cannot melt, except where there is a decrease in this pressure.

Studies at the surface of the earth have revealed areas where great heat flow comes from the mantle. Near the center of the Atlantic Ocean, the basaltic ocean bottom has split; the two sides are being pulled away from each other as Europe and Africa move away from the American continents. At this split, a decrease in pressure allows the hot mantle rock to melt and well upward, filling the gap between the dividing ocean bottoms. Thus the new ocean basin is made of material directly from the mantle. Within the mantle are large, slow-moving convection currents where hot mantle rock moves upward, cools off, and slowly sinks. These currents are believed to be the driving forces of continental drift.

At the bottom of the mantle, beneath the Gutenberg Discontinuity, is the earth's core. Seismic studies have shown that the outer core, which extends from roughly 1,800 to 3,100 miles beneath the earth's surface, is not a perfect sphere. The core rises in areas where hot mantle rock is moving upward and is depressed where cooler mantle rock is moving downward. The density of the core is much greater than that of the mantle. It is believed that this core is made of molten nickel and iron and that its motion generates the earth's magnetic field and aurora borealis.

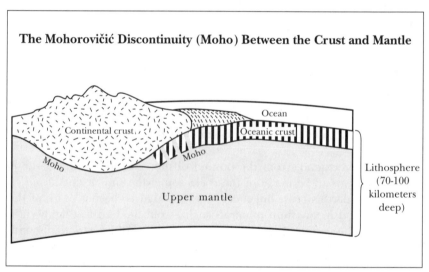

The Mohorovičić Discontinuity (Moho) Between the Crust and Mantle

In 1936, Danish seismologist Inge Lehman discovered evidence for a solid core within the molten center of the earth by detecting seismic waves that had been deflected back to the surface from within the core. When she realized that these waves, though very weak, travel faster through this most-central part of the earth than through the rest of the molten core, she was able to infer that this inner core was composed of solid material completely surrounded by the molten outer core. This most central layer of the earth extends from 3,100 miles beneath the earth's surface to the center of the earth, 4,000 miles down.

Seismograms have shown that the earth is composed of four major layers: the crust, mantle, outer core, and inner core. The crust is the only layer of which scientists have any direct evidence. In 1957, a project was conceived to drill a hole through the thin oceanic crust down past the Mohorovičić Discontinuity to bring up rock from the mantle. Although the "Mohole" project was approved and funded by the National Science Foundation, funds for it were cut off by the U.S. Congress in 1966.

Methods of Study

Although the deep interior of the earth cannot be seen or examined directly, earthquake waves can give an indirect picture of the earth's interior. These waves, formed by an earthquake or when artificial explosions are set off, go through the earth at speeds determined by the type of rock through which they pass. Like sound and light waves, earthquake waves may be reflected and refracted when they move from one medium to another if the media are of different densities.

The seismoscope is an ancient instrument that shows earth movements. A Chinese seismoscope of the second century A.D. had eight dragon figures each with a ball in its mouth. When the earth trembled, a ball would fall from the dragon's mouth into the mouth of a frog figure underneath it. European seismoscopes often used bowls of water that would spill when agitated. In 1853, Luigi Palmieri designed a seismometer that used mercury-filled tubes that would close an electric circuit and prompt a recording device to start moving when the earth vibrated.

In 1880, British seismologist John Milne invented the first modern seismograph, which employed a heavy mass suspended from a horizontal bar. When the earth would quake, the bar would move, and that movement would be recorded on light-sensitive paper beneath. Most seismographs employ a pendulum, which, because of inertia, remains still as the earth moves underneath it.

When seismographs measure shock waves from nearby earthquakes, they first receive the P waves, which vibrate in the direction in which the waves are moving. S waves, which vibrate perpendicular to the direction in which the waves are moving, then arrive, followed by surface waves. When seismographs record more distant earthquakes, the results are complicated by the reflection and refraction of seismic waves resulting from the various discontinuities underground. As the complications were deciphered, seismologists realized that the recordings described the rock layers below and between the quake and the seismograph.

Once geologists realized that they could learn about the earth by examining seismograms, some researchers became impatient when they wanted to study a particular area but had to wait for an earthquake to occur. This became particularly difficult in areas where earthquakes did not occur frequently. Milne solved the problem by dropping a one-ton weight from a height of about 25 feet. The impact of this weight on the ground generated seismic waves that were weaker than, but similar to, those generated by earthquakes. To create stronger waves, seismologists explode charges of dynamite. These artificially induced shock waves have enough energy to reach deep into the planet. Since the 1970's pistons on large trucks have been used to strike the earth and create artificial seismic waves.

When charges are exploded and the vibrations recorded by several nearby seismographs, a detailed description of rock layers can be detected. Since 1923, when a seismograph was first used to locate a large underground pool of petroleum, seismology has played a large part in the oil and gas industries. Earthquake waves artificially produced by explosions are also able to determine the location of underground geologic structures that may contain mineral deposits.

With the advent of the space age, seismographs connected to radio transmitters have been placed on the surfaces of the moon and Mars. There are more than a thousand seismographs in constant operation gathering seismic data around the world. Data from the National Earthquake Information Center is updated daily and is available on the Internet.

Context

The same technology that has indicated the location of discontinuities deep within the earth has also provided a greater knowledge about the crust. Whereas ancient civilizations feared earthquakes as manifestations of angry gods, quakes are now seen as results of energy released when plates of the earth's crust move past one another.

Although earthquakes do occur in many places on the earth's crust, they are most common in certain areas such as the "Ring of Fire" around the Pacific Ocean. Most earthquakes are linked directly to the movement of the earth "plates," or sections of the crust. The Pacific plate and the North American plate meet along the San Andreas fault, which runs from western Mexico through California to the Pacific Ocean. The two plates are moving past each other along this fault. Each time there is movement along the fault, tremendous amounts of energy are released, and the earth quakes. Quakes along this fault and others have caused untold damage.

One of the primary goals of seismologists is to determine a way to predict exactly when earthquakes will occur. If this information were known in advance, people could prepare for quakes, and far fewer deaths would occur. Many phenomena have been observed before quakes, such as increased strains upon bedrock, changes in the earth's magnetic field, changes in seismic wave velocity, strange movements of animals, changes in groundwater levels, increased concentrations of rare gases in well water, geoelectric phenomena, and changes in ground elevation. However, none of these dependably occurs before every quake, and thus these signs have not become reliable indicators.

Seismologists cannot prevent earthquakes from occurring, nor can they yet predict the exact time of a major quake, but they can predict where earthquakes are likely to occur. It is believed that certain active faults where there has been no earthquake activity for thirty years or so are about ready for an earthquake. With this information, urban and regional planners can provide for quake-resistant roads, bridges, and buildings.

Seismology and the search for minor discontinuities play a great part in the search for oil, gas, and mineral resources. Since much petroleum is retrieved from off-shore locations where the crust of the earth is thinner, knowing the location of the Mohorovičić Discontinuity sets the lower boundary for exploration.

Seismic studies are used regularly to assist in the search for oil and gas reservoirs. Natural gas and petroleum both can become trapped under some geologic formations. Seismologists routinely create a survey of an area before drilling to find minor discontinuities or boundaries between two different rock types, such as shale and sandstone. These surveys are made by measuring the reflection of seismic waves from the underlying rock layers. Geologic structures that can contain petroleum, natural gas, or mineral deposits can be identified from these surveys. Seismic surveys can show the distance and direction to these structures.

Bibliography

Calder, Nigel. *The Restless Earth: A Report on the New Geology.* New York: Viking Press, 1972. A companion book to the television program "The Restless Earth," this book emphasizes how geologists came to the conclusion that the continents are moving. Illustrated with black-and-white and color photographs and diagrams. Indexed.

Cromie, William J. *Why the Mohole?* Boston: Little, Brown, 1964. A 1960's view of the never-finished American and Soviet plans to drill holes through the entire crust of the earth in order to reach the mantle. The author was public-information officer for the American project. Several diagrams and photographs, a bibliography, and an index.

Emiliani, Cesare. *Planet Earth: Cosmology, Geology, and the Evolution of Life and Environment.* Cambridge, England: Cambridge University Press, 1992. A large, comprehensive book containing basic information about matter and energy, many aspects of the physical and historical earth, and a large section about the earth's relationship to the universe. The last section is a brief history of the earth sciences.

Erickson, Jon. *Rock Formations and Unusual Geologic Structures.* New York: Facts on File, 1993. An easy-to-read description of the earth's crust, including the creation, deformation, and erosion of rock. Clear black-and-white photographs, diagrams, and maps along with a large glossary, bibliography, and index.

Lambert, D., and the Diagram Group. *Field Guide to Geology.* New York: Facts on File, 1988. A profusely illustrated book about the earth, its seasons, rocks, erosional forces, and geological history. Contains a list of "great" geologists (including Mohorovičić) and a list of geologic museums, mines, and spectacular geologic features. Indexed.

Miller, Russell. *Continents in Collision.* Alexandria, Virginia: Time-Life, 1983. A thorough text describing how earth motions have created geologic features. Profusely illustrated with color and black-and-white illustrations. Bibliography and index.

Tarling, D., and M. Tarling. *Continental Drift: A Study of the Earth's Moving Surface.* Garden City, N.J.: Anchor Press, 1971. A small paperback book with black-and-white photographs and diagrams that help the reader to understand the principles of earth structure and plate tectonics.

Vogt, Gregory. *Predicting Earthquakes.* New York: Franklin Watts, 1989. A good text on the earth's interior and on how earthquakes are generated, detected, and measured. The last chapter discusses the prediction of earthquakes and efforts to control their effects. Black-and-white photographs and diagrams, glossary, and index.

Weiner, Jonathan. *Planet Earth.* New York: Bantam Books, 1986. A companion volume to the television series "Planet Earth," this book is well illustrated with both black-and-white and color pictures and diagrams. No glossary, but a comprehensive bibliography and index.

Kenneth J. Schoon

Cross-References

Earth's Core, 171; Earth's Crust, 179; Earth's Mantle, 195; Earth's Structure, 224; Plate Tectonics, 505.

Drainage Basins

Drainage basins reflect the operation of physical laws affecting water flow over the earth's surface and through rocks. They define natural units that concentrate flow into rivers, which, in turn, remove both water and sediment, the latter eroded from the surface of the basin. They become a focus for human activities whenever water is exploited as a resource.

Field of study: Sedimentology

Principal terms

ALTITUDE: the height (in meters or feet) above mean sea level

BASIN ORDER: an approximate measure of the size of a stream basin, based on a numbering scheme applied to river channels as they join together in their progress downstream

CHANNEL: a linear depression on the earth's surface caused and enlarged by the concentrated flow of water

EROSION: the removal of sediment (particles of various sizes) from the earth's surface

GROUNDWATER: water that sinks below the earth's surface and that slowly flows through the rock (ground) toward river channels; it keeps rivers flowing long after rainstorms

HYDROLOGICAL: relating to the systematic flow of water in accordance with physical laws at or close to the earth's surface

LIMESTONE: a rock composed primarily of the mineral calcium or dolomite, which can be dissolved by water that is acidic

Summary

A drainage basin is an area of the earth's surface that collects water, which accumulates on the surface from rain or snow; its slopes deliver the water to either a channel or a lake. Normally, the channel that collects the water leads eventually to the ocean. In this case, the drainage basin is defined as the entire area upstream whose slopes deliver water to that channel or to other channels tributary to it. Thus, strictly speaking, drainage basins are defined as natural units only when streams enter bodies of water such as lakes or the ocean or when two streams join; one speaks of the Mississippi River drainage basin or the Ohio River drainage basin—this latter basin lying above the Ohio River's confluence with the Mississippi.

141

Less often there is no exit to the ocean: This type of basin is called a basin of inland or interior drainage. Notable examples are the basins containing the Great Salt Lake in Utah, the Dead Sea in Israel and Jordan, and the Caspian Sea in Asia. The Basin and Range province in the Rockies (an area of about 1 million square kilometers extending from southern Idaho and Oregon through most of Nevada, western Utah, eastern California, western and southern Arizona, southwestern New Mexico, and northern Mexico), has at least 141 basins of inland drainage. The center of these basins is usually marked by a playa—a level area of fine-grained sediments, often rich in salts left behind as inflowing waters evaporate. At certain times in the geological past, when the annual rainfall regime was wetter, some of these basins completely filled with water to the point of overflowing, at which point the drainage system may either connect to another interior basin or connect to a river system which drains to the sea. The basin now containing the Great Salt Lake (known to geologists as Lake Bonneville) overflowed at Red Rock Pass about 15,000 years ago. The overflowing waters discharged into the Snake River system, and thus to the Columbia River and the Pacific. It is clear, therefore, that drainage basins may change in character over relatively short periods of geological time. There is some evidence that the entire Mediterranean Sea was a basin of inland drainage for a period about 3-5 million years ago: Substantial salt deposits are found on its bed, and traces of meandering rivers have been seen in certain geological sections.

Although the term "drainage basin" is normally thought of as applying to the surface, a very important component of the basin as a hydrological unit is the rock beneath the surface. Much of the water that arrives on the surface sinks into the soil and the underlying rocks, where it is stored as soil water and as groundwater. Soil water either sinks farther to become groundwater, or it may flow through soil and back out onto the surface downslope when and where the soil is saturated. Groundwater moves very slowly through the rock (millimeters to centimeters per day), but it eventually seeps into stream channels and so keeps water flowing in rivers long after rain has finished falling.

This characteristic of groundwater can lead to a circumstance that alters the definition of a drainage basin when the rocks are primarily composed of limestone or any rocks susceptible to solution. Because limestone is soluble in acidic water (natural rain is slightly acid; acid rain accentuates the acidity), over long periods of time (thousand of years), percolating ground-water dissolves substantial volumes of rock and causes a system of under-ground channels to develop, which may sometimes be enlarged to caverns. Yugoslavia is famous for its underground cave systems; in the United States, Kentucky, Florida, and New Mexico (with its famous Carlsbad Caverns) are well known for their limestone terrain and underground drainage systems. In this case, the route of the water underground may bear little or no relation to the pattern of channels and slopes seen on the surface, some or all of which may have become completely inactive. The determination of the drainage basin is then very difficult as various types of tracer (colored dyes or tracer chemicals) have to be placed in the water in order to track it, and the pattern of water flow to any given site will also depend on the location of the storm waters causing the flow. The very slow solution of limestone to form underground river systems and caves emphasizes the fact that water moving through the basin removes solid rock. As the rivers dissolve their way downward, they leave some caves "high and dry" above the general level of the underground water (the water table), which points up the fact that rivers work down through the rock with the passage of time.

This aspect of drainage basins—that solid rock is slowly removed—is harder to observe in areas of less soluble rock, even though water flowing out of the drainage basin carried sediment (small particles of soil and rock) and has been doing so for long periods of geological time. Thus, in the long run, the surface of the earth is lowered, and even in the short run, enormous amounts of sediment may be removed from the basin every year. The Mississippi removes a total of about 296 million metric tons per year, or 91 tons per square kilometer; yet this is small compared to the Ganges, which takes out 1,450 million metric tons per year, or 1,520 tons per square kilometer. If there were no corresponding uplift of the drainage basins or other interference, this removal would lead to the leveling of entire basins within 10 or 50 million years, depending on the lowering rate and the mean altitude of the basin. If the Ganges basin has a mean relief of 4,500 meters, it would be completely lowered in about 7 million years at present rates. If the Mississippi has an average relief of 1,000 meters, the time needed is on the order of 40 million years. These figures, however, are unrealistic because other processes cause compensating uplift, but they do indicate that drainage basins are being actively eroded within reasonable spans of geologic time.

The process of erosion proceeding at different rates in adjacent basins may cause the drainage divide (the line separating different flow directions for surface waters) to migrate toward the basin with the lower erosion rate. This is most common in geologically "new" terrain when stream systems are not deeply incised into the rocks. Drainage diversions may be simulated, as in the Snowy Mountain diversion in Australia, where waters are diverted across a divide by major engineering works in order to provide irrigation water for the Murray Darling river basin and also for hydroelectric power generation.

When winter snow melts (the late spring and early summer peak in Mississippi River flow is caused by melting snow in the Rocky Mountains) or severe storms bring heavy rain to large areas, the water which falls flows over the surface to channels and generates floods in the rivers. Floods are not abnormal; they are an expectable occurrence in drainage basins. It is easy to understand that when basin relief is high and slopes are steep (in the Rockies or the Appalachians), floods tend to generate higher flood peaks than when slopes are very gentle. A basin that is round in shape tends to concentrate floodwater quickly because the streams tend to converge in the

143

middle, whereas a long, narrow stream has the effect of attenuating the flow peak, even when the total amount of water falling on the basin may be the same. Similarly, a forest tends to attenuate flood peaks and to promote higher river flows between flood peaks than does open farmland. With the latter, there is a tendency for water to flow very rapidly off the surface into channels, whereas with a forest much water is intercepted by the leaves of trees and the impact of rain on the surface is weaker, in part because leaf cover protects the soil. Because the soil is not so well protected, sediment loss from the surface into streams is greater from farmland than from forests and is higher again from land disturbed by major building projects.

Methods of Study

Various methods have been devised to classify basins according to size, and the most common method depends on a numbering system applied to the streams that drain them. All the fingertip streams are labeled with 1. When two of these tributaries meet the channel, it is termed a second order channel and is labeled with 2. Subsequently, however, the order of a stream increases by one only when two streams of equal order join. Otherwise, if two streams of unequal order join, the order given to the downstream segment is that of the larger of the two orders. The order of the drainage basin is then the order of the stream in the basin. In this type of numbering system (called Horton/Strahler ordering), the Mississippi drainage basin is an eleventh or twelfth order basin; the exact number depends on the detail (map scale) with which the fingertip streams are defined. The larger orders are rare because if the Mississippi basin is twelfth order, it would take

CHARACTERISTICS OF SELECTED MAJOR DRAINAGE BASINS

River	Outflow	Length	Area	Average Annual Suspended Load
Amazon	180.0	6,300	5,800	360
Congo	39.0	4,700	3,700	—
Yangtze	22.0	5,800	1,900	500
Mississippi	18.0	6,000	3,300	296
Irawaddy	14.0	2,300	430	300
Brahmaputra	12.0	2,900	670	730
Ganges	12.0	2,500	960	1,450
Mekong	11.0	4,200	800	170
Nile	2.8	6,700	3,000	110
Colorado	0.2	2,300	640	140
Ching	0.06	320	57	410

Note: Rivers are ordered by outflow; outflow is multiplied by 1,000 cumecs (a cumec is 1 cubic meter of water per second); length is measured in kilometers; area is measured in square kilometers multiplied by 1,000; average annual suspended load is measured in millions of metric tons.

another river of roughly similar magnitude to join it to make a thirteenth order basin.

Basin order may be used as a relatively natural basis for the collection of other data about the basin (see table). The simplest measure is of the area in square kilometers. In addition, one may record the basin relief, the height difference between the lowest and the highest points, and the mean relief, or the average height of the basin above the outlet. The most precise method of recording basin relief is by computing the hypsometric (height) curve for the basin, which requires an accurate topographic map. When constructed, it shows, for any altitude, the proportion of the basin area above that particular altitude and, for comparative purposes, it may be produced in a dimensionless form by dividing both the height and the area measures by their maximum values or by the difference between the maximum and minimum heights if zero is not the minimum height.

Basin shape and basin dimensions (length and width) may also be recorded, although the notion of basin shape suffers from the problem that no completely unambiguous numerical measure exists that can be used to define the shape of an area in the plane (that is, on a map), and the problem is especially intractable if there are indentations in the edge of the basin. All measures are dependent to a considerable degree on the accuracy of source maps, and in mountainous terrain, such maps may often be much less than perfect, if they exist at all. Even with automated drafting aids and digitizers (which automatically record positions on maps and save them for a data file), the measurement of basin properties is a tedious and time-consuming process. Unless there are pressing reasons for a new analysis, it is common to rely on data tabulations made by hydrological or environmental agencies whenever possible.

Measurements are made of drainage basin properties because they are often used in statistical analyses together with the known flow of the few gauged rivers in order to predict flow characteristics for rivers that have not been metered. The direct measurement of stream flow, while straightforward in principle, is time-consuming, especially in the early stages, and a flow record is not very useful for predictive purposes until it has recorded at least twenty years of flow (preferably much more). Because of the high capital and maintenance costs involved in collecting river records, there has been an understandable emphasis on records for large rivers, the economic benefits from prediction (and eventual control) of the flow are more obvious, and measured flow records can sometimes be supplemented by anecdotal evidence of historic large floods—those flows which are often of most interest in land-use planning (for example, in the zonation of land for residential use). In recent years, it has been acknowledged that the hydrological behavior of low order basins is less well understood, and more information has been collected on them, especially for urban areas where the routing of the large quantities of water that run off from impermeable surfaces in the city (roofs and roadways) has been recognized as a serious planning problem, especially when the flow systems connect with urban sewage systems.

Context

The control of water outflow from drainage basins is necessary in some regions in order to promote irrigation, to supply domestic and industrial water, to generate power, and to implement flood control. The Hoover Dam on the Colorado was originally conceived as a control dam, but hydroelectrical generators were also built in order to help defray costs by selling power. There are nineteen major dams in the Colorado basin. Difficulties (aside from the legal technicalities of water ownership and redistribution) arise from the fact that to control substantial amounts of water, large areas of the basin have to be regulated and, in addition, there are economies of scale in large projects, particularly in the construction of large dams and reservoirs. A single control dam strategically placed may regulate flow downstream for hundreds of miles, whereas it would require hundreds of small dams on first- and second-order streams to achieve the same effect.

Large control dams do generate problems. The reservoirs trap sediment coming from upstream (in addition to water), which will eventually fill them, at which point they will be useless, and small dams may fill within a few years. An original estimate for the Hoover Dam suggested that it would take only four hundred years to fill Lake Mead; after only fourteen years, surveys revealed that the water capacity had been reduced by 5 percent and that sediment in the lake bottom reached a maximum of 82 meters where the upstream river entered the still waters of the lake. Downstream of a dam, the reduced sediment content and the regulated water flow often seriously affect riparian environments, and there may be a variety of channel responses, often unpredictable, to the interference in the river regime caused by the dam. The stream may cut into its bed; it may change the dimensions of its channel; or it may even aggrade its bed. In the case of the Hoover Dam, the water downstream, deprived of its sediment by the dam, had an increased ability to remove fine sediment from the river bed but left coarser rocks behind because the flood peaks that would remove them normally were now controlled—that is, much reduced. The result is an "armouring" of the stream bed with coarse rocks, an effect that extends 100 kilometers downstream in the case of the Colorado River below the Hoover Dam. In the Colorado system as a whole, the net effect of controlling flood peaks has caused rapids to stabilize and to increase in size as sediment becomes trapped in them. A corollary of the "winnowing" of fine material has been the disappearance of river beaches and an increased propensity to pollution as sediment becomes much less mobile and more concentrated in space.

Bibliography

Chorley, Richard J. "The Drainage Basin as the Fundamental Geomorphic Unit." In *Water, Earth, and Man,* edited by Richard J. Chorley.

London: Methuen, 1969. Treats the modern geologic and geometric approaches to the measurement of the physical characteristics of the basin: stream numbering and ordering techniques, relief measures, and the relations of basin size, shape, and relief with stream flow behavior. Excellent bibliography.

Graf, William L. *The Colorado River: Instability and Basin Management.* Washington, D.C.: Association of American Geographers, 1985. An excellent, well-written study of the particular management problems and practices associated with this large and famous river. Focuses on the way the river has adjusted to a variety of changes caused by climatic change, rangeland management, the building of large dams, and the extraction of water for irrigation. Easily understood by the layperson; sound bibliogrpahy.

Gregory, K. J., and D. E. Walling. *Drainage Basin Form and Process: A Geomorphological Approach.* London: Edward Arnold, 1973. A comprehensive academic textbook aimed at the serious undergraduate or a well-prepared, scientifically minded layperson. Examples from around the world; well-illustrated with photographs, maps, and diagrams. Detailed information on instrumentation, and in chapters 5 and 6, the implications for socioeconomic management are carefully considered. Extensive bibliography.

Montgomery, David R., and Tim B. Abbe. "Distribution of Bedrock and Alluvial Channels in Forested Mountain Drainage Basins." *Nature* 381, no. 6583 (June 13, 1996): 587-590.

More, Rosemary J. "The Basin Hydrological Cycle." In *Water, Earth, and Man,* edited by Richard J. Chorley. London: Methuen, 1969. Lays out in considerable diagrammatic detail how water circulates in and through the basin and the theoretical framework possible for modeling and studying it, particularly with a view to interfering with basin flows in an optimal manner. Excellent bibliography.

Smith, C. T. "The Drainage Basin as an Historical Basis for Human Activity." In *Water, Earth, and Man,* edited by Richard J. Chorley. London: Methuen, 1969. Explains how the drainage basin has long been a natural unit for the focus of human economic activity, with examples from China, Europe, and the Americas. The importance of the basin declined somewhat as the Industrial Revolution progressed, but the need for large-scale planning of water use may be reversing this trend. A helpful guide to the historical perspective.

Keith J. Tinkler

Cross-References

Floodplains, 241; River Flow, 526; River Valleys, 534; Water-Rock Interactions, 692.

Earthquakes

An earthquake is the sudden movement of the ground caused by the rapid release of energy that has accumulated along fault zones in the earth's crust. The earth's fundamental structure and composition are revealed by earthquakes through the study of waves that are both reflected and refracted from the interior of the earth.

Field of study: Geophysics

Principal terms

CRUST: the uppermost 5-40 kilometers of the earth

DEFORMATION: a change in the shape of a rock

ELASTIC REBOUND: the process whereby rocks snap back to their original shape after they have been broken along a fault as a result of an applied stress

LITHOSPHERE: the solid part of the upper mantle and the crust where earthquakes occur

MANTLE: the thick layer under the crust that contains convection currents that move the crustal plates

STRAIN: the percent deformation resulting from a given stress

STRESS: a force per unit area

Summary

Earthquakes are sudden vibrational movements of the earth's crust and are caused by a rapid release of energy within the earth. They are of critical importance to humans, first, because they reveal much about the interior of the earth and, second, because they are potentially one of the most destructive naturally occurring forces found on earth.

The outermost skin of the earth, called the crust, is in constant motion as a result of large convection cells within the upper mantle that circulate heat from the interior of the earth toward the surface. The crust of the earth is about 5 kilometers thick in the oceanic basins and about 40 kilometers thick in the continental masses, while the upper mantle is about 700 kilometers thick. Because the crust is relatively thin compared to the upper mantle, the crust is broken up into several plates that float along the top of each convection cell in the upper mantle. Most earthquakes occur along the boundaries separating the individual plates and are represented by faults that may be thousands of kilometers long and tens of kilometers deep.

Although the vast majority of earthquakes occur along these plate boundaries, some also occur within the plate interior. The rocks on either side of the fault fit tightly together and produce great resistance to movement. As the blocks of rock attempt to move against one another, the resistance of movement causes stress, which is a force per unit area, to build up along the fault. As the stress continues to build, the rocks in the immediate vicinity slowly deform, or bend until the strength of the rock is exceeded at some point along the fault. Suddenly, the rocks break violently and return to their underformed state, much as a rubber band snaps to its original shape when it breaks. This rapid release of stress is called elastic rebound. The point at which the stress is released is called the focus of an earthquake, and that point at the earth's surface directly above the focus is called the epicenter.

The release of energy associated with elastic rebound manifests itself as waves propagating away from the focus. When these waves of energy reach the surface of the earth, the land will oscillate, causing an earthquake. These waves move through the earth in two ways. P (primary) waves move in a back-and-forth motion in which the motion of the rock is in the same direction as the direction of energy propagation. This type of wave motion is analogous to placing a spring in a tube and pushing on one end of the spring. The motion of the spring in the tube is in the same direction as is the motion of the energy. These waves are called primary because they move through the earth faster than do other waves—up to about 25 kilometers per second. Thus, P waves are the first waves to be received at a seismic recording station. Because the individual atoms in a rock move back and forth along the direction of energy movement, P waves can move through solids and liquids and, for this reason, do not tell geologists much about the state (solid or liquid) of a given rock at depth. In contrast to P waves, for S waves, the rock motion is perpendicular to the direction of energy propagation. Guitar strings vibrate in a similar manner: Each part of the guitar string moves back and forth while the energy moves along the string to the ends. S waves are the second waves to be received at a seismic recording station and derive their name from this fact. Unlike P waves, S waves cannot move through liquids but can move through solids. Thus, when a P wave is received by a seismic station but is not followed by an S wave, seismologists know that a liquid layer is between the focus of the earthquake and the receiving seismic station. Both S and P waves are bent, or refracted, as they move in the earth's interior. This refraction occurs as the result of the increase in density of rocks at greater depths. Furthermore, both types of waves are reflected off sharp boundaries, representing a change in rock type located within the earth. Thus, by using these properties of S and P waves, geologists have mapped the interior of the earth and know whether a given region is solid or liquid.

Although S and P waves represent the way seismic energy moves through the earth, once this energy reaches the earth's surface, much of it is converted to another type of wave. L (Love) waves move in the same manner

149

as do S waves, but they are restricted to surface propagation of energy. L waves have a longer wavelength and are usually restricted to within a few kilometers of the epicenter of an earthquake. These waves cause more damage to structures than do P and S waves because the longer wavelength causes larger vibrations of the earth's surface.

The amount of energy released by an earthquake is of vital importance to humans. Many active fault zones, such as the famous San Andreas fault in California, produce earthquakes on an almost daily basis, although most of these earthquakes are not felt and cause no damage to human-made structures. These minor earthquakes indicate that the stress that is accumulating along some portion of a fault is continuously being released. It is only when the stresses accumulate without continual release that large devastating earthquakes occur. The intensity of an earthquake is dependent not only on the energy released by the earthquake but also on the nature of rocks or sediments at the earth's surface. Softer sediments such as the thick muds that underlie Mexico City will vibrate with a greater magnitude than will the very rigid rocks, such as granites, found in other parts of the world. Thus, the great earthquake that devastated Mexico City in 1985 was in part the result of the nature of the sediments upon which the city is built.

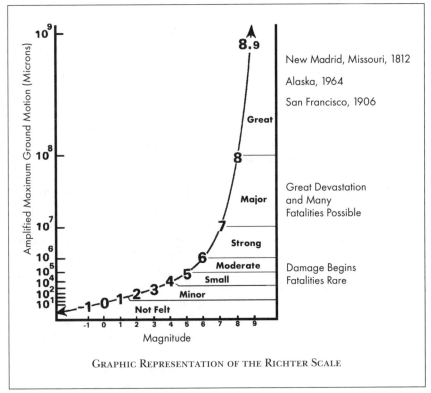

GRAPHIC REPRESENTATION OF THE RICHTER SCALE

For a given locality, earthquakes occur in cycles. Stress accumulates over a period of time until the forces exceed the strength of the rocks, causing an increase in minor earthquake activity. Shortly thereafter, several foreshocks, or small earthquakes, occur immediately before a large earthquake. When a large earthquake occurs, it is usually followed by many aftershocks, which may also be rather intense. These aftershocks occur as the surrounding rocks along the fault plane readjust to the release of stress by the major earthquake. The cycle then repeats itself with a renewed increase in stress along the fault. Although seismologists can usually tell what part of the seismic cycle a region is experiencing, it is difficult to predict the duration of each of these cycles; thus, it is impossible to predict precisely when an earthquake will occur.

Methods of Study

Seismographs are the primary instruments used to study earthquakes. All seismographs consist of five fundamental elements: a support structure, a pivot, an inertial mass, a recording device, and a clock. The support structure for a seismograph is always solidly attached to the ground in such a fashion that it will oscillate with the earth during an earthquake. A pivot, consisting of a bar attached to the support structure via a low-friction hinge, separates a large mass from the rest of the seismograph. This pivot allows the inertial mass to remain stationary during an earthquake while the rest of the instrument moves with the ground. The recording device consists of a pen attached to the inertial mass and a roll of paper that is attached to the support structure. Finally, the clock records the exact time on the paper so that the time of arrival of each wave type is noted. When an earthquake wave arrives at a seismic station, the support structure moves with the ground. The inertial mass and the pen, however, remain stationary. As the paper is unrolled, usually by a very accurate motor, the wave is recorded on the paper by the stationary pen. Modern seismographs, however complex in design, always contain these basic elements. The clock, which each minute places a small tick mark on the recording, is calibrated on a daily basis by a technician using international time signals from atomic clocks. The recording pen often consists of an electromagnet that converts movement of the inertial mass relative to the support structure to an electrical current that drives a light pen. The light pen emits a narrow beam of light onto long strips of photographic film that are developed at a later date.

Seismologists have adopted two widely used scales, which are called the Richter and Mercalli scales, to measure the energy released by an earthquake. The Richter magnitude scale is based on the amplitude of seismic waves that are recorded at seismic stations. Because seismic stations are rarely located at the epicenter of earthquakes, the amplitude of the seismic wave must be corrected for the amount of energy lost over the distance that the wave traveled. Thus, the Richter magnitude reported by any seismic station for a given earthquake will be approximately the same. Richter magnitudes are open-ended, meaning that any amount of seismic energy can be calculated. The weakest earthquakes have Richter magnitudes less than 3.0 and release energy less than 10^{14} ergs. These earthquakes are not usually felt but are recorded by seismic stations. Earthquakes between magnitudes 4.0 and 5.5 are felt but usually cause no damage to structures; they release energy between 10^{15} and 10^{16} ergs. Earthquakes that have magnitudes between 5.5 and 7.0 cause slight to considerable damage to buildings and release energy between 10^{18} and 10^{24} ergs. Earthquakes that are greater than 7.5 on the Richter scale generate energy up to 10^{25} ergs—as much as a small nuclear bomb. The Mercalli intensity scale is not based on the energy released by an earthquake but rather on the amount of shaking that is felt on the ground; it rates earthquakes from Roman numerals I to XII. Unlike the Richter scale, the Mercalli scale provides descriptions of sensations felt by observers and of the amount of damage that results from an earthquake. Thus, an earthquake of Mercalli intensity I is felt only by very few persons, while an earthquake of intensity XII causes total destruction of virtually all buildings.

Both the Mercalli and Richter scales have advantages and disadvantages. The Mercalli scale provides the public with a more descriptive understanding of the intensity of an earthquake than does the Richter scale. The damage caused by an earthquake is a function not only of the energy released by such an event but also of the nature of the sediments or rocks upon which the buildings in the vicinity are constructed. The Richter scale is best used to study specifically the amount of energy released by an earthquake. Finally, the Richter scale, which is purely quantitative, does not rely on subjective observations such as those required by the Mercalli scale.

The exact location of an earthquake epicenter can be deduced from three seismographic stations using triangulation techniques. Because the P and S waves travel at different velocities in the earth, seismologists can determine the distance from the station to the epicenter. They calculate the difference in time between the first arrival of the P and S waves, respectively, at the station. They then divide this time difference by the difference in wave velocities to obtain the distance to the epicenter. The earthquake must have occurred along a circle whose radius is the distance so calculated and whose center is the seismographic station; any three stations that record the event can be used to draw three such circles, which will intersect at a single point. This point is the epicenter.

Context

Earthquakes are one of the most important processes that occur within the earth because they have such a profound effect on how and where people should develop cities. Geologists understand how and where earthquakes occur yet, despite their best efforts, they still cannot accurately determine when an earthquake will happen. They are merely able to predict that a large earthquake will occur in a particular region "in the near future." Very great earthquakes of magnitude 8 or greater, such as the San Francisco earthquake of 1906, occur about every five to ten years throughout the world. Industrialized societies, such as Japan, the United States, and many European countries, have developed buildings that are capable of withstanding devastating seismic catastrophes, but other countries are not as fortunate. Furthermore, some great earthquakes occur in regions that are not considered seismically active. The great Charleston, South Carolina, earthquake of 1886 or the Tangshan, China, earthquake of 1976 are examples of seismic events that could not have been easily predicted using modern technology. In such regions, buildings are not designed to withstand devastating earthquakes. Finally, many regions of the world do not experience earthquakes on a daily basis and, thus, their governments lack the motivation to plan adequately for such potential catastrophic events.

Bibliography

Bolt, Bruce A. *Earthquakes: A Primer.* San Francisco: W. H. Freeman, 1988. This book is an excellent introduction to earthquakes and is written at a level that most laypersons can understand.

Hodgson, John H. *Earthquakes and Earth Structure.* Englewood Cliffs, N.J.: Prentice-Hall, 1964. This source provides the reader with an understanding of how earthquakes are used to determine the structure and composition of the interior of the earth.

McKenzie, D. P. "The Earth's Mantle." *Scientific American* 249 (September, 1983): 66-78. This article, written at the college undergraduate level, is a very complete description of scientific understanding of the interior of the earth.

Matthews, Robert. "Predicting the Big Quake." *Focus: Technology, Life, and Outrageous Science,* July, 1998, 18-24.

Nichols, D. R., and J. M. Buchanan-Banks. *Seismic Hazards and Land-Use Planning.* U.S. Geological Survey Circular 690. Washington, D.C.: Government Printing Office, 1974. The effect of earthquakes on human-made structures is discussed in this short bulletin. Written explicitly for the layperson by the United States government, it provides additional sources of information for land-use planning.

Press, Frank. "Earthquake Prediction." *Scientific American* 232 (May, 1975): 14-23. Press's article details geologists' understanding of earthquake prediction. Also provides a discussion of the methods by which earthquakes can be predicted. Written at the college undergraduate level.

Press, Frank, and R. Siever. *Earth.* 4th ed. San Francisco: W. H. Freeman, 1985. This text includes one of the most complete descriptions of the causes of earthquakes, their measurement, where they occur, how they can be predicted, and how they affect humans. Places earthquakes into the scope of the science of geology. An excellent bibliography, index, and a short geologic dictionary. Written for the freshman college student.

United States Department of the Interior. *Earthquake Information Bulletin.* Washington, D.C.: Government Printing Office. This bimonthly bulletin provides the reader with a concise understanding of where earthquakes occur in the United States and which regions are likely to be affected in the future. Also lists other sources of information on earthquakes. For general and specialized readers.

A. Kem Fronabarger

Cross-References

Earth's Structure, 224; Plate Tectonics, 505; Stress and Strain, 607.

Earth's Age

Determining the age of the earth is one of the great achievements of science. Until the eighteenth century, all geological phenomena were believed to have been produced by historical catastrophes such as great floods and earthquakes. The new geology showed that the earth was billions of years old, rather than thousands as many had previously believed, and that the earth had the form it did because of slow uniform processes rather than catastrophes.

Field of study: Geochronology and paleontology

Principal terms

CATASTROPHISM: the theory that the large-scale features of the earth were created suddenly by catastrophes in the past; the opposite of uniformitarianism

GEOCHRONOLOGY: the study of the time scale of the earth; it attempts to develop methods that allow the scientist to reconstruct the past by dating events such as the formation of rocks

ISOTOPE: atoms with the same number of protons in the nucleus but with differing numbers of neutrons; a particular element will generally have several different isotopes occurring naturally

RADIOACTIVITY: the process by which an unstable atomic nucleus spontaneously emits a particle (or particles) and changes into another atom

SEDIMENTARY: rocks that are formed by a layering process that is generally easily visible in a cross section of the rock

UNIFORMITARIANISM: the theory that processes currently operating in nature have always been operating; it suggests that the large-scale features of the earth were developed very slowly over vast periods of time

Summary

In the middle of the seventeenth century, Joseph Barber Lightfoot of the prestigious University of Cambridge in England penned the following words: "Heaven and earth, center and circumference, were made in the same instant of time, and clouds full of water, and man was created by the Trinity on the 26th of October 4004 B.C. at 9 o'clock in the morning." At the time that Lightfoot wrote those words, this statement expressed the most informed opinion on the age of the earth—namely, that it could be calculated by adding up the ages of the people recorded in the Old Testament and assuming that Adam and Eve were created at about the same time as was the earth. This was the method that most scientists—including Nicolaus Copernicus, Johannes Kepler, and Sir Isaac Newton—used to date the earth, and much effort was expended analyzing the first few books of the Old Testament "scientifically."

A little over a century later, a Scottish geologist named James Hutton suggested that there was a better way to determine the past history of the earth than by poring over biblical genealogies. Hutton believed that processes currently operating in nature could be extrapolated back in time to shed light on the historical development of the earth. This idea—that historical processes are essentially the same as present processes—is called uniformitarianism. In 1785, he presented his new views on geology in a paper entitled "Theory of the Earth: Or, An Investigation of the Laws Observable in the Composition, Dissolution, and Restoration of Land upon the Globe." Uniformitarianism became the foundation of the newly developing science of historical geology. Charles Lyell, who was born in the year of Hutton's death, extended these new ideas and laid the foundation for what was to become a powerful new science. The major argument was over the age of the earth. Was it really billions of years old, as suggested by new discoveries and theories, or was it only a few thousand years old, as everyone had previously believed? The materials from which the earth is constructed are certainly very old. Many of the atoms in the earth date from the beginning of the universe, 15 to 20 billion years ago. The establishment of criteria by which the age of anything will be determined is guided by the need for that age to be a meaningful physical quantity. The conventional definition of age for a person (number of years since birth) is meaningful; the number of years since the origin of the atoms in a person's body would not be meaningful, because it is not relevant to that particular person's duration of existence as

Charles Lyell played a key role in the early debates over earth's age. *(Library of Congress)*

that person. A meaningful definition for the age of the earth can thus be formulated as follows: The age of the earth is the time since its composite materials acquired an organization that could be identified with the present earth.

Current theories of the formation of the earth suggest that the atoms of the earth and all the other members of the solar system formed a cloud of interstellar material that existed in a corner of the Milky Way galaxy several billion years ago. Under the influence of gravity, this cloud of material began to condense in those regions where the concentration of material was sufficiently higher than average. This nebular cloud, as it is called, gave birth to the earth, the sun, and the planets. As the material from which the earth was forming condensed, a number of events occurred: The density increased to the point where the mutual repulsion of the particles balanced the gravity from the newly formed "planet"; the planet became hotter as friction from the now-dense material became a significant source of energy; and energy given off by materials inside the planet was unable to escape into space and was absorbed, further increasing the temperature. The early earth was therefore very hot and existed in a molten state for many years.

There is thus no unique age for the earth. Rather, there is a time period that can realistically be described as the "birth" of the earth. This time period was millions of years long, and any dates given for the age of the earth must necessarily reflect this ambiguity. Fortunately, the age of the earth is measured in billions of years, so the uncertainties surrounding the exact time of its birth do not significantly affect measurements of its age.

Since the initial formation of the earth, many processes have been taking place: Unstable (radioactive) materials have been decaying into other elements; the initial rotation rate has been declining as friction from the tides and the moon has worked to slow the rotation of the earth; mountains have been rising under the influence of global tectonics, and rivers have been formed from the ceaseless activities of erosion; and evolution has been transforming the planet, changing sterile compounds into organic, and barren wasteland into ecological congestion as the phenomenon of life has manifested itself over the face of the globe. As these various physical processes traverse the earth, they leave footprints as evidence of their passing. When these footprints are studied, the history of the earth can be reconstructed. In some cases, this reconstruction can lead all the way back to the origin of the earth, thus providing an answer to the question "How old is the earth?"

Methods of Study

Current estimates put the age of the earth at about 4.6 billion years. This figure is firmly supported by a number of measurements—some very direct and straightforward and some rather subtle. Life itself can be used as a clock. For example, trees add distinguishable layers of growth at a rate of one a year; these are the familiar "rings" that can be counted on a stump of wood.

Counting these rings provides a very accurate clock for determining the age of the tree. Giant sequoias in California are regularly dated at about three thousand years old, and the bristlecone pine has been dated at almost five thousand years. Samples of sedimentary rock, which form yearly layers called varves, can extend back as far as twenty thousand years. Unfortunately, all these annual processes that provide a direct year-by-year chronicle of earth history provide no useful data beyond a few tens of thousands of years.

There are other, less direct, uniformitarian processes, however, that perform somewhat better in this regard. Measurements of erosion, the salinity of the ocean, the strength and direction of the earth's magnetic field, and the internal heat of the earth can all yield values for the "age" of the earth, measured in millions rather than thousands of years. The validity of each of these indirect measurements requires a strict uniformitarian character for the nature of the process; this assumption, however, is not legitimate for most of these processes, which explains why the ages determined from their application are so discordant and unreliable.

The most consistent geological chronometer is based on radioactive decay, an atomic/nuclear phenomenon. All atoms consist of a densely packed nucleus housing a number of protons, which have a positive charge, and neutrons, which have no charge. Because the protons are all positively charged, they repel one another; an atomic nucleus would immediately explode if it were not for a different nuclear force, called the strong force, that holds them together. Every nucleus exists in a state of dynamic tension as the electrical force tries to blow it apart and the strong nuclear force tries to hold it together. Certain nuclei are frequently unstable; that is, they have a tendency to disintegrate spontaneously into other, more stable, nuclei. This disintegration is initiated by yet another nuclear force, the weak force.

Usually the protons in the nucleus of an atom are paired with a particular number of neutrons in such a way that the nucleus will be stable. For the first few elements on the periodic table, the neutron/proton ratio is equal to one, but for larger atoms, the ratio increases as the neutrons start to outnumber the protons. For almost all the elements, there are certain nuclear combinations of protons and neutrons that are stable. By definition, members of the same atomic species have the same number of protons in the nucleus and thus the same atomic number. Atoms with differing numbers of neutrons are called isotopes of that element. Carbon, for example, normally has twelve particles in the nucleus—six protons and six neutrons—and is therefore designated carbon 12. A common isotope, however, has two extra neutrons and is designated carbon 14.

The detailed structure of a particular nucleus determines its long-term stability. Most of the nuclear configurations found in nature, such as hydrogen and helium, are stable indefinitely, or at least for a time that is much longer than the age of the universe (about 20 billion years). Unstable nuclei, on the other hand, are stable for only a finite period of time, which can be either very short (a fraction of a second) or very long (billions of years), depending on the composition of the particular nucleus.

The period of stability for an unstable nucleus is known as its half-life. A half-life is defined to be the time period during which one-half of the nuclei of a given sample will spontaneously decay into another nuclear species. The half-life of carbon 14, for example, is about 5,730 years. This means that in 5,730 years, one-half of an original carbon 14 nucleus, called the parent, will spontaneously decay into another element, nitrogen 14, called the "daughter" element. Over time, the parent element will gradually transform into the daughter. The ratio of daughter to parent can be used to determine how long the parent has been decaying and thus how old the material containing the parent is. It is important to note that the assumption of uniformitarianism for radioactive decay rates is considered very reasonable. Unlike the other processes mentioned above, there seem to be very few mechanisms in nature that can disturb the constancy of the radioactive "clock."

A number of radioactive materials are found in nature, all with differing half-lives. Each can be used to find the ages consistent with their half-lives; that is, a material with a long half-life, such as uranium 238 (whose half-life is almost 5 billion years), can be used to date objects that are billions of years old, and carbon 14 can be used to date objects that are thousands of years old.

Radioactive dating has been applied to many rocks found on the earth. The oldest rocks believed to have formed on the earth are from a volcano in western Greenland and have been dated at about 3.8 billion years, using uranium 238. It is difficult to find very ancient rocks on the surface of the earth, because most of the earth's surface has been rebuilt many times since the earth was born. There are probably older rocks in the deep interior of the earth.

The currently accepted age for the earth, 4.6 billion years, was obtained by dating meteorites that fall to earth from space. These meteorites are believed to have been formed at the same time as was the earth and to have existed in the vacuum of space until they were captured by the gravity from the earth. Similar dates have been obtained from the rocks brought back from the moon, which is believed to have formed at about the same time as the earth.

While many questions remain about the details of the formation of the earth, two facts seem clear: First, the earth owes its origin to the same processes that brought the solar system into existence; second, those processes can be dated with a high degree of confidence at between 4 and 5 billion years ago.

Context

The problem of the age of the earth is part of a much larger scientific question, which exists at the interface between the very practical study of the earth and its various properties and the more esoteric question of the origin and evolution of the universe as a whole. On the practical side, knowledge of the earth's various and occasionally delicate properties is important for the future of the human race. By knowing how long the earth has been in existence, scientists are better able to understand the processes that have shaped the surface of the earth into the form that it has today. Predicting earthquakes, hunting for oil, monitoring the spread of the sea floor—all these practical questions require knowledge of large-scale planetary pro-cesses, the same kind of knowledge that illuminates the question of the age of the earth. Furthermore, knowing that the earth is billions of years old and can easily survive for billions more should encourage human societies to take better care of the planet.

From a more esoteric or speculative point of view, the age of the earth is important because it speaks to the most fundamental questions that are asked about the place of human beings in the universe. How old is this planet? How was it formed? In the century or so since geological science overthrew the seventeenth century notion of a much younger earth, people have struggled with finding a new place in the universe. The argument that began centuries ago is still heard in courtrooms across the United States as "creation science" once again argues that the earth is thousands, not bil-lions, of years old. Legal battles rage over the issue of whether high schools across the country should teach geochronology that is based on religious dogma rather than on scientific research. Research is still being done on this very important scientific question and no doubt will continue into the foreseeable future as the human mind strives to learn more about the earth. The growing awareness of how dependent humans are on the continued health of the earth is a powerful incentive to learn more about their planetary home.

Bibliography

Haber, Frances C. *The Age of the World: Moses to Darwin.* Baltimore: Johns Hopkins University Press, 1959. Reprint. Westport, Conn.: Greenwood Press, 1978. This interesting book does not focus on current estimates of the age of the earth but rather on the historical controversy that emerged when nonbiblical values for the age of the earth began to be accepted. Provides insight into the conflict between science and dogma.

Hurley, Patrick M. *How Old Is the Earth?* Garden City, N.Y.: Doubleday, 1959. One of the few full-length books on geochronology for the layperson. Even though published many years ago, it is still valid, as most of the material relevant to the age of the earth has not changed appreciably since its publication.

Ozima, Minoru. *The Earth: Its Birth and Growth.* Cambridge, England: Cambridge University Press, 1981. A translation of a Japanese book that was written by a scientist whose specialty is geochronology. Written at an introductory level.

Stearn, Colin W., et al. *Geological Evolution of North America*. New York: John Wiley & Sons, 1979. Several excellent chapters discussing the age of the earth. Contains an excellent chapter on geological time and the various ways it can be measured.

Stokes, William Lee. *Essentials of Earth History: An Introduction to Historical Geology*. 4th ed. Englewood Cliffs, N.J.: Prentice-Hall, 1982. A standard introductory text on historical geology. All the various methods for determining the age of the earth are discussed in the first few chapters.

_____, et al. *Introduction to Geology: Physical and Historical*. Englewood Cliffs, N.J.: Prentice-Hall, 1978. Textbook similar to Stokes's other book in terms of its discussion of geochronology.

Thackray, John. *The Age of the Earth*. New York: Cambridge University Press, 1989. A very short publication, about forty pages long, published by a British geological museum. Contains more pictures than text, but the pictures, most in color, are helpful and make this an interesting source.

Karl Giberson

Cross-References

Earth's Composition

Understanding the processes that have evolved the earth can help to unify various earth and biological sciences. The theories about earth's evolution are speculative, and much of the earth's earliest history is unknown. Using meteorites and some of the oldest-known crustal rocks, geochemists are trying to unravel the mysteries of the early earth's composition.

Field of study: Geochemistry

Principal terms

ACCRETION: the process by which small bodies called planetesimals are attracted by mutual gravitation to form larger bodies called proto-planets

ARCHEAN EON: the older of a two-part division of the Precambrian, also known as the Archeozoic

CHONDRITES: stony meteorites that contain rounded silicate inclusion grains called chondrules; they are believed to have formed by crystallization of liquid silicate droplets and volatiles

DIFFERENTIATION: the process by which a planet is divided into zones as heavy elements (metals) sink to the core, while lighter elements collect near the surface

ISOTOPE: atoms of an element that have the same number of protons in the nucleus, the same atomic number, and the same chemical properties but that have different atomic masses because they have different numbers of neutrons in the nucleus

MAFIC AND ULTRAMAFIC: rock-forming magmas that are high in dense, refractory elements such as iron and magnesium; oceanic basalts are examples of mafic rocks

REFRACTORY (SIDEROPHILE) ELEMENTS: elements least likely to be driven off by heating; the last elements to be melted as a rock is heated to form magma

VOLATILE ELEMENTS: elements most likely to be driven off by heating; those that are first to melt or be driven off as gas in a heated rock

ZIRCONS: mineral inclusions found in granitic rocks, zircons are often the only evidence left of early crustal rocks

Summary

About 4.5 billion years ago, scientists believe, a massive star exploded in a supernova event that shined as brightly as a whole galaxy of stars. Shock waves from the celestial fireworks overtook a cloud of gas and dust a few light-years away and triggered its contraction, simultaneously seeding the nebula with heavy elements (those heavier than iron on the periodic table). The solar nebula's collapse led to the formation of the sun (which swept up most of the matter), and the planets formed by the accretion of small bodies called planetesimals. As the planetesimals grew into protoplanets, their gravitational fields increased, so they continued to sweep up material not garnered by the protosun. The innermost planets, Mercury, Venus, Earth, and Mars, contained the dense metals and rocks, while the outer planets were mostly made of gases and volatile ices. During the protoearth's initial accretion process, small, cold bodies collided to form a large mass of homogeneously heterogeneous composition. By the process of differentiation, the heavier metallic elements, such as iron and nickel, migrated to the core of the early earth, while the lighter elements migrated to the outer portions of the contracting planet.

Meteorites offer clues to the composition of the earth. Extraterrestrial pieces of an asteroid or planetesimal, meteorites are remnants of the earliest period of planetary formation and come in three basic types. Stony meteorites make up the most abundant group and are composed of silica-associated, or lithophile, elements such as those found in the earth's crustal materials. Stony-iron meteorites are composed of roughly equal parts of rock (typically olivine) suspended in a matrix of iron. Iron meteorites are composed of siderophile elements, iron being the major constituent, along with (perhaps) 10-20 percent nickel. Iron meteorites are of particular interest to scientists attempting to model the composition of the earth's core. The mean density of the earth is about 5.5 grams per cubic centimeter. The mean density of crustal rocks, however, is only about 2.7 grams per cubic centimeter (water is conveniently 1 gram per cubic centimeter), which indicates a core density of ten to twelve times that of water. The only known objects approaching these densities are the iron meteorites.

After the initial accretion of the planetesimal materials and just prior to differentiation of the lithophile and siderophile elements, the earth's thermal history began through the process of radioactive decay. During this early thermal period, short-lived radioactive nuclides (atoms of a specific isotope, distinguished by their atomic and mass numbers) produced heating seven times greater than that of today's molten core. Most of the heating was attributable to the decay of potassium 40 as well as of the short half-lived elements such as aluminum 26. After about 100,000 years, the planet separated into the iron-nickel core and magnesium-iron-silicate lower mantle. Over a longer time scale (probably more than 10 million years), the high-volatility compounds, such as lead, mercury, thallium, and bismuth, along with the noble gases, water in hydrated silicates, and carbon-based organic compounds, all condensed. This volatile-rich material migrated to the surface, where it was melted into magmas in a continuous period of crustal reprocessing that lasted for about 1 billion years.

The earth's original inventory of gases appears to have been lost, based on the relative present abundances of the rare gases (helium, neon, argon, krypton, xenon, and radon) compared to the present silicon content of the earth. Later periods of volcanic outgassing and perhaps impacts with volatile-rich cosmic objects such as carbonaceous chondritic meteorites and comets may also have played a role in the evolution of the atmosphere and oceans. Separated into three main layers—the crust, mantle, and core—the earth is an active body, its internal heat far from exhausted. The complexity of the chemical composition increases as one examines each successive outward layer. This generalized model gives a starting point from which to examine the complex nature of earth materials. Earth's wide range of pressure and temperature regimes helps explain why there are more than two thousand distinct minerals and numerous different combinations of minerals in rock types.

The core is actually composed of two basic parts: the solid inner core, with a density equal to twelve times that of water and a radius of 1,300 kilometers, and a molten outer core, 2,200 kilometers thick, with a density of about 10 grams per cubic centimeter. This model consists of an essentially iron-nickel inner core at high pressure and a metallic outer core that also contains iron sulfide and light elements such as silicon, carbon, and oxygen. As a whole, the core unit comprises about 32 percent of the earth's mass. Comprising the outer 68 percent of the earth's bulk, the mantle makes the crust, atmosphere, and oceans insignificant by comparison. The mantle is rich in dense, or mafic, rocks such as olivine and pyroxene (which comes in two basic types, calcium-rich or calcium-poor), with olivine the dominant mineral.

Basic earth materials are derived via reaction series from mafic magmas melting and settling out in the mantle's upper regions. As the temperatures drop in the melt zone, a discontinuous series (a set of discrete reactions) occurs. Magnetite, an oxide of iron and titanium, is the first to settle out, with the highest melting point at about 1,400 degrees Celsius. Olivine, a mineral whose silicate structure is a simple tetrahedron, is the next to solidify out of the melt, with a density of 3.2-4.4, followed by the single chain structure pyroxene, with a density of 3.2-3.6. As temperatures in the magma drop to near 1,000 degrees Celsius, the amphibole group forms with a lesser density, 2.9-3.2. As the cooling progresses, the structures increase in complexity with the micas—biotite and muscovite, which form in planar sheets. Next in the cooling sequence would be orthoclase, or potassium feldspar, and plagioclase, or calcium feldspar, and, finally, quartz, which are all distinguished by their characteristic three-dimensional diamond shapes and varying colors. The calcic through sodic plagioclase to muscovite, biotite, and quartz occurs in a smooth, or continuous, transition rather than the

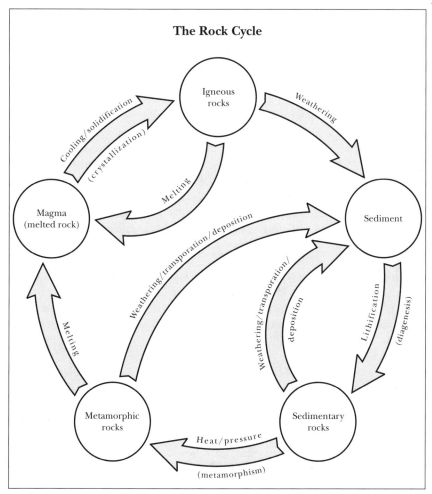

The Rock Cycle

The rock cycle, the basic geochemical cycle, operates on a time scale of hundreds of millions to billions of years. It includes subcycles such as the oceanic cycle and the biological cycle, which could be called parts of the "atmospheric-hydrologic-biological-sedimentary" cycle.

stepwise, or discontinuous, reactions that characterize the formation of olivine through biotite.

An estimate of the crustal elemental composition of the earth indicates that only a handful of elements (oxygen, silicon, aluminum, iron, magnesium, calcium, sodium, and titanium) make up more than 99 percent of the earth's crust. The simple oxide quartz is the most common of the silicate minerals, which account for 95 percent of the crust. With these facts in mind, one can start to hypothesize about how the continents evolved. About 700 million years after the initial formation of the earth through accretion and differentiation, the first rocks of the Archean eon formed. They are composed of olivine, pyroxene, and anorthite (calcium-rich plagioclase feldspar), which settled out of the basaltic magma. The lighter plagioclase would rise to the surface to form a hardened crust of anorthosite, the same material that makes up the moon's ancient highlands, which are about 3.8 billion years old.

The anorthosite formed a thick sheet that was fractured into pieces and subjected to further heating through radioactive decay, leading to an essentially granitic rock layer 10-15 kilometers thick. Extensive volcanic activity and high surface temperatures gradually diminished until the hydrosphere (water cycle) was established. The earth's crust is divided into two main types: the dense, or mafic, oceanic crust and the lighter, or sialic (silica-aluminum), continental crust. Archean rocks (up to 3.5 billion years old) found in the stable interiors of the continents contain massive anorthosite inclusions and may be viewed as the nuclei of the continents.

About the time of the formation of the continental nuclei, or cratons (relatively stable portions of crust), the oldest-known sedimentary rocks accumulated as the rock cycle began, eroding the parent igneous rocks into secondary types of rock. This occurrence may coincide with the beginning of plate tectonics, as the lithosphere (rock crust) of the earth broke into plates and began its hallmark active motion. Life is thought to have arisen at about the same time, with primitive blue-green algae found in strata 2.8 billion years old. With the oceans growing in volume and salinity and the development of oxygen-releasing blue-green algae, earth's geochemistry became more complex. Chemically precipitated rocks of calcium carbonate, commonly known as limestones, are an example of the evolving rock cycle.

Life forms shaped the earth's chemical composition. By the end of Precambrian time, oxygen levels had reached 1 percent of its present value. Multicelled animals in the oceans scrubbed the carbon dioxide from the atmosphere and locked it up in the carbonate rocks, forming biochemically precipitated limestones. By the late Paleozoic era, coal formations grew as a result of the first land forests being periodically inundated by ocean transgressions.

Methods of Study

Perhaps no other earth science is as speculative as that of early earth history and the geochemical evolution of the earth. Varying models for crustal development are advanced and overturned annually. Despite the problems of extrapolating back to a time before there were solid rocks, the established models are based on some solid lines of evidence as well as on conjecture. In 1873, American geologist James D. Dana made one of the initial advances in the study of the earth's internal chemical composition when he suggested that analogies could be drawn from the study of meteorites. Believed to be pieces of differentiated bodies that were later disintegrated into smaller pieces, meteorites come in differing types that are analogous to the earth's interior. Because meteorite types approximate elemental distribution in the earth, they are valuable samples for laboratory examination by scientists. Geochemists studying meteorites have derived radiometric dates of 4.6 billion years—corresponding to the initial time of accretion and differentiation of the planets.

Geophysicists use seismic waves to study the earth's interior. Changes in velocity and deflections of the waves passing through the earth have revealed a differentiated earth with a very dense core, less dense mantle, and a light crust "floating" on top. The well-established theory of plate tectonics has shown that the crust is broken into pieces, or plates, that are moving, driven by convection currents in the upper mantle. Some of the major challenges confronting earth scientists are the questions about how the earth's crust formed and about when plate movement began.

During the 1960's, interest in Archean crustal evolution was aroused by the discovery of Archean era magnesium-oxygen-rich lavas similar to those found in the early Precambrian. Called komatiites, these rocks date back to 3.7 billion years ago and represent ultramafic lavas that form at 1,100 degrees Celsius. Komatiites are generally found around greenstone belts, an agglomeration of Archean basaltic, andesitic, and rhyolitic volcanics, along with their weathering and erosion derived sediments. One hundred million years older than any previously known rocks, the finds led to further exploration of Archean formations by field geologists in West Greenland-Labrador, Zimbabwe, Transvaal-Swaziland, Ontario-Quebec, southern India, Western Australia, and more recently, China and Brazil.

Important work by field geologists in these regions launched a new era in Precambrian geology. The primary targets for study are the greenstone belts and granitic-gneiss associations. An important twentieth century find included detrital zircon, discovered in Australia. An age of 4.2 billion years for the zircons was determined using precise ion microprobe analysis. The zircon find is significant because it places an approximate birth date for the continental crust, as zircon is a mineral constituent of granite (recall that oceanic crust is composed of mafic and ultramafic rocks while continental crust is granitic).

The drive to study Archean rocks was further fueled by the United States' Apollo missions to the moon, which returned rocks of slightly older age from the lunar surface. At the same time, geochemists were able to refine their study of these ancient rocks with more sophisticated methods to determine ratios of isotopes in the samples. Instruments common in the geochemical lab today are X-ray diffraction and gamma-ray spectrometers, which probe the nuclei of atoms to determine the spectral fingerprint of elements and their various isotopes. Isotopic ratios in rocks are of particular interest to geochemists because they provide clues as to chemical cycles in nature, such as the sulfur, chemical, nitrogen, and oxygen cycles. The equilibria of these cycles, as indicated by the isotopic ratios, offer insights into volcanic, oceanic, biological, and atmospheric cycles and conditions in the past.

Context

It is generally accepted by most earth scientists that crustal formation and heat flow were substantially greater in Archean times. The question is whether this crust was broken into moving lithospheric plates as it has been for the past 900 million years. The question of plate motion during this early period has generated debate among scientists and has led to two general theories of early crustal evolution. If plate tectonics was occurring 4 billion years ago, one would expect to find formations of arc deposits and complexes similar to the Franciscan formation in California's coast range. Oölite and arc deposits are terranes that accumulate near zones of subduction, where dense mafic rocks are recycled into the mantle. Such formations have not been found to date—geologic evidence arguing against rapid plate motion.

If crustal rock production was great and yet plate tectonics minimal, what process shaped the early earth? An answer may have emerged from one of the earth's sister planets. Shrouded in clouds, Venus did not give up the mysteries of its geology until the radar maps generated by Soviet and American spacecraft. Like Mars, with its giant volcanoes in the Tharsis region, Venus appears to have great shield volcanoes and continent-like regions the size of Africa and Australia. Hot-spot volcanism, in which plumes of magma rising from the planetary interior erupt to form shield volcanoes at the surface, may indeed be the key to understanding incipient plate tectonics on the early earth.

Perhaps no other area of scientific study is as intriguing and controversial as that of the origin and evolution of the earth. Geochemists have been at the forefront of the quest for understanding the earth's present geology in terms of its past. Before the 1960's, little was known of the earth's history during early Precambrian times. This lack is significant when one realizes that the Precambrian comprises about 87 percent of the geologic time scale.

It is likely that new techniques used to analyze rocks and minerals in the laboratory will lead to a better understanding of the formation of the earth's crustal materials and the evolution of moving crustal plates. Precise dating of zircons from ancient rocks, isotope analysis, and high-resolution seismic data will help scientists to comprehend the relationships between the granite-greenstones and gneiss terranes (crustal blocks) that typify Archean formations.

Bibliography

Burchfiel, B. Clark, et al. *Physical Geology*. Westerville, Ohio: Charles E. Merrill, 1982. An excellent and comprehensive textbook covering all aspects of geology suitable for the layperson or liberal studies college student. Of special interest are chapter 2 on mineralogy, chapter 7 on the earth's interior, chapter 9 on crustal materials and mountain building, and chapter 10 on the origin and differentiation of the earth and early geologic time.

Fyfe, W. S. *Geochemistry.* Oxford, England: Clarendon Press, 1974. Part of the Oxford Chemistry series, this work was written for lower-division college chemistry students. Although in some respects dated, it is nevertheless a brief (about one-hundred-page) and excellent introduction to the science of geochemistry. Of special interest is chapter 9, "Evolution of the Earth." The book has a bibliography, glossary, and index.

Gregor, C. Bryan, et al. *Chemical Cycles in the Evolution of the Earth.* New York: John Wiley & Sons, 1988. A systems approach to geochemistry, this book is suitable for the serious college student. Although filled with graphs, tables, and chemical equations, sections are very readable for the layperson. Discussions of mineralogical, oceanic, atmospheric, and other important chemical cycles are extensive, and the work is well referenced.

Kroner, A., G. N. Hanson, and A. M. Goodwin, eds. *Archaean Geochemistry: The Origin and Evolution of the Archaean Continental Crust.* Berlin: Springer-Verlag, 1984. A collection of reports by the world's leading geochemists studying the geochemistry of the world's oldest rocks. Although many of the articles are technical in nature, the abstracts, introductions, and summaries are accessible to a college-level reader interested in the work of top international scientists.

Levin, Harold L. *The Earth Through Time.* 3d ed. Philadelphia: Saunders College, 1988. An excellent and very readable text dealing with historical geology. Filled with illustrations, photographs, and figures, this book is suitable for the layperson. Chapters on planetary beginnings, origin and evolution of the early earth, and plate tectonics are of special interest. Contains an excellent glossary and index.

McCall, Gerald J. H., ed. *The Archean: Search for the Beginning.* Stroudsburg, Pa.: Dowden, Hutchinson and Ross, 1977. A superb collection of thirty-eight papers by outstanding geologists, arranged under topical headings. The papers are at times technical, but the editor provides an introduction and integrating commentary that helps bridge the gap for the nontechnical reader. Contains a subject index.

Ponnamperuma, Cyril, ed. *Chemical Evolution of the Early Precambrian.* New York: Academic Press, 1977. A collection of papers from the second colloquium of the Laboratory of Chemical Evolution of the University of Maryland, held in 1975. Written by experts in the field, the papers are still, for the most part, accessible to the nontechnical reader. The volume contains a subject index.

Salop, Lazarus J. *Geological Evolution of the Earth During the Precambrian.* Berlin: Springer-Verlag, 1983. A top Soviet geologist conducts an exhaustive survey of Precambrian geology. Suitable for a college-level reader with a serious interest in the subject. Contains numerous graphs and tables, with extensive references.

Tarling, D. H. *Evolution of the Earth's Crust.* New York: Academic Press, 1978. Written for the undergraduate-level college reader with some background in geology, this volume is an excellent collection of nontechnical, well-written essays covering the origin and evolution of the earth's crust and plate tectonics. Contains references and an index.

Wedepohl, Karl H. *Geochemistry.* New York: Holt, Rinehart and Winston, 1971. An accessible and brief introduction to geochemistry fundamentals. Contains an excellent chapter on meteorites and cosmic abundances of the elements. Suitable for the nontechnical reader, with index and references. A good starting point for those unfamiliar with mineral formation.

Wetherill, George W., A. L. Albee, and F. G. Stehli, eds. *Annual Review of Earth and Planetary Sciences.* Vol. 13. Palo Alto, Calif.: Annual Reviews, 1985. Three articles of interest to the earth history student are "Evolution of the Archean Crust," by Alfred Kroner, and "Oxidation States of the Mantle: Past, Present, and Future" and "The Magma Ocean Concept and Lunar Evolution," by Richard Arculus. Kroner's article is particularly readable for the college-level audience, with an excellent overview of the historical views on Precambrian geology. References at the end of each article.

David M. Schlom

Cross-References

Earth's Age, 155; Earth's Core, 171; Earth's Crust, 179; Earth's Mantle, 195; Earth's Oldest Rocks, 202; Earth's Origin, 210; Earth's Structure, 224; Magmas, 383; Ultramafic Rocks, 655.

Earth's Core

The core is the earth's densest, hottest region and its fundamental source of internal heat. The thermal energy released by the core's continuous cooling stirs the overlying mantle into slow, convective motions that eventually reach the surface to move continents, build mountains, and produce earthquakes.

Field of study: Geophysics

Principal terms

CONVECTION: the process in liquids and gases by which hot, less dense materials rise upward to be replaced by cold, sinking fluids

CORE: the spherical, mostly liquid mass located 2,900 kilometers below the earth's surface; the central, solid part is known as the inner core

MAGNETIC FIELD: a force field, generated in the core, that pervades the earth and resembles that of a bar magnet

P WAVES: seismic waves transmitted by alternating pulses of compression and expansion; they pass through solids, liquids, and gases

S WAVES: seismic waves transmitted by an alternating series of sideways movements in a solid; they cannot be transmitted through liquids or gases

SEISMIC WAVES: elastic oscillatory disturbances spreading outward from an earthquake or human-made explosion; they provide the most important data about the earth's interior

Summary

The earth's core extends from a depth of 2,900 kilometers to the center of the earth, 6,371 kilometers below the surface. The core is largely liquid, although toward the center, it becomes solid. The liquid part is known as the outer core; the solid part, the inner core. Ambient pressures inside the core range from 1 million to nearly 4 million atmospheres, and temperatures probably reach more than 5,000 degrees Celsius at the earth's center.

Being almost twice as dense as the rest of the planet, the core contains one-third of the earth's mass but occupies a mere one-seventh of its volume. Surrounding the core is the mantle. The boundary between the solid mantle and the underlying liquid core is the core-mantle boundary (CMB), a surface that demarcates the most fundamental compositional discontinuity in the earth's interior. Below it, the core is mostly made of iron-nickel oxides.

Above it, and all the way to the surface, the mantle is made of silicates (rock-forming minerals). The solid inner core contains 1.7 percent of the earth's mass, and its composition may simply be a frozen version of the liquid core. The boundary between the liquid and the solid cores is known as the inner core boundary (ICB); it appears sharp to seismic waves, which easily reflect off it.

The core has lower wave-transmission velocities and higher densities than the mantle, a consequence of its being of a different chemical composition. The core is probably composed of 80 to 90 percent (by weight) iron or iron-nickel alloy and 20 to 10 percent sulfur, silicon, and oxygen; it therefore must be a good electrical and thermal conductor. The mantle, in contrast, is composed mainly of crystalline silicates of magnesium and iron and is therefore a poor conductor of electricity and a good thermal insulator.

This sharp contrast in physical properties is a major end product of the way in which the earth evolved thermally, gravitationally, and chemically. It is difficult, however, to tell whether the earth's core formed first and the earth was accreted from the infall of meteorites and other gravitationally bound materials or, alternatively, the core differentiated out of an already formed protoearth, in which silicates and iron were separated after a cataclysmic "iron catastrophe." This event may have occurred when iron, slowly heated by radioactivity, suddenly melted and sank by gravity toward the earth's center, forming the core. Unfortunately, the two scenarios are equally likely, and both give the same end result; moreover, there probably are other scenarios. Calculations show, however, that iron sinking to the core must have released great amounts of energy that would have eventually heated and melted the entire earth. Cooling of the outer parts proceeded rapidly, by convection, but as the silicate mantle solidified, it created a thermal barrier for the iron-rich core, which, not being able to cool down as readily, remained molten. The inner core began then to form at the earth's center, where the pressure was greatest and solidification was (barely) possible.

The most tangible consequence of the existence of a fluid, electrically conducting core is the presence of a magnetic field in the earth that has existed for at least 3.5 billion years with a strength not very different from what it has today. The process that generates and maintains the geomagnetic field is attributable to a self-exciting dynamo mechanism—that is, an electromagnetic induction process that transforms the motions of the conducting fluid into electric currents, which in turn induce a magnetic field that strengthens the existing field. (For the system to get started, at least a small magnetic field must be present to initiate the generation of electric currents.) The increased magnetic field in turn induces stronger currents, which further strengthen the field, and so on. As the magnetic field increases beyond a certain high value, it begins to affect the fluid flow; there is a mechanical force, known as the Lorentz force, that is induced in a conductor as it moves across a magnetic field. The stronger the magnetic field, the stronger the Lorentz force becomes and the more it will tend to modify the motion of the fluid so as to oppose the growth of the magnetic field. The result is a self-regulating mechanism which, over time, will attain a steady state.

A dynamo mechanism is needed to explain the geomagnetic field, because there can be no permanently magnetized substances inside the earth. Magnetic substances lose their magnetism as their temperature increases above the so-called Curie temperature (around 500 degrees Celsius for most magnetic substances), and most of the mantle below the depth of 30 kilometers and all the core is at temperatures well above the Curie point. The basic problem, then, is to find a source of energy that can maintain the steady regime of flow in the core against decay by somehow maintaining the fluid currents that induce the field. A favored view is that the necessary energy to maintain the flow is provided by the growth of the inner core as it is fed by the liquid core. According to some researchers, this process would provide enough gravitational energy to stir the core throughout. Thermally and compositionally driven flows can also be invoked as possible models of core fluid dynamics, but there is still no evidence that decides the question.

A most extraordinary feature of the core-generated magnetic field is that, at least over the past few hundred million years, it has reversed its polarity with irregular frequency. For example, it is known that at times the field has reversed as frequently as three times every million years, but in other cases, more than 20 million years went by without a noticeable reversal. A reversed geomagnetic field means simply that the magnetic needle of a compass would point in the opposite direction as it does today. (For convenience, the present orientation of the needle is considered normal.) The important point is that the rocks that form (for example, lavas that cool below the Curie point as they become solid rock) during either a reverse or a normal period acquire and preserve that magnetism. Unlike the swinging compass needle, a rock keeps the magnetic field direction that existed at the time of its formation forever frozen in its iron-bearing minerals. Therefore, rocks formed throughout geologic time have recorded the alternating rhythms of normal and reverse earth magnetism. This sequence of magnetic reversals contains the clue to the core's nature.

Geophysicists are anxious to learn whether the core is vigorously convecting as a consequence of the inner core's growth. If that were the case, the core would be delivering a great amount of heat to the mantle, whose low thermal conductivity would create a barrier to the upcoming heat. As a result, the local temperature gradient at the base of the mantle would probably be very high, so that a layer 100 kilometers thick, say, at the base of the mantle, would be gravitationally unstable. From this layer, thermal inhomogeneities would rise through the mantle in the form of plumes of buoyant, hot, lower-mantle material. Several such plumes might reach the upper mantle or set the entire mantle into convection. These convection currents would be responsible for the motion of the tectonic plates on the earth's surface and, consequently, for the uplifting of mountain ranges, the formation of oceanic basins, and the occurrence of volcanic eruptions and earthquakes. Continental drift and plate tectonics, the most visible effects of the internal cooling of the earth, would thus be linked to the growth of the inner core and to earth's earliest history. This view of the earth is very speculative, but it is favored by many geoscientists, who recognize its beauty and simplicity.

Methods of Study

Knowledge of the structure, physical properties, and composition of the core is entirely based on indirect evidence gathered mostly from analyses of seismic waves, the study of the earth's gravitational and magnetic fields, and laboratory experiments on the behavior of rocks and minerals at high pressures and temperatures. The first evidence for the existence of the core was presented in a paper suggestively entitled "The Constitution of the Interior of the Earth, As Revealed by Earthquakes," published in 1906 by Richard D. Oldham of the geological survey of India. Thirty years later, Inge Lehmann, from the Copenhagen seismological observatory, presented seismic evidence for the existence of the inner core. In the past few decades, with the advent of high-speed computers and technological advances in seismometry, seismologists have developed increasingly sensitive instrumentation to record seismic waves worldwide and sophisticated mathematical theories that allow them to construct models of the core that explain the observed data.

Seismic waves provide the most important data about the core. Earthquakes or large explosions generate elastic waves that propagate throughout the earth. These seismic waves may penetrate deep in the earth and, after being reflected or transmitted through major discontinuities such as the CMB and ICB, travel back to the surface to be recorded at the seismic stations of the global network. The most direct information that seismic or elastic waves carry is their travel time. Knowing the time it takes for elastic waves to traverse some region of the earth's interior allows the calculation of their velocity of propagation in that region. The velocity of seismic waves strongly depends on the density and rigidity, or stiffness, of the material through which they propagate, so estimates of the mechanical properties of the earth can in principle be derived from seismic travel time analyses. Seismic waves that propagate through the deep interior of the earth are of two types: compressional waves (also called P waves) and shear waves (also called S waves). Compressional waves produce volume changes in the elastic medium; shear waves produce shape distortion without volume change. If the medium has some rigidity, both P and S waves can be transmitted. If the medium has no rigidity, it offers no resistance to a change in shape; no elastic connection exists that can communicate shearing motions from a point in the medium to its neighbors, so S waves cannot propagate, although P waves can.

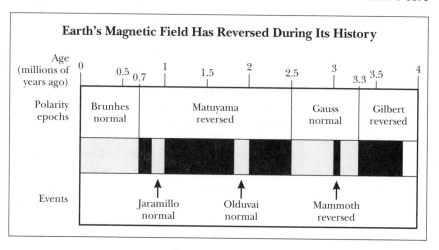

After many years of careful observations, it has been determined that S waves are not transmitted through the outer core. Therefore, the outer core material has no rigidity, but behaves as a fluid would. Similar observations suggest that the inner core is solid; the actual rigidity of the inner core is very difficult to estimate, however, since shear waves inside the inner core are isolated from the mantle by the outer core and can only travel through it as P waves converted from S waves at the ICB. Nevertheless, when the whole earth is set into vibration by a very large earthquake, the average rigidity of the inner core can be estimated by comparing the observed frequencies of oscillation with those theoretically computed for models of the earth that include a solid inner core. Model studies have indicated that the inner core is indeed solid, because a totally liquid core model does not satisfy the observations.

The average velocity of P waves in the earth is about 10 kilometers per second, whereas the average P-wave velocity in the rocks accessible to measurement at the earth's surface is 4 to 5 kilometers per second. The S-wave velocity is nearly half that of P waves in solids and zero in perfect fluids.

The velocity of P waves drops abruptly from 13.7 kilometers per second at the base of the mantle to 8.06 kilometers per second across the CMB, at the top of the core. From this point down, the velocity steadily increases to 10.35 kilometers per second at the ICB, where it jumps discontinuously to 11.03 kilometers per second at the top of the inner core. From there to the center of the earth, the velocity of P waves increases slowly to reach 11.3 kilometers per second. The S-wave velocity increases from zero at the ICB to around 3.6 kilometers per second at the earth's center. The core's density abruptly increases from 5,500 kilograms per cubic meter at the base of the mantle to nearly 10,000 kilograms per cubic meter just underneath the CMB. From there, the density increases slowly to nearly 13,100 kilograms per cubic meter at the earth's center. In comparison, the density of mercury at room temperature and ambient pressure is 13,600 kilograms per cubic meter.

That the core is mostly iron is consistent with iron's being cosmically more abundant than other heavy elements and with the high electrical conductivity the core needs to have in order to generate the earth's magnetic field. The fluidity of the outer core has been demonstrated by measurements not only of seismic wave transmission but also of the oscillation period of gravitational waves in the core excited by the lunisolar tides. The existence of a sustained, steady magnetic field is also consistent with a fluid outer core.

Seismic data can probe the inner core only partially from the earth's surface, unless the source of the seismic waves and the receivers are located antipodally to each other. Such an arrangement would allow scientists to measure seismic waves that had penetrated the center of the earth. It would be possible to construct a global experiment to investigate the inner core by deploying an array of highly sensitive seismic sensors antipodal to either a seismically active region or an underground nuclear explosion testing ground. Despite the wealth of unique data that would be obtained from such an experiment, it would be a very expensive endeavor, and not devoid of risk.

New views of the earth's interior are produced, sometimes unexpectedly, by the analyses of data collected by satellite missions. Data from orbiting satellites that measure tiny variations of the earth's gravitational field, combined with computer-aided seismic tomography of the earth's interior, have revealed large-density anomalies at the base of the mantle and a large relief of more than 2 kilometers on the CMB. Seismic tomography uses earthquake-generated waves that penetrate the mantle in a multitude of directions to map the three-dimensional structure of its deep interior, just as computerized medical tomography uses multiple X-ray images to create a three-dimensional view of internal organs of the body. Essential to the success of these studies, however, is the installation of dense networks of seismic sensors all over the surface of the earth; this installation, however, is another very expensive procedure.

Context

Any study of the earth's physical environment is likely to provide insight into the nature and future of the planet and, consequently, the future of humankind. If geophysicists come to understand how the earth's core works, they will be able to predict the geomagnetic field's activity for years to come. Thus, they will be able to predict an upcoming reversal. According to the best estimates, a reversal does not occur suddenly but takes about ten thousand years. That means that during a reversal, there is a time of very small or even zero field intensity. Under such conditions, the magnetic shielding that prevents the highly energetic solar wind particles from reaching the earth's surface will disappear, leaving the earth directly exposed to lethal radiations.

The inner core has not yet been sufficiently explored with seismic waves.

One reason is that it is the remotest region of the earth and therefore the most difficult to reach; another is that it is hidden beneath the "seismic noise" created by the crust, mantle, and outer core. The inner core, however, holds the key to the understanding of the earth's early history and its subsequent development as a planet.

Bibliography

Bolt, Bruce A. *Inside the Earth: Evidence from Earthquakes.* San Francisco: W. H. Freeman, 1982. An elementary treatment of what is known about the earth's interior, mostly through the study of seismic waves, the author's major field of research. The book contains abundant diagrams that illustrate accurately important results of the investigation of the core and mantle. For readers with some knowledge of mathematics, the book includes brief derivations of important formulas, separated by "boxes" from the main text. It is well written and includes anecdotal descriptions of great scientific discoveries along with personal views of the history and development of seismology. Illustrated.

Clark, Sydney P. *Structure of the Earth.* Englewood Cliffs, N.J.: Prentice-Hall, 1971. Although slightly out of date, this short review of the earth's structure and composition is an excellent first reading to gain a global perspective on geology and geophysics. Illustrations are abundant and very clear. The text is simply written, yet the author manages to convey complex concepts about tectonics, wave propagation, and ray theory with ease. The chapter dedicated to seismology is the best and most carefully written section of the book.

Hamblin, W. Kenneth. *The Earth's Dynamic Systems: A Textbook in Physical Geology.* New York: Macmillan, 1989. This geology textbook offers an integrated view of the earth's interior not common in books of this type. The illustrations, diagrams, and charts are superb. Includes a glossary and laboratory guide. Suitable for high school readers.

Jacobs, J. A. *The Earth's Core.* 2d ed. New York: Academic Press, 1987. This is a highly technical text, but it is perhaps the best reference for a detailed description of the most accepted core models. The tables—which give the numerical values of the density, temperature, rigidity, and wave velocity distributions within the earth—are of interest to anyone wanting a quantitative description of the core. A long list of research articles is included.

Jeanloz, Raymond. "The Earth's Core." *Scientific American* 249 (September, 1983): 56-65. The best elementary treatment of the structure and composition of the core. Jeanloz is a leading expert in the field. In this article, the origin, evolution, and present state and composition of the core are discussed in detail. The language is precise but not too specialized. The entire issue is dedicated to the earth and earth dynamics, so it should be of great interest to some readers.

Okuchi, Takuo. "Hydrogen Partitioning into Molten Iron at High Pressure: Implications for Earth's Core." *Science* 278 (December 5, 1997): 1781-1785.

Press, Frank, and Raymond Siever. *Earth*. 3d ed. San Francisco: W. H. Freeman, 1985. The most geophysical and probably the best written of all elementary geology textbooks. It includes an intriguing description of the evolution of the earth, the iron catastrophe, and the formation of the atmosphere. Well illustrated. Includes a glossary.

J. A. Rial

Cross-References

Earth's Differentiation, 188; Earth's Mantle, 195; Earth's Origin, 210; Earth's Shape, 218; Earth's Structure, 224.

Earth's Crust

Humankind's existence and modern society depend upon the crust of the earth. The dynamic changes involved in the creation and destruction of crustal rock also liberate gases and water that form oceans and the atmosphere, cause earthquakes, and create mineral deposits essential to society.

Field of study: Geophysics

Principal terms

ANDESITE: a volcanic igneous rock type intermediate in composition and density between granite and basalt

BASALT: a dark-colored igneous rock rich in iron and magnesium and composed primarily of the mineral compounds calcium feldspar (anorthite) and pyroxene

DENSITY: the mass per unit of volume (grams per cubic centimeter) of a solid, liquid, or gas

GRANITE: a silica-rich igneous rock light in color, composed primarily of the mineral compounds quartz and potassium- and sodium-rich feldspars

ISOSTASY: the concept that suggests that the crust of the earth is in or is trying to achieve flotational equilibrium by buoyantly floating on denser mantle rocks beneath

P WAVE: the fastest elastic wave generated by an earthquake or artificial energy source; basically an acoustic or shock wave that compresses and stretches solid material in its path

PLATE TECTONICS: the theory that the crust and upper mantle of the earth are divided into a number of moving plates about 100 kilometers thick that meet at trench sites and separate at oceanic ridges

REFLECTION: the bounce of wave energy off a boundary that marks a change in density of material

REFRACTION: the change in direction of a wave path upon crossing a boundary resulting from a change in density and thus seismic velocity of the materials

SNELL'S LAW: a statement of the fact that refraction of seismic waves across a boundary will occur such that the ratio of the two velocities of the material on either side of the boundary is equal to the size of the two angles on either side of the boundary formed by the ray path and a line perpendicular to the boundary

179

Summary

The crust of the earth is the outermost layer of rock material of the earth. It is distinct from the region of rocks lying beneath it, called the mantle, in that the rock materials that compose the crust are of a different composition and a lower density. Density may be described as the weight per unit of volume of solid materials. Therefore, if a cubic centimeter of granite, which makes up much of the crust of the earth of continents such as North America, could be weighed, it would total 2.7 grams. Deeper crustal rocks under continents have higher densities, some approaching the 3.3 grams per cubic centimeter characteristic of upper mantle rocks. A sample of crustal rock underlying the ocean basins would reveal that it is a rock type known as basalt, with a density of about 2.9 grams per cubic centimeter.

Compared with the rocks of the mantle, the rocks of the earth's crust are quite varied. The rocks of the crust can be classified as belonging to one of three broad groups: igneous, sedimentary, and metamorphic. Both granite and basalt are igneous rocks. Such rocks are formed by cooling and crystallization from a high-temperature state called magma or lava. Other igneous rock types of the earth's crust that are intermediate in rock composition and density between granite and basalt include andesite and granodiorite. Igneous rocks may form by melting of other igneous and metamorphic rocks in the crust or upper mantle, or by melting of sedimentary rocks.

Metamorphic rocks are formed from other rocks that have been subjected to pressures and temperatures high enough to cause the rock to

CHEMICAL COMPOSITION OF EARTH'S CRUST

Element	Weight (%)	Volume (%)
Oxygen (O)	46.59	94.24
Silicon (Si)	27.72	0.51
Aluminum (Al)	8.13	0.44
Iron (Fe)	5.01	0.37
Calcium (Ca)	3.63	1.04
Sodium (Na)	2.85	1.21
Potassium (K)	2.60	1.88
Magnesium (Mg)	2.09	0.28
Titanium (Ti)	0.62	0.03
Hydrogen (H)	0.14	—

Source: Data are from William C. Putnam, *Geology*, 2d ed., revised by Ann Bradley Bassett, 1971, and Sybil P. Parker, ed., *McGraw-Hill Concise Encyclopedia of Science and Technology*, 2d ed. 1989.

respond by change in the crystalline structure of the rock materials. These temperatures are not high enough to melt the rock. Such changes often occur in the deep parts of the crust, where heat is trapped and great pressure occurs from the weight of the overlying rock. As a consequence of this high pressure, densities of metamorphic rocks of the lower crust average about 2.9 grams per cubic centimeter.

Sedimentary rocks of the earth's crust are formed by chemical change and physical breakdown into fragments of other rocks exposed to the atmosphere and water of the earth's surface. The density of sedimentary rocks is generally less than that of igneous rocks, ranging from about 2.2 to as high as 2.7 grams per cubic centimeter.

The boundary between the rocks of the crust and the mantle is known as the Mohorovičić discontinuity, or simply Moho. The nature of this boundary varies from place to place. Under parts of the crust that have recently been stretched or compressed by mountain-building forces, such as under the great desert basins of the western United States, the Moho is a very sharp, distinct boundary. Elsewhere, in the interior of continents that have not been deformed for long time periods, the Moho appears to be an area of gradual density change with increasing depth rather than a distinct boundary. The position of the Moho, and thus the thickness of the crust, varies widely. The crust is thickest under the continents, reaching a maximum of 70 kilometers beneath young mountain chains such as the Himalaya. Under the ocean basins, the crustal thickness varies from 5 to 15 kilometers.

Thickness of the crust is directly related to its formation and evolution through geologic time. Only within the last twenty-five years have geoscientists understood this relationship. The crustal rocks of the earth are constantly being created, deformed, and destroyed by a process known as plate tectonics. Plate tectonics is a theory that suggests that the crust and upper mantle of the earth are divided into a number of separate rock layers that resemble giant plates. These plates are in motion, driven by heat from the earth's interior. Where the heat reaches the surface along boundaries between plates on the ocean floor, new rocks are formed by rising lava, creating new ocean basin crust. Because new crust is being created, crust must be consumed or destroyed elsewhere so that the earth's volume will remain constant. The sites where crust is consumed also lie on the ocean floor. Topographically, such sites are deep trenches where the crust bends down into the mantle to be heated and remelted. Such a process of recycling ocean-basin crust means, first of all, that ocean-basin crust is never geologically very old. The oldest sea-floor crust in the western Pacific is 175 million years old as compared to about 4.5 billion years for the age of the earth. Second, it suggests that since ocean-basin crust goes through a geologically short life and uncomplicated history, it has a rather uniform thickness of about 5 to 15 kilometers, unchanged between the time it is born and the time it is destroyed.

Continental crust has a much longer life and a more complicated history, reflected in a highly variable crustal thickness. It is initially created at the sites where oceanic crust is consumed, also known as subduction zones. As the crust and upper mantle, or lithospheric plate, is bent back down into the earth, it is heated up. Eventually, melting of part of this rock material occurs, creating volcanoes near trench sites. Such volcanoes have lavas rich in elements such as calcium, potassium, and sodium. When these lavas cool to form rock, the rock type that results is an andesite, named for volcanic rocks abundant in the Andes of South America. These continental volcanic rocks are less dense than basalts and, once created, remain on the top of a lithospheric plate, where they are carried along by the motion of the sea floor and the underlying lithospheric plate as it moves away from the ocean ridge boundaries. Eventually, the sea-floor motion may cause pieces of this continental crust to collide and weld together, forming larger pieces of continental crust. Thus, continents grow with time by two processes: volcanism above subduction zones and collision. The process of collision causes rocks to pile up like a throw rug pushed against a wall, creating high mountains that also extend downward with roots that increase crustal thickness. Continents thus grow along their edges where young mountain belts, called orogenic belts, are found, such as the Andes and the mountain systems of the western United States. The crust is relatively thick under young mountain belts, piling upward and sinking downward simultaneously to form a thick wedge of rock. In this sense, it is much like a buoyant iceberg, with the majority of its mass below the water or, in this case, below sea level. The buoyancy of the lighter continental rocks above the denser mantle rocks is known as the principle of isostasy, or flotational equilibrium. Just as the iceberg must reach a flotational level by displacing a volume of water equal to its mass, so must the continental crustal rocks displace a volume of denser mantle rocks to reach their buoyancy level. Thus, under higher mountainous terrain thicker crust is found, whereas at lower elevations, such as under the ocean basin, the thinnest crust is found.

Toward the center of continental landmasses are core areas of older rocks known as cratons. The age of rocks found in the cratons range from about 500 million to an extreme of 3.8 billion years. The cratons of the world compose about one-half of the area of the continental crust and have been free of deformation and mountain-building forces for long periods of time. Consequently, their surfaces tend to be relatively flat as a result of surface processes such as weathering and stream-cutting acting on the exposed rocks over a geologically long period of time. The thickness of the continental crust in cratonic areas is variable, which is a reflection of their long and complex histories. These areas were at one time thickened because of mountain-building activity, but long and varying periods of stability have caused them to lose some crustal thickness as well. Figures for central Canada and the United States show a range of from 30 to 50 kilometers for thickness of the craton.

PRIMARY ROCKS AND MINERALS IN EARTH'S CRUST

Rocks	% Volume of Crust	Minerals	% Volume of Crust
Sedimentary		Quartz	12
Sands	1.7	Alkali feldspar	12
Clays and shales	4.2	Plagioclase	39
Carbonates (including		Micas	5
salt-bearing deposits)	2.0	Amphiboles	5
		Pyroxenes	11
Igneous		Olivines	3
Granites	10.4	Clay minerals (and	
Granodiorites, diorites	11.2	chlorites)	4.6
Syenites	0.4	Calcite (and aragonite)	1.5
Basalts, gabbros,		Dolomite	0.5
amphibolites, eclogites	42.5	Magnetite (and	
Dunites, peridotites	0.2	titanomagnetite)	1.5
		Others (garnets, kyanite,	
Metamorphic		andalusite, sillimanite,	
Gneisses	21.4	apatite, etc.)	4.9
Schists	5.1		
Marbles	0.9	**Totals**	
		Quartz and feldspars	63
Totals		Pyroxene and olivine	14
Sedimentary	7.9	Hydrated silicates	14.6
Igneous	64.7	Carbonates	2.0
Metamorphic	27.4	Others	6.4

Source: Michael H. Carr et al., *The Geology of the Terrestrial Planets,* NASA SP-469, 1984. Data are from A. B. Ronov and A. A. Yaroshevsky, "Chemical Composition of the Earth's Crust," American Geophysical Union Monograph 13.

Methods of Study

Elastic waves are created by both earthquakes and artificial sources and may be used to study the crust of the earth. This branch of earth science is called seismology. When energy is released in rock by a source, the rock is set in motion with an up-and-down or back-and-forth wavelike motion. These waves force the rock to respond like a rubber band, stretching and compressing it without permanently deforming it. Such a response is called elastic. This response can be used as a key to studying the physical properties of the rocks because of a wave generated by a seismic relationship between velocity and rock properties. The fastest elastic wave is the primary, or P, wave. This wave is basically an acoustic wave or sound wave traveling in rock, compressing and stretching rock materials in its path. Therefore, the density, rigidity, and compressibility of a material determine wave velocity. A

183

simple example would be to compare the velocity of an acoustic wave in air, called the speed of sound, to that in rock. The result is that in air near sea level, an acoustic wave travels at about 0.3 kilometer per second, whereas in rocks near the earth's surface, the same kind of wave travels at about 5 kilometers per second. Air is much less dense than rock and has no rigidity (no permanent shape).

Rocks of the lower crust beneath the continents have P-wave velocities of between 6.8 and 7.0 kilometers per second. It can be shown in the laboratory that metamorphic rocks known as granulites, when placed under the pressures and temperatures of the lower crustal depths, have velocities in this range. Other rocks under the same pressures and temperatures (600 to 900 degrees Celsius, 5,000 to 10,000 atmospheres) may have similar velocities. Samples of lower crustal rocks known as xenoliths, however, exist at the surface, having been brought up by volcanic activity. These samples also suggest that granulite is a good choice.

Rocks of the upper crust in continental areas have P-wave velocities of around 6.2 kilometers per second. Here, rocks at the surface of a granite to granodiorite composition suggest the appropriate choice of rock. When such rocks are velocity-tested in the laboratory under the appropriate range of pressures and temperatures, there is a good match between rock type and velocity.

The composition of the oceanic crust is well known. Here, basalts yield a velocity of around 6.7 kilometers per second, reflecting the rather uniform composition of the geologically simpler oceanic crust. Finally, part of the continental crust is mantled by sedimentary rocks. This material has among the lowest velocities, ranging from less than 2 kilometers per second up to an extreme of about 6 kilometers per second and reflecting a wide range of compositions as well as the presence of open space and fluids contained therein.

The thickness of continental crust has been determined by the study of seismic waves that bounce off (reflect) or bend (refract) when they cross the Moho. The density contrast and resulting change in velocity of seismic waves when they cross the Moho from crust to mantle cause the wave path to change angle or bend. The same phenomenon occurs when light crosses from air to liquid in a glass. This can be shown by placing a straight straw in the liquid and gazing along its length. The straw will appear bent even though it is actually the light wave that has bent.

Waves that leave the source at one critical angle will cross the boundary and travel along beneath it, radiating energy back to the surface at the same angle. This is the critically refracted ray path, and the sine of the critical angle can be predicted from Snell's law of refraction to be the ratio of the crust and mantle velocities. The geometry of this wave path is determined by two factors, the ratio of the crust and mantle velocities, and the thickness of the crust. The thicker the crust, the longer the travel time of the wave for a particular pair of velocities. Using critically refracted

P waves, thicknesses have been estimated for much of the crust.

It has been possible in many areas to check the crustal thickness determined by critically refracted waves by using information from reflected waves. This has been applied with particular success to the study of earth's crust in continental areas. The technique is similar to that of depth sounding in ships, in which an acoustic wave is sent down from a ship, bounces off the bottom, and returns. The depth is proportional to the time of travel of the wave, also called the two-way time. The depth can be found by multiplying water velocity by travel time. The same basic procedure has been used under the continents with artificial acoustic wave sources such as explosives and vibrator trucks.

Context

An understanding of the geometry, evolution, and composition of the earth's crust increases humankind's knowledge of the nature of the world. It is easy to show that humankind's very existence, as well as the material wealth of modern societies, is totally dependent on the crust of the earth. The crustal state is one of dynamic evolution, with rock materials being created, deformed, and destroyed at plate tectonic boundaries. The process that makes creation and destruction of rocks possible is that of crystallization and melting of the mineral compounds that compose rock, a process known as volcanism. Volcanic activity over the billions of years of the earth's existence has, by the expulsion of gases trapped in lavas that reach the surface, provided the water vapor and other gases necessary to form the oceans and atmosphere, which are necessary to support life.

An understanding of volcanoes, of the how, why, and where they occur, requires an understanding of the earth's crust and of crustal dynamics. Certainly, this can be important as viewed from the perspective of Mount St. Helens and other volcanoes of the northwestern United States. Mount St. Helens is a volcano formed by the remelting of part of the oceanic crust that is slowly being taken back into the interior of the earth. As this process of subduction and remelting of the oceanic crust will continue into the future for millions of years, so will eruptions continue to occur at Mount St. Helens, as well as at other volcanoes of the Cascade Mountains. Thus, an understanding of the crust of the earth shows that the disastrous May 18, 1980, eruption of Mount St. Helens was not a onetime event.

The dynamic evolution of the earth's crust is also accompanied by the movement of large plates of the crust and upper mantle, up to 100 kilometers thick, against one another. The San Andreas fault of California is one place where two of these plates of crustal material rub against each other. The forces created by this motion are released as energy in large earthquakes, posing a threat to life and property. Eventually, knowledge of how crustal rocks change and respond to these forces before an impending earthquake may allow their prediction.

Exploration for important economic minerals is guided by knowledge about the evolution and composition of the crust. The creation of valuable metal deposits, such as gold and copper, during volcanic activity at ocean ridge sites where new oceanic crustal rocks are also being created is occurring in the Red Sea between Africa and Asia. Consequently, exploration efforts for such metallic ores can be directed toward identifying ancient ridge site deposits. The formation of continental sedimentary rocks in the Gulf of Mexico is trapping organic materials that will be turned into oil and natural gas. Looking for similar types of sedimentary rocks in the appropriate crustal environment would be worthwhile for explorers in the petroleum and natural gas industries.

Bibliography

Bally, A. W. *Seismic Expression of Structural Styles.* Tulsa, Okla.: American Association of Petroleum Geologists, 1983. An excellent visual treatment of the structure and layering of primarily the upper crust throughout the world. Sections into the crust of offshore Scotland and northwest Germany show the Moho. Suitable for a broad audience from general readers to scientific specialists.

Bott, M. H. P. *The Interior of the Earth.* New York: Elsevier, 1982. This book was intended for undergraduate and graduate students of geology and geophysics as well as for other scientists interested in the topic. The plate tectonic framework of the outer part of the earth is strongly emphasized.

Brown, G. C., and A. E. Mussett. *The Inaccessible Earth.* London: Allen & Unwin, 1981. A good general introduction geared toward the undergraduate college student. The primary topics are the internal state and composition of the earth. Included is background material on seismology and three chapters discussing the earth's crust.

Phillips, O. M. *The Heart of the Earth.* San Francisco: Freeman, Copper, 1968. An excellent and well-written book intended for a general college and noncollege audience with no background in geophysics. The book has an excellent chapter on seismology and the way in which earthquake waves are used to determine physical properties from velocity and to infer crustal structure by refracted waves.

Smith, David G., ed. *The Cambridge Encyclopedia of Earth Sciences.* Cambridge, England: Cambridge University Press, 1981. This general reference provides an excellent overview of the earth sciences. Chapter 10 is an extensive discussion of the earth's crust, including useful illustrations and diagrams. Contains a glossary, an index, and recommendations for further reading.

Taylor, Stuart R., and Scott M. McLennan. *The Continental Crust: Its Composition and Evolution.* Boston: Blackwell Scientific, 1985. A text aimed at undergraduate and graduate geology and geophysics students as well as general earth scientists. It is clearly written and up-to-date and has excellent, well-rounded scientific references.

Yardley, Bruce W. D., and John W. Valley. "How Wet Is the Earth's Crust?" *Nature* 371, no. 6494 (Sept. 15, 1994): 205-207.

David S. Brumbaugh

Cross-References

Continental Crust, 78; Continental Growth, 102; Earthquakes, 148; Earth's Composition, 162; Earth's Structure, 224; Igneous Rocks, 315; Plate Tectonics, 505.

Earth's Differentiation

Earth's differentiation describes the formation of "layers" within the early earth when it originated more than 4 billion years ago. A core surrounded by a mantle overlain with a crust, on which humans now live, was created by chemical and physical processes as the earth cooled from a hot molten sphere.

Field of study: Geophysics

Principal terms

CRUSTAL DIFFERENTIATION: origin of continental and oceanic crust through remelting of original, heavier crust

DIFFERENTIATION: origin of layers (core, mantle, and crust) through differential settling of material in the molten earth, based on densities

HOMOGENEOUS ACCRETION THEORY: one of two major theories on earth's differentiation: that differentiation occurred after the earth had formed through accretion of debris

INHOMOGENEOUS ACCRETION THEORY: one of two major theories on earth's differentiation: that differentiation occurred while the earth was accreting debris because denser debris formed first

PARTIAL MELTING: melting of rock that results in a magma concentrated in some minerals and depleted in others as compared to the original unmelted rock

SEISMIC WAVES: vibrational waves caused by earthquakes that are a major tool in analyzing the layers of rock within the earth

Summary

The earth today is a spherelike body composed of layers arranged according to density. The highest-density ("heaviest") material, mainly nickel and iron, is at the core, which is about 3,400 kilometers (2,100 miles) thick with an average density of about 10-13 grams per cubic centimeter. The lightest elements are dominant in the crust, the outermost layer at the surface of the earth. This layer is very thin—only 5-60 kilometers (3-37 miles) thick. The average density of the crust is about 2.8 grams per cubic centimeter. The common elements of the crust include silicon, aluminum, calcium, potassium, and sodium. They combine to form various silicate minerals, especially the feldspars and quartz. These minerals are major components of granites and other abundant rocks. Between the crust and the core is the mantle, which is intermediate in density (4.5 grams per cubic centimeter). It is very thick, about 2,900 kilometers (1,800 miles).

There are two major theories on how the earth became differentiated in this way: the homogeneous accretion theory and the inhomogeneous accretion theory. The homogeneous accretion theory states that the earth formed by randomly sweeping up debris (meteors, dust) in its orbit around the sun. This process, which occurred about 4.6 billion years ago, caused the debris to be added on (accreted) to the early earth in a sort of "snowball" effect that made the earth progressively larger. The bigger it became, the more its gravity increased and the more debris it accumulated. According to this theory, because the debris was accreted at random, the earth at this time was undifferentiated; that is, it was homogeneous, meaning it was roughly the same throughout. At some time (scientists are uncertain exactly when) during or after the accretion, differentiation occurred when this body became molten. The liquid state allowed heavier (denser) elements to sink to the core so that it became enriched in iron and nickel. The lighter elements rose to the surface to form the crust, while elements of intermediate density stayed below them. This process would also explain why radioactive elements such as uranium and thorium are common in the crust, because they would have tended to combine with the low-density crustal minerals at this time.

For the homogeneous accretion theory to be correct, there must have been some process which heated up the originally solid body, causing it to become molten liquid. Scientists agree that there were probably three processes that could have caused this heating. First, radioactive isotopes were much more common in the early earth; radioactive decay of these isotopes would have produced much heat. This heat source has greatly decreased through time as the isotopes have decayed to more stable states. Second, during accretion of debris, the energy released as the debris impacted the earth was converted to heat energy. Third, much heat is created by gravitational compression. As more and more material was added to the earth, progressively greater temperatures were generated at the core. Some scientists have estimated that the combination of these three processes would have raised the temperature of the earth by as much as 1,200 degrees Celsius. While this theory has many supporters, it is far from conclusively proven. Current research based on computer models of physical laws indicates that even with all three of these processes operating, there might not have been enough heat generated to warm a cold planetary body to a molten state.

To provide an alternative to the problems of the homogeneous accretion theory, the inhomogeneous accretion theory was proposed. The main difference is that this latter theory states that the differentiation of layers

189

occurred as accretion was occurring. Instead of accreting as an originally homogeneous body, which then became layered through density gradients, the earth is thought to have accreted the layers in sequence. First, a naked core developed from dense matter in the debris of the orbit. Later, a less dense mantle and an even lighter crust accreted around the core as lighter debris in the orbit was encountered. Yet why should progressively lighter matter be encountered in such a cloud of debris? Calculations show that in a cooling cloud of hot gas and dust, such as that in the early earth's orbit, iron and nickel would condense first to form the core. As the cloud cooled further, silicates of progressively lighter elements would condense so that they would be accreted in that order.

In spite of their differences, the two theories agree on a number of major points about the earth's differentiation, the most basic being the assumption that the earth originated from condensation of a large cloud of dust and debris around the sun. The theories also agree that the differentiation must have been well under way by 3.8 billion years ago. The oldest rocks on earth date to this age, and they contain remanent magnetism, which indicates that the earth had a magnetic field by then; therefore, scientists know that the core had formed. Rotation of the core produces the field. Another major point of agreement is that the different melting points of the various elements played a key role in the origin of layering. In both theories, undifferentiated matter formed separate layers because the denser materials tended to solidify first. Thus, nickel and lead condense early in the homogeneous accretion theory and then sink to the core, or condense early in the dust cloud in the inhomogeneous theory and are accreted first as the core.

Scientists are currently debating the evidence for the two models and, presumably, a consensus will someday be reached. Whichever of the two models is correct, it must be noted that the differentiation of the earth did not end with the formation of merely these three layers. The early crust almost certainly consisted mainly of ultramafic minerals, which are relatively heavy minerals high in magnesium and iron. Nevertheless, today's crust, composing the continents, contains much lighter sialic minerals, which are high in silicon and aluminum. Experimental evidence shows that in order for this continental crust to differentiate, the ultramafic crust would need to undergo remelting, which would cause the lighter minerals to separate from the heavier ones. There are two theories as to how this remelting came about. One relies on convection currents within the very hot early earth itself. For millions of years after it formed, the earth stayed extremely hot compared to today's interior temperatures. Therefore, as the crust began to harden and surface cooling began, there was much volcanism as hot magma from the interior sought release through the crust. Volcanoes created some local density differences in the crust and also led to erosion as some of the higher areas were exposed to weathering. It is thought by many scientists that these density differences together with accumulating sediment along the volcano margins led to remelting as convection in the underlying magma carried up great amounts of heat. In addition, convection currents return to deeper depths when they cool off, so they may have carried ultramafic crust and sediment down with them to be remelted. This recycling of rock by convection cells in hot magma was probably the beginning of plate tectonics. This same recycling process continues, although at a much slower pace because the convection currents are driven by heat and the earth is much cooler today. The second theory about the differentiation of the crust states that meteorite impacts were involved. Scientists know that the early earth was (with all the planets) heavily bombarded with debris. This theory says that large meteors penetrated the original ultramafic crust, remelting some of the crust and causing a large rim to form around the crater. The remelting, along with the rim formation, might have led to a continental "nucleus" that formed the center onto which later continental material was accreted. This theory is not exclusive of the first theory because the nuclei created by the impacts would then participate in convection recycling, helping to create still more continental crust. Whether impacts were involved or not, it is clear that the amount of continental crust has continued to increase since the original differentiation to form the land masses as they are known today. At the same time, the amount of ultramafic crust has diminished because it is readily destroyed by the erosion and remelting process. Instead there remains only oceanic crust, underlying the ocean basins. The oceanic crust is more mafic than the continental crust but not as high in magnesium and iron as the original ultramafic crust. It consists of crust that has not been as completely recycled as is continental crust and is therefore somewhat denser. Today, the crust is composed entirely of the lighter continental crust, with a density of about 2.7 grams per cubic centimeter, and the denser oceanic crust, with a density of about 3.0 grams per cubic centimeter.

Methods of Study

It is not possible to study the differentiation of the earth directly because the process occurred at least 4 billion years ago. If technology ever becomes adequate, it may be possible to study the differentiation of other planets in other solar systems. Astronomers have recently discovered debris around other stars that appears to represent a solar system in the process of formation. Until then, there are two less direct methods of studying earth's differentiation: observations of the current internal structure of the earth to see what changes continue to occur and laboratory and field study of the earth and meteorites to make inferences about those processes involved in differentiation.

The core, mantle, and even the deeper crust are far too deep to be reached by conventional drilling; therefore, no one has ever seen rocks from those parts of the earth. Instead, all scientists know of them comes from the study of seismic waves produced by earthquakes, which provide a kind of X ray of the interior. For example, if an earthquake occurs on one side of the planet, the seismic vibrations traveling to stations on more distant parts of the planet not only travel rapidly but follow refracted (curved) paths. This high speed indicates that the earth's interior becomes denser with depth, because vibrations travel faster in denser material—which re- fraction confirms, because the changing density will cause waves to travel at different speeds depending on where they are. Even more telling is the behavior of different kinds of seismic waves. P (primary) waves move back and forth in the direction of travel and will go through solid or liquid material. S (secondary) waves, which move at right angles to the travel direction, will go through solids but not through liquids. Observations show that S waves are not received at locations directly on the other side of the earth to an earthquake. This "shadow zone" indicates that the earth has a core that is composed of liquid. On the other hand, P waves are received on the opposite side of the earth, but they are refracted in some parts but not in others. Calculations show that this phenomenon occurs because the core is divided into two layers: a solid inner core and a liquid outer core. Similar refractions are used to locate the top of the mantle, where "seismic discontinuities" cause waves to change speed beween the mantle and crust.

Laboratory studies used to model differentiation show that iron and nickel will differentially separate from silicate minerals at an early stage in cooling. This finding is confirmed by the study of metallic meteorites, which are composed mostly of nickel and iron. These meteorites appear to have formed under extremely high temperatures and pressures such as exist in planetary cores. In fact, such meteors are thought to have originated from early planets that were shattered in collisions shortly after the solar system formed. The next major group of minerals to separate are olivine and pyroxene. These are silicates that are high in iron and magnesium and that form the rock called peridotite. This observation, plus density estimates of the mantle itself using seismic velocity data, indicates that the mantle is composed largely of that rock. (The rate of seismic wave travel can be related to what the medium is composed of chemically.) Laboratory studies of peridotite also indicate how oceanic and continental crust would differenti- ate from it. As peridotite rises up from the mantle at divergent plate boundaries, the pressure and temperature of the magma's environment begin to decrease dramatically. This temperature decrease leads to partial melting of the magma, which produces minerals of basaltic composition. These minerals become accreted onto the oceanic basin floor to become oceanic crust. At convergent plate boundaries where subduction is occur- ring, the oceanic crust, along with many sediments on it, is being pushed underneath the other plate and is being remelted. Laboratory studies of magmas of this composition show that its partial melting will create a new magma that is rich in silica and other minerals found in granites and in other common rocks making up continental crust.

Context

It is difficult to appreciate the importance of an event that occurred at least 4 billion years ago, but there are good reasons to do so. The process of Earth's differentiation very much shaped the earth as humans now live on it; the better those events are understood, the better the planet will be understood and its resources used wisely. For example, humankind lives on only a tiny fraction of the crustal thickness, yet the crust itself is only a small fraction of the total planet. How far down will mineral resources be found? A project that had been planned to drill to the mantle to sample it directly was abandoned because it was too expensive. At present, industrialized society is running low on many materials, such as chromium, which is mined from ores naturally enriched by differentiation. If it were not for such differentiation, there would be no civilization, because it costs too much (uses too much energy) to separate the usable minerals in rocks that have not undergone natural enrichment. For example, any average rock (such as a granite) contains many valuable elements, such as gold. They are so dilute, however, that the cost of extracting them is too great to be economical. Is petroleum formed by magmatic processes of differentiation as some (only a few) geologists say? It is not known for certain. Of particular relevance is the origin of earthquakes. The processes that formed the mantle and crust played a major role in establishing the processes of plate tectonics. Scientists now know that most earthquakes are directly associated with plate tectonism. Some earthquakes originate at great depths, while others begin nearer to the surface.

On a longer time frame, our knowledge of earth's differentiation will make it much easier to understand the formation and to utilize the resources of other planets. The other terrestrial planets—Mercury, Venus, and Mars—also underwent planetary differentiation at about the same time as did the earth. Mars is the most habitable by humans, but all of them, especially Mercury, are rich in minerals and in other materials that humans can use. Comparative planetology can also tell scientists about the geological future of the earth. It is important to note that earth's differentiation is not truly complete; plate tectonism means that the crust continues to evolve. Furthermore, physiochemical changes continue to occur in the core and mantle as the interior slowly cools. By looking at other worlds that have already gone through these changes, scientists can draw conclusions about earth's fate. Mercury, for example, because of its small size (about that of earth's moon), lost its internal heat billions of years ago, and any movement of crust or volcanic activity has long since ceased.

Bibliography

Head, James W., Charles A. Wood, and Thomas A. Mutch. "Geologic Evolution of the Terrestrial Planets." *American Scientist* 65 (January/February, 1977): 21. Comparative evolution of Mars, Venus, Mercury, and Earth in a widely read review article. Technical but very informative to the motivated layperson or student.

Kaufmann, William J., III. *Planets and Moons.* New York: W. H. Freeman, 1979. One of the standard and best-respected texts on comparative planetology, this book includes excellent discussions of planetary origins and differentiation. Some parts of this college text are highly advanced, but much is suitable for the interested layperson and the advanced high school student because chapters begin with the basics.

Levin, H. *The Earth Through Time.* New York: Saunders, 1988. A summary of the earth's growth and differentiation is found in this highly respected and widely used basic freshman text. Very well illustrated and clearly written; an excellent introduction to the subject. Technical references for further research.

Moorbath, Stephen. "The Oldest Rocks and the Growth of Continents." *Scientific American* 236 (March, 1977): 92. A well-illustrated discussion of differentiation processes of the continental and oceanic crusts, including the role of plate tectonism. Especially interesting discussion of earliest known rocks. Suitable for the interested layperson or the advanced high school student.

Ozima, Minoru. *The Earth: Its Birth and Growth.* Translated by J. F. Wakabayashi. New York: Cambridge University Press, 1981. An excellent overview of earth's differentiation from the beginning of planetary condensation to the present. Suitable for interested laypersons and advanced high school students. Technical in parts, but many basic concepts, too.

Short, Nicholas M. *Planetary Geology.* Englewood Cliffs, N.J.: Prentice-Hall, 1975. A complete introduction to planetary evolution, with detailed description on an elementary level. Comprehensible to interested laypersons and high school students.

Wetherill, George W. "The Formation of the Earth from Planetesimals." *Scientific American* 244 (June, 1981): 162. A well-illustrated account of earth's origin and subsequent differentiation. Very readable by the motivated layperson or high school student.

Wicander, R., and J. Monroe. *Historical Geology.* St. Paul, Minn.: West, 1989. An up-to-date survey of earth history, with a good summary discussion of earth's differentiation. A basic college-level text, but readable for the layperson and the advanced high school student.

Michael L. McKinney

Cross-References

Continental Growth, 102; Earth's Composition, 162; Earth's Core, 171; Earth's Crust, 179; Earth's Mantle, 195; Earth's Origin, 210; Earth's Structure, 224; Plate Tectonics, 505.

Earth's Mantle

The mantle is that portion of the inner earth that lies between the crust and the outer core. It is composed of rocks that are of greater density than those of the crust. The mantle contains a zone in which the rock is under such great temperature and pressure that it exists in a plastic state. It is upon this zone that the major plates of the earth's crust slide.

Field of study: Geophysics

Principal terms

DIFFERENTIATION: layering within rock that results from differences in density; the lighter material rises to the surface while the heaviest material sinks to the bottom of a mixture of substances

DISCONTINUITY: a rapid change in the properties of rock with increasing depth

FOCUS: the point within the earth that is the center of an earthquake and the point of origin of seismic waves

IGNEOUS ROCKS: a family of rocks which have solidified from the molten state

POLYMORPHISM: the characteristic of a mineral to crystallize into more than one form

PRIMARY WAVE: a compressional type of earthquake wave, which will travel in any medium and is the fastest wave

SECONDARY WAVE: a transverse type of earthquake wave, slower than a primary wave, which will not travel in a liquid

Summary

The mantle is that portion of the interior of the earth that extends from the base of the crust to the boundary of the outer core. This distance is approximately 2,900 kilometers, roughly 45 percent of the radius of the earth. Since the thickness of the earth's crust is not uniform, the distance from the ground surface to the upper boundary of the mantle varies significantly. It has been determined that the thickness of the crust in continental areas is approximately 40 kilometers, while in the ocean basins the thickness is only some 5 kilometers.

Evidence for the existence of differentiation, layering within the earth caused by density differences, was first observed from the study of earth-

quake waves. In 1906 a seismologist, Andrija Mohorovičić studied the records of an earthquake that had taken place in Yugoslavia. At a certain
distance from the actual focus of the earthquake, two types of earthquake
waves were received. Those types were the primary waves, P waves, and the
secondary waves, S waves. Although there was only a single shock, a short
time later another set of P and S waves were received by the same seismograph. Mohorovičić concluded that the second set of waves were actually
reflections of the original waves. When the rock was stressed and broken at
the focus, P and S waves were sent out in all directions. The P and S waves
initially received by the seismograph traveled by the most direct route. Waves
directed downward into the earth were reflected back from a surface and
were recorded. This reflecting surface is called a discontinuity. A discontinuity is a rapid change in the properties of rock with increased depth.
Knowing the velocities of seismic waves in the rocks nearer to the surface,
Mohorovičić calculated the distance from the surface to the discontinuity.
This boundary between the crust and the mantle is known as the Mohorovičić discontinuity, or Moho, named in his honor.

Unlike rocks of the crust, rocks of the mantle have never been directly
observed, and therefore only indirect evidence of their nature or composition exists. At one time in the mid-1960's, a project to drill down through
the earth's crust to the mantle was begun. The undertaking was appropriately named Project Mohole. Unfortunately, because of lack of funding, the
idea was abandoned.

The greatest source of information on the nature of mantle materials
comes from the study of earthquake waves. Since the velocity of seismic
waves through the mantle is known, the types of rock that conduct waves at
this known velocity are primary candidates for being mantle materials.
These rocks are peridotite and eclogite , which both occur to some extent
in the crust. Peridotite is a heavy, dark-green rock from the igneous rock
family. Igneous rocks are those that have cooled and solidified from a
molten state. Peridotite consists of the elements magnesium, oxygen, and
silicon. The second possibility, eclogite, is composed of the minerals garnet
and jadeite. Eclogite is very similar chemically to basalt, which is a lava
commonly associated with worldwide volcanic activity. Since the source of
volcanic activity is believed to be in the mantle, eclogite might well be the
material that is transformed into basalt as pressures are reduced. Since the
crust is far less dense than the mantle, as the molten material moves upward
toward the surface, the lithostatic pressure exerted by overlying rock layers
would be significantly less.

There is a third type of material that is believed to originate in the
mantle. This is the rare substance known as kimberlite. Kimberlite occurs
in pipe-shaped deposits and is mined extensively for diamonds. Diamonds
are a form of carbon that has been placed under great pressure. These
pressures have been calculated, and it has been concluded that pressures
of this magnitude could occur only 100 kilometers or more within the

earth. The diamond pipes must have originated within the mantle.

By the use of seismic wave information, it has been shown that the mantle is not uniform throughout. It is assumed that greater depths in mantle rock would produce greater pressures and, therefore, greater rock density. If this is true, seismic wave velocity would also increase. Primary wave velocities in the upper mantle are approximately 8 kilometers per second and gradually increase with depth to a velocity of roughly 8.3 kilometers per second. At this depth, the velocity of the waves begins to drop to a value of somewhat less than 8 kilometers per second. It has been concluded that the rock composition at that depth does not change but that its physical state does. Because of the geothermal gradient, temperatures at this depth and pressure have risen to near the partial melting point of the rock. The material then assumes plastic or flow properties. This low-velocity layer was first identified by Beno Gutenberg in 1926.

According to the theory of continental drift, the earth's surface consists of pieces called tectonic plates. These eighteen or so lithospheric plates slide over a plastic zone in the mantle. Apparently the low-velocity layer of the mantle is the asthenosphere, the plastic zone upon which the plates move. The asthenosphere has been found to vary significantly in depth from the surface of the earth. It has been found to be as close as 20 kilometers in depth near an ocean ridge; the asthenosphere averages some 100 kilometers in depth under continents.

Beneath the asthenosphere the wave velocity begins to increase again. Sharp increases at depths of 400 kilometers and 650 kilometers have been noted. Scientists believe that these increased velocities are caused by polymorphism. Polymorphism is a term that means "many different forms." When a rock or a mineral is subjected to increasing temperature and pressure, it may rearrange its internal structure to compensate for this added stress. As a result, the density of the substance is increased, and therefore the velocity of the seismic waves passing through it is also increased. From the 650-kilometer anomaly, the wave velocities gradually increase until reaching the boundary of the outer core, where the S wave is no longer conducted.

Critical to the modern explanation of how the continents move is the topic of heat flow in the mantle. Although the mechanism of heat flow in the mantle is not completely understood, it is known that the earth's interior heat was left over both from its original formation and from the decay of radioactive elements. It is believed that this heat causes rock to rise in the form of a current from the depths. It cools nearer to the surface and then plunges back deep within the mantle. This process is known as convection and is easily explained with the heating of a beaker of water or a container of a gas. A convection current is the density flow of a liquid or a gas. The hot material is less dense and rises, cooler material moves in to replace the rising material. As surface material cools, it plunges below to be replaced by more rising material. This concept of a convection current is the modern explanation of heat flow in the mantle and of the mechanism that drives continental movements.

It has been known since the nineteenth century that the age of the Hawaiian islands increases from southeast to northwest. In the early 1960's, when the theory of continental drift was becoming more acceptable to earth scientists, it was suggested that these volcanic islands recorded the movement of the sea floor. It was postulated that there existed a magma, molten rock, source deep within the mantle. This hot spot, or plume, was a long-lived source of magma. During an eruption of the hot spot, volcanic material would be extruded out onto the sea floor. Eventually the material would break the surface of the ocean and an island would be formed. Since the earth plates were in constant motion, the newly created island would then move away from the hot spot. Further eruption would create new islands and therefore a chain of islands like Hawaii. More than one hundred hot spots have been found worldwide.

Methods of Study

Although the earth's mantle cannot be directly observed, it can be studied indirectly by various techniques. The primary method of study is by use of seismic waves. These waves may be generated by an earthquake or a large explosion such as that produced by a nuclear test. At the point of rock fracture, energy is released in the form of several types of waves traveling outward in all directions and at velocities that depend upon the density of

Diamonds are produced from carbon by the tremendous pressures in the earth's mantle.
(Reuters/The Smithsonian/Archive Photos)

the conducting medium. The waves that travel deep into the earth increase in velocity as they encounter denser material. Since the mantle is much denser than the crust, wave velocities in the mantle are higher than those velocities in the crust. The study of velocities of seismic waves helps scientists determine the type of rock through which the waves are traveling. When calculated velocities of waves are compared with known velocities in various types of rock, the subsurface material can be identified. The study of wave velocities can also be used to identify discontinuities in the subsurface.

When an earthquake or a large explosion occurs, energy waves travel outward from the point of energy release in all directions. These energy waves are of three main types: P waves, S waves, and L waves (surface waves). The P waves are similar to sound waves in that they are compressional in nature. The particles in a compressional wave vibrate back and forth parallel to the direction in which the wave is traveling. Primary waves will pass through any type of material. The secondary waves are transverse types of waves similar to the wave form of electromagnetic radiation. The particles of matter that make up an S wave travel perpendicular to the direction of wave propagation. The S waves are considerably slower than P waves, so at a recording station the P waves always arrive first. The L waves are also transverse waves that travel along the surface; these types are the slowest of the three waves.

As these waves travel through rock, their velocity depends on the density of the material. Since pressure increases with depth, and increased pressure results in rocks of greater density, the velocity of seismic waves is in general greater with depth. As waves encounter boundaries between rock layers of different density or composition, some of the waves are reflected back toward the surface. It is this reflection of waves that allows scientists to determine the depth of various parts of the earth's interior, including the mantle.

Not all waves are reflected at a discontinuity. Some waves are refracted into the newly encountered material. If the material is denser, the velocities are higher. Waves reaching the surface after being refracted through an area of greater density may arrive in a shorter time interval than those that traveled a shorter distance but through a less dense medium.

Another modern method of studying the subsurface was first used in the field of medicine. The CAT (computerized axial tomography) scan is a composite image of X rays taken from a number of different angles. The computer assembles these images into a three-dimensional representation of the particular organ under study. The earth science equivalent of the CAT scan is known as seismic tomography. Seismic data from all over the world are analyzed. These data provide pictures of the earth from many different angles. The goal of seismic tomography is to assemble these pictures into a three-dimensional image of the interior of the earth. Although this technology is in its infancy, it holds promise for future studies.

Context

By studying the change in velocities of waves, scientists are able to determine the nature and the composition of the earth's mantle. The behavior of the material of the mantle has a direct bearing on much of the activity of the earth's surface; consideration of the role of the mantle is essential in studying such concerns as earthquakes, volcanism, movement of the continents, and even diamond mining.

A low velocity zone in the mantle is referred to as the asthenosphere. It is upon the plastic rock of this zone that the tectonic plates of the earth's crust move. It is these moving plates that make the active fault zones and areas of extensive volcanism that exist on the earth's surface. For example, Southern California is prone to earthquakes because it lies upon two different plates. The city of Los Angeles lies on the Pacific plate while the city of San Francisco lies upon the North American continental plate. As the Pacific plate moves to the north, it rubs against the boundary of the continental plate. The result is an earthquake.

In areas where one plate is moving below another plate, volcanism is common. The plate that is being subducted into the mantle undergoes remelting. This molten material then finds its way up to the surface through cracks and fissures. The result is a volcano. The eruption of Mount St. Helens is an example of this type of action.

The mechanism that causes the movement of the continents upon the asthenosphere is heat. Scientists believe that a convection current operates within the mantle. Hot rock rises into zones of less temperature and pressure, gives off heat, and then plunges below to be re-heated again. This endless motion of rock is a "conveyor belt"-like action and causes the movement of the continents.

Of economic importance is the mining of diamonds. Diamonds are a polymorphic form of carbon and are found in deposits of kimberlite, an igneous rock that originates deep within the mantle. As molten material it works its way to the surface through cracks and fissures, thus, knowledge of the activity of the mantle and its action on the crust aids in finding likely areas for diamond mining.

Bibliography

Cailleux, André. *Anatomy of the Earth.* Translated by J. Moody Stuart. New York: McGraw-Hill, 1968. A complete, well-illustrated volume describing the earth's interior and how it is studied. The book also treats such topics as the origin of the earth and continental drift. The book is for general readers.

Compton, R. R. *Interpreting the Earth.* New York: Harcourt Brace Jovanovich, 1977. This well-illustrated volume discusses the geology of the earth's surface. It also offers chapters on tectonics and continental drift. The book is suitable for general readers.

Jacobs, John A., Richard D. Russell, and J. T. Wilson. *Physics and Geology*. 2d ed. New York: McGraw-Hill, 1974. A technical volume covering such topics as composition of the earth, geochronology, isotope geology, thermal history of the earth, magnetism, and seismic studies. The text is intended for college-level students of geology or physics. Some differential equations are used in the book.

Phillips, Owen M. *The Heart of the Earth*. San Francisco: Freeman Cooper, 1968. A technical volume covering various topics in geophysics, such as gravitation, mass, earthquakes and seismic waves, volcanism, continental drift, and earth magnetism. The volume is well illustrated with drawings and numerical tables. The reader should have a working knowledge of college algebra.

Skinner, B. J., and S. C. Porter. *The Dynamic Earth*. New York: John Wiley & Sons, 1989. A well-written, well-illustrated, and very colorful volume on the geology of the earth. It would be suitable for the college student beginning geology.

Tennissen, A. C. *The Nature of Earth Materials*. 2d ed. Englewood Cliffs, N.J.: Prentice-Hall, 1983. A complete, well-illustrated volume covering the nature and structure of rocks and minerals. The volume would be suitable for the college student of geology, mineralogy, or petrology.

Vinnik, Lev, et al. "Anisotropic Structures at the Base of the Earth's Mantle." *Nature* 393, no. 6685 (June 11, 1998): 564-568.

Weiner, Jonathan. *Planet Earth*. New York: Bantam Books, 1986. A colorful, well-illustrated, well-written book describing the earth and how it is studied. This volume is the companion to the PBS television series of the same name. It is suitable for general readers.

David W. Maguire

Cross-References

Continental Growth, 102; Earth's Core, 171; Earth's Crust, 179; Earth's Structure, 224; Magmas, 383; The Oceanic Crust, 471; Plate Tectonics, 505; Rocks: Physical Properties, 550; Subduction and Orogeny, 615.

Earth's Oldest Rocks

The oldest-known rocks on earth have absolute (radiometric) ages approaching 3.8 billion years. Although earth apparently has no rocks resulting from the first 7,600 million years of its history, rocks with ages ranging back to the earliest age for the terrestrial planets, about 4.56 billion years, occur for many meteorites and are closely approached in age by some rocks from the moon.

Field of study: Geochronology and paleontology

Principal terms

ABSOLUTE DATE/AGE: the numerical timing, in years or millions of years, of a geologic event, as contrasted with relative (stratigraphic) timing

GEOCHRONOLOGY: the study of the absolute ages of geologic samples and events

HALF-LIFE: the time required for a radioactive isotope to decay by half of its original weight

ISOCHRON: the line connecting points representing samples of equal age on a radioactive isotope (parent) versus radiogenic isotope (daughter) diagram

ISOTOPES: species of an element that have the same number of protons but with differing numbers of neutrons, and therefore different atomic weights

MASS SPECTROMETRY: the measurement of isotope abundances of elements, commonly separated by mass and charge in an evacuated electromagnetic field

NUCLIDE: any observable association of protons and neutrons

RADIOACTIVE DECAY: a natural process by which an unstable (radioactive) isotope transforms into a stable (radiogenic) isotope, yielding energy and subatomic particles

RADIOGENIC ISOTOPE: an isotope resulting from radioactive decay of a radioactive isotope

Summary

Present knowledge of the oldest rocks on earth developed slowly and descriptively until the 1950's, when it became possible to measure the absolute (quantitative) ages of minerals and rocks by radiometric means. These means involve the instrumental (commonly, mass spectrometric)

measurement of unstable (radioactive) and stable (radiogenic) isotopes, or species of elements that differ only in their masses.

Prior to the ability of physicists, chemists, and geologists to make absolute age determinations, the oldest rocks on earth were known only through field relations. The main field relation used is stratigraphic sequence, which involves an application of the principle of superposition: In a sequence of undisturbed layered rocks such as sedimentary layers and lava flows, the oldest rock unit—that is, the first to be deposited—is at the bottom of the sequence. Another important field principle is the manner in which one rock is cut or cuts another rock unit. The obvious chronological conclusion is that the structure or rock that transects must be younger than the structure or rock that is transected. Through a combination of these stratigraphic and cross-cutting relationships, rock units studied and mapped can be assigned to a relative chronologic order and the geologic history of the mapped area worked out.

Accompanying the development of classical geologic principles was the understanding of the time dependence of biological evolutionary characteristics displayed by fossils found in the enclosing—primarily sedimentary—layers. Although it was early understood that fossil morphology changed through time from simpler to more complex forms, the time required for such evolutionary change could only be guessed. It is a tribute to early geologists and paleontologists that, before a quantitative measure of evolutionary scale was available, it was realized that enormous amounts of time probably were required between the deposition of rocks containing, for example, fossil collections of extinct marine animals such as trilobites and those of horses.

The geologist's most important document, the stratigraphic column (the geologic time scale), was developed over the past several hundred years through the cumulative observations of field relations, paleontologic studies, and absolute dating methods. Its refinement will continue to be an important result of geologic endeavor. Correlation, the principal activity of the geologic study of stratigraphy, whereby rock units are related through their temporal and physical characteristics, enabled scientists to have some sense of the earth's oldest rocks, long before numbers of years could be assigned to paleontologic and physical geologic phenomena. The Precambrian era, however—the vast period of geologic time that comprises more than 85 percent of the known age of the earth—was not known to have harbored life or to provide fossils until the past few decades; the discovery of widespread bacterial and stromatolitic fossils in Precambrian rocks has reversed this conclusion. Prior to these discoveries, the ages of the earliest rocks thus were surmised only through field relations and not through fossils.

A misunderstanding of old rocks on earth occurred because of the reasoning that the older the rock, the more opportunity for it to have altered, such as by weathering, tectonism (as in mountain building), or especially metamorphism. Thus, it was expected that the oldest rocks should be highly metamorphosed, as in much of the Precambrian terrain of Canada and other, commonly central continental areas of Precambrian rock (cratons or shields), and that essentially unaltered sediments and sedimentary rocks must be geologically young. Miscalculations in geologic age of billions of years occurred, owing to the incorrect correlation of rocks of similar petrology and metamorphic grade. Absolute age determinations, while not negating the essential premise of this theory, have nevertheless shown that some of the oldest rocks on earth are not highly altered and that many young rocks may be highestgrade metamorphic and tectonized.

A major advance in geochronology has developed since radiometric ages were attached to points of the stratigraphic time scale and a quantitative framework for major fossil assemblages was established. Once the ages of characteristic, representative fossils are quantified through absolute age determinations, the ages of sedimentary rocks containing chronologically diagnostic fossils can be assigned by comparison of these "guide" fossils with points on the stratigraphic time scale. Thus, the field geologist may establish the approximate age of sediments or sedimentary rocks in his or her area of interest (and, through field relations, the qualitative ages of associated igneous and metamorphic rocks and of geologic structures), simply through fossil identification. Although fossils, especially diagnostic fossils, are rare in many sedimentary rocks (especially those sedimentary rocks that formed prior to about 600 million years) and are absent in most igneous and metamorphic rocks, the use of paleontology as a chronologic tool is routine and in most cases quicker and less expensive than are geochemical (radiometric) age determinations.

Methods of Study

Antoine-Henri Becquerel presented his discovery of the phenomenon of "radioactivity" to the scientific community in Paris about one hundred years ago, laying the cornerstone for scientists' present understanding of earth's oldest rocks. The finding was followed rapidly by the seminal work of Marie Curie in radioactivity, a term she first used. Her discovery of the intensely radioactive radium as well as plutonium led Ernest Rutherford to distinguish three kinds of radioactivity—alpha, beta, and gamma—and, in 1900, with Frederick Soddy, to develop a theory of radioactive decay. Soddy later proposed the probability of isotopes, the existence of which was demonstrated on early mass spectrographs and mass spectrometers.

Rutherford and Soddy's theory of the time dependence of radioactive decay, followed by breakthroughs in instrumentation for the measurement of these unstable species and their radiogenic daughter nuclides by Francis William Aston, Arthur Jeffrey Dempster, and Alfred Otto Carl Nier, among others, caught the rapt attention of early geochronologists and had a revolutionary effect on the study of geology. In 1904, Rutherford proposed that geologic time might be measured by the breakdown of uranium in U-bearing minerals and, a few years later, Bertram Boltwood published the "absolute" ages of three samples of uranium minerals. The ages, of about half a billion years, indicated the antiquity of some earth materials, a finding enthusiastically developed by Arthur Holmes in his classic *The Age of the Earth.* Holmes's early time scale for earth and his enthusiasm for the developing study of radioactive decay, although not met with instant acceptance by most contemporary geologists, helped to set the stage for the acceptance of absolute ages as the prime quantitative components in the study of geology and its many subdisciplines.

After the early study of the isotopes of uranium came the discovery of other unstable isotopes and the formulation of the radioactive decay schemes that have become the workhorses of geochronology, such

Ernest Rutherford, who pioneered the use of radioactive decay to measure geologic time. *(The Nobel Foundation)*

as the rubidium-strontium, samarium-neodymium, potassium-argon, uranium-thorium-lead, and fission track methods. The formulation of the theory of radioactive decay of the parent, unstable nuclide (or the growth of the stable daughter nuclide), developed in the early 1900's, is the basis for the measurement of time, including geologic time, for any of these parent-daughter schemes used by geochronologists. Although each of the dating techniques is based on the formulation, differences occur in the kind of measurement and in the geochemical behavior of the several parent and daughter species. Thus, the geological interpretation of the data obtained is very different for the several chronometric schemes. These techniques for establishing absolute ages for minerals and rocks have been applied to the study of earth's oldest rocks since the early 1900's and, with the development of modern mass spectrometry, more intensely since the 1950's.

Although not indigenous to earth, the oldest rocks found on earth (and also seen to fall to earth) are meteorites, many of which yield radiometric ages near 4.56 billion years, the accepted time of formation of many solar system materials. With respect to the oldest rocks indigenous to earth, these rocks have the most likely chance of being destroyed by ongoing geologic processes such as erosion, metamorphism, and subduction. It is no surprise that fewer and fewer outcrops are found as ages become older back into Precambrian time. The oldest rocks are most commonly found in continental, cratonic regions, in many cases because of geologic preservative features such as their protective superjacent rocks, their location in tectonically stable continental interiors, and their low density and thus lower propensity for subduction than the more common basaltic rocks.

Histograms of rock ages thus show fewer and fewer data the further back in time. Such figures also show a feature whose significance has not been immediately apparent: the clustering of ages in rather discrete groupings. These groupings correlate with regionally defined rock/tectonic units such as those of the Grenville and Superior provinces of the Canadian Shield and indicate the intense geologic activity that resulted in these Precambrian rocks. Many scientists believe that the "magic numbers" that mark the groupings represent geologic periodicity, perhaps a result of major, discrete plate tectonic episodes. Others, however, point out that many radiometric dates fall outside these groupings and that the picture is incomplete and thus misleading. Although certainly incomplete, the available data indicate to many that there is some patterning in both the chemical and chronologic analyses of these rocks.

Early radiometric results showed some ages far back in time, near 3 billion years, and further analyses confirmed their antiquity. The oldest rock was thought to be the Morton gneiss in Minnesota, at about 3.2 billion years and questionably older, until several cratons yielded rocks with ages near 3.5 billion years. One such exposure, at North Pole, Australia, is of special significance, because of the concurrence of its age by several chronometric schemes (3.5 billion years) and especially because of its well-preserved bacterial and stromatolitic fossil assemblage, the earliest known. (Equivocal chemical evidence for organic life in even older rocks has been described; the existence of well-developed life at 3.5 billion years presupposes the existence of earlier life.) The oldest rocks, however, appear to be the well-studied Amîtsoq gneiss and contiguous, related rocks in the Godthaab area of western Greenland. Although there is incomplete agreement as to the exact range and significance of these earliest ages, several are close to or perhaps slightly greater than 3.8 billion years. Some of the disagreement with respect to these rocks, as well as for similar rocks around the world, results from incompletely known and undoubtedly variable diffusive and "freezing-in" behavior of the parent/daughter nuclides of the several chronometric systems. This varying behavior commonly results in different "ages" (dates) for the same analyzed rock specimen. A further uncertainty is whether the several isotopic systems can be completely reset, on the whole-rock scale, in metamorphic terrains that have been metamorphosed to lower physicochemical conditions.

Although not all scientists may agree, minerals of even older ages have been analyzed from Archean sandstones of Australia. Zircons (residual mineral phases from the final stages of crystallization of igneous rocks, especially granites) were separated from this stratigraphic unit and analyzed by uranium-thorium-lead dating using an innovative technique, the ion probe mass spectrometer. Although many of these zircons have been analyzed, only a few have exceptionally old ages, but these ages, ranging back to almost 4.3 billion years, are especially important. Because they are detrital (fragmental) in their host sandstone, they must have eroded from even more ancient rocks, perhaps granites or granitic gneisses, whose age, composition, and petrologic features are important to an understanding of the development of earth's earliest crust. So far, their provenance (parental rock) has not been found; apparently they have been completely eroded, altered, or buried by younger rock. If one accepts these earliest ages, crustal rocks existed on earth less than 300 million years after earth accreted from the solar nebula.

It is useful to place earth's oldest rocks within the framework of the ages of other available solar system materials, especially meteorites. Although the formation of the earth—that is, earth's time of accretion—is accepted by most scientists as having occurred about 4.56 billion years ago, it is obvious from the discussion above that no terrestrial rocks have ages this old. Earth's absolute age, therefore, as well as that of other solid materials of the solar system except for the sun, is known only by analogy with meteorites. Many of the meteorites have been dated by the techniques discussed above and give formational ages near 4.56 billion years; some are thought to represent the oldest and most primitive material in the solar system with the possible exception of cometary material and cosmic dust. A few apparently unprocessed (primitive) meteorites yield radiometric age and initial isotopic composition data that suggest formational ages slightly older than 4.56 billion years. The terrestrial planets (Earth, Mercury, Venus, and Mars) are thought to have originated at the same time as did the meteorites.

A few meteorites have ages significantly younger than 4.56 billion years. These rocks are considered to have originated from parent bodies that were large enough to have maintained internal heat, and, therefore, igneous processes significantly after 4.56 billion years, as did earth, with its continuing volcanism and other geologic processes that have resulted in rocks of all ages from 4.56 billion years to the present. Several of these exotic rocks, collected from ice fields in Antarctica, were recognized almost immediately as pieces of the moon, owing to scientists' familiarity with the Apollo mission's lunar rock collections. Even more spectacularly, a small collection of meteorites, long known to be different from the main collection of meteorites, was found to have crystallization ages of about 1.3 billion years, much younger than the accepted accretion age for solar system materials. These rocks must have originated from a body large enough to have maintained geologic processes between 4.56 and 1.3 billion years, unlike the moon, which has so far yielded no rocks younger than about 3.0 billion years. This parent body is widely assumed by scientists

to be Mars, a theory that is much strengthened by the compositional similarity of gases dissolved in glass from these meteorites and atmospheric
compositions of present Mars, as measured from the Viking lander a de-
cade ago.

Rocks returned from the moon by the U.S. and the Soviet space programs
yield ages from about 3.0 to 4.5 billion years. Although the moon is thought
to have originated at the same time as earth, it was not massive enough to
have provided an internal heat source capable of the continuation of
geologic processes and thus the formation of new rock to the present time.
Instead, significant igneous processes may have terminated about 3.0 billion
years ago, though there is speculation that some younger volcanic rocks may
exist. (A current and popular theory is that the moon originated from a
giant impact on early earth; the impact threw out earth mantle material into
near-earth space, which coalesced to form the moon. If this theory is true,
the moon may have an age that is measurably younger than that of earth,
and the process may also be recognizable in the chemistry and age of earth
rocks, as well.)

A widespread though not fully accepted theory for the early moon is that
it underwent a massive, perhaps global melting not long after formation
(whether by nebular accretion or earth impact). Upon cooling, plagioclase
feldspar crystallized, floated, and formed the earliest lunar crust (anorthosite), which thus dates from some time after moon accretion. If this theory
is correct, no rocks older than the anorthosite will be found; this rock has
yielded ages of 4.44 billion years and, arguably, somewhat older. If the moon
underwent significant or complete melting, it is possible or likely that the
earth experienced the same event, in which case there also will be no earth
rocks representing its earliest history. Finally, owing to earth's continuing
history of constructive and destructive geologic processes, it seems unlikely
that significant amounts of rock will be found that date from earth's first 200
million years.

Context

The absolute dating of geologic materials and events has had unprece-
dented influence on the evolution and understanding of geologic events
on earth, including earth's origin and its oldest rocks, as well as other
ancient minerals and rocks of the solar system. The ability for the scientist
to establish events in terms of actual years, rather than in relative terms
such as "older than" or "younger than," has led to a realistic knowledge
of earth's origin and its oldest rocks and has led to calibrated time scales
for major geologic processes such as organic evolution. Owing to their
usefulness in the precise determination of the ages of very old rocks,
dating methods such as uranium-thorium-lead, rubidium-strontium, and
samarium-neodymium will continue to be of major use in refining the
sequence and meaning of earth's oldest rocks and extraterrestrial materials.

Bibliography

Ashwal, L. D., ed. *Workshop on Early Crustal Genesis: The World's Oldest Rocks.* Technical Report 86-04. Houston, Tex.: Lunar and Planetary Institute, 1986. Technical but interesting and valuable introduction to the earth's oldest rocks, especially those of western Greenland. Suitable for college-level readers.

_____. *Workshop on the Growth of Continental Crust.* Technical Report 88-02. Houston, Tex.: Lunar and Planetary Institute, 1988. A technical but interesting series of articles that bear directly on earth's oldest rocks and related material. Suitable for college-level readers.

Faure, Gunter. *Principles of Isotope Geology.* 2d ed. New York: John Wiley & Sons, 1986. This textbook is an excellent though technical introduction to geochronology and the use of radioactive isotopes in geology and includes a thorough treatment of several dating techniques. Well illustrated and indexed. Suitable for college-level readers.

Taylor, S. R., and S. M. McLennan. *The Continental Crust: Its Composition and Evolution.* Oxford, England: Blackwell Scientific, 1985. An up-to-date though technical review of processes contributing to the formation of earth's oldest rocks. Suitable for college-level readers.

York, Derek, and Ronald M. Farquhar. *The Earth's Age and Geochronology.* Reprint. Oxford, England: Pergamon Press, 1975. Contains good accounts of the chronologic techniques required to date rocks and earth's age but does not include the more recent work on the oldest rocks. Technical but suitable for college-level readers.

E. Julius Dasch

Cross-References

Earth's Age, 155; The Geologic Time Scale, 272; Igneous Rock Bodies, 307.

Earth's Origin

The earth's early evolution, its subsequent internal differentiation, and its external weathering have left little substantive evidence intact for direct study. Much about the formative processes involved in the planet's origin can be learned, however, from the study of seismology, space exploration, meteoritics, and geomagnetics.

Field of study: Geochronology and paleontology

Principal terms

ACCRETION: the process of building an object by accumulation; the planets were so formed within vortices in the nebular disk around the protosun

GRAVITATIONAL DIFFERENTIATION: the separation of minerals, elements, or both as a result of the influence of a gravitational field wherein heavy phases sink or light phases rise through a melt

METEORITICS: the study of the naturally occurring masses of matter that have fallen to the earth's surface from outer space

NEBULAR HYPOTHESIS: the concept that the solar system and all of its parts are the result of the contraction of a gaseous nebula

OUTGASSING: the process of releasing gases that are trapped in molten rock

PROTOPLANET: an early phase of a planet's formation in which accretion is still a major contributor to the final product

RADIOACTIVE DECAY: the conversion of one element into another by the emission of charged particles from an atom's nucleus

SOLAR NEBULA: the disk-shaped accumulation of gas and particulate matter from which the sun was formed

SOLAR WIND: the high-speed emission of charged particles from the sun's atmosphere

Summary

In order to understand the origins of the earth, it is necessary to be aware of the sources of the materials from which it is made. It must also be kept in mind that the earth is a geologically active body and, as such, is still evolving; the final product of the origin processes has yet to be reached. In the cores of stars, hydrogen nuclei are fused together to form the heavier elements. The first product of this fusion, in addition to energy release, is helium. As the amount of helium increases in the star's core, opportunities to form the heavier elements, such as lithium, beryllium, and carbon, arise. For average, sunlike stars, carbon is usually the last major element to be formed before the star flickers out to end its life as a white dwarf and ultimately as a black cinder body. For the more massive stars, the demise is more spectacular. As the carbon state is reached, the core collapses with such force that the star explodes in what is known as a supernova. It is during this time that the heavier elements are produced and thrown out into interstellar space. The expanding shell of gas and other debris from the supernova forms a planetary nebula. This material continues to expand and mix with interstellar hydrogen, eventually forming a large, irregular nebula many light-years across.

According to the nebular hypothesis, the sun, planets, and other bodies of the solar system formed as the result of the contracting and cooling of such a nebula. The hypothesis suggests that this cloud was composed primarily of hydrogen and helium along with a small percentage of the other naturally occurring, heavier elements. The forces responsible for the initiation of the nebula's contraction, some 5 billion years ago, are not well understood. It is generally agreed that the gravitational effects within the nebula or that of nearby stars must have been instrumental in the process. A rotational component was established that increased its speed with the compaction of the nebula as a result of the conservation of angular momentum. (The same effect is seen on spinning figure skaters. As their arms are drawn in, their speed of rotation increases.) The rotation caused the nebula first to become oblate and eventually to form a flattened disk known as the solar nebula. Most of the solar nebula's mass concentrated at the center of the disk, which, when heated by gravitational pressure, formed the protosun. It has been shown by some studies that perhaps there was a compositional gradient across the nebula that may have given rise to the different planet types.

Not very long after the protosun's formation, temperatures within the remaining disk were significantly decreased. With a decrease in temperature, fractional condensation began—a process in which substances with the highest melting points solidify first into small sand-sized particles. Iron and nickel were the initial elements to solidify, followed by substances, such as the silicates, that form rocky materials. These solidified particles collided with one another and accreted in the small eddylike concentrations of the disk. As the accretion process continued, the resulting masses increased in size so that, within the span of a few tens of millions of years, the protoplanets, their satellites, and other smaller bodies of the solar system were formed.

As the protoplanets and other large mass objects in the accretion disk continued to grow, their gravitational influence grew as well. They could attract greater amounts of disk materials, thus accelerating their expansion while at the same time sweeping the surrounding interplanetary space

Scientists theorize that the Sun, Earth, and other planets were created out of a swirling nebula of interstellar matter. *(PhotoDisc)*

clean. Solar radiation could then penetrate the distances between the sun and the planets, bringing light and heat to their still-evolving surfaces. It was during this time that the planets started to evolve in different ways. The earth, like its inner solar system neighbors, had a relatively weak gravitational field, which, coupled with its now high surface temperature and exposure to the solar wind, caused it to lose significant amounts of the lighter gaseous elements. These gases, such as hydrogen and helium, were vaporized from the planet's surface and blown away from the inner parts of the solar system by the sun's emanations. The outer planets, beyond Mars, were not so affected because of their cold temperatures, greater masses, and the lessened influence of the solar wind. As a result, they are now recognized as the low-density, giant, gaseous planets with small, rocky cores.

As the earth continued to grow by accretion, certain processes were under way that would lead to the melting of its interior. Radioactivity levels were much higher in the formative periods of the earth's history as a result of the fact that many of the radioactive elements with the shorter half-lives had not yet decayed. When the protoplanet Earth was smaller and not as consolidated, the energy released during the spontaneous breakdown of the radioactive elements could escape into outer space. A larger earth would tend to insulate itself, making it more difficult for the energy released deep within it to reach its surface and to escape. Thus, the insulation effect helped in increasing the growing planet's internal temperature. Another source of heat energy for the forming planet Earth was meteoritic impact. As the earth attracted more and more material, the influx rate increased tremendously. As each colliding body struck the earth's surface, its energy of motion was converted into heat energy. The impact pressures of the larger masses were also transmitted into the earth, generating heat not only at the surface but also in its depths. There was yet another major source of heat for the earth's interior: pressure. As the accretion continued, the crushing weight of the overburden on the inner materials caused molecular friction and heat. The combined effect of all these thermal sources was the melting of the earth's interior. Once the materials of the earth were in a liquid form, they became mobile, and the process of differentiation could proceed.

The heavier elements, particularly iron and nickel, sank to form the core, while the lighter rock-forming components floated up toward the planet's surface. Also occuring at this time was the release of vapors either trapped in the solid materials or held under pressure in the molten phases or through simple sublimation or vaporization of the frozen and liquid gases. These processes were thus responsible for the generation of the earth's first atmosphere, which was primarily composed of hydrogen, helium, methane, and ammonia. The fate of this first atmosphere was quickly evident. The hydrogen and helium were swiftly lost to outer space because of insufficient gravity, while the other gases were chemically altered in the harsh environment of the rapidly evolving planet's surface. The segregation process is an ongoing one for the earth, but on a greatly reduced scale. As a result of differentiation, the interior structure of the earth consists of concentric spheres, or shells, made of materials possessing different physical and chemical properties. It has also been proposed that the earth went through a two-stage accretion process in which first 80 percent of the planet accumulated and the metallic core differentiated. During the second stage, the remaining nebular material cooled further, and the outer 20 percent of the earth accreted to form the lighter element-enriched mantle.

As time went on, the gases that were released rose to the planet's surface in a process known as outgassing. They were expelled at the surface through volcanoes, fissures (large cracks in the crust), and fumaroles (vents expelling gases of steam and sulfurous vapors). Ninety-five percent of the released gas was composed of steam and carbon dioxide and a small percentage of nitrogen. These emissions supplanted the original atmospheric gases already on the earth's surface, giving the earth its second-generation atmosphere. As several hundreds of millions of years passed, the atmosphere gradually cooled. The outpouring of gas from the earth's interior continued, however, and provided an abundant supply of water vapor that would eventually condense to form rain clouds. As the rains fell, the rivers, lakes, and seas were formed. All this water caused large amounts of carbon dioxide to be absorbed from the atmosphere by dissolving. As more gaseous material was taken up by the water, nitrogen remained, eventually becoming the major atmospheric constituent.

Thus all the criteria had been established to bring the earth through its formative stage to maturity. A phase in its existence is now in place where life as it is now known thrives, where rocks are formed and weathered, and where an evolving sea and atmosphere are present to keep the planet dynamically alive on its surface as well as in its depths.

Methods of Study

Much of what is known about the origins of the earth is derived from studies of meteorites and comets, space exploration programs, seismology, and geomagnetics. In meteorites and space exploration, scientists have two pristine, unadulterated data sources—early solar system materials to observe and study. Meteorites, moon rocks, photographs, and other remote-sensing data are being derived from sources that have not been exposed to the destructive forces of chemical and physical weathering that are at work on the earth's surface. These materials give investigators a glimpse of the composition and element distribution of the ancient solar nebula and of the various types of bodies that were formed from it. Seismology and geomagnetics give researchers clues about the internal structure of the planet. Thus, between the heavenly sources and the plutonic sources, scientists have been able to piece together an acceptable hypothesis of the earth's origin.

Meteorites have been described as miniature asteroids that have survived the rigors of an encounter with the earth. Not all meteorites fit into this description; some have been recovered that suggest lunar and Martian origins. All meteorites, however, can be classified into two major categories based on composition: The iron-nickel types have a density of about 7.5 and comprise 25 percent of all the falls; the story types have a density average of 3.5 and comprise 75 percent of all the falls. When the combined total composition of all meteorites is calculated, the resulting values compare quite favorably with those for the terrestrial planets. The difference in composition lends support to one interpretation of the nebular hypothesis: that all protoplanets differentiated approximately at the same time. It is believed that the iron-nickel meteorites are representatives of a metallic core, while the stony types were derived from the rocky mantle of some embryonic planet that was torn apart early in its formative period.

Comets provide evidence of the more volatile components of the early solar system. They are composed of ice and gravel, remnants of materials blown out from the inner solar system some 4-5 billion years ago by the solar wind but not before a portion was incorporated into the accreting proto-planets.

Data obtained from seismological studies led to the discovery that the earth's interior is divided into several distinct zones. Observations of the discontinuous transmission of earthquake shock-waves near the earth's surface led to the discovery of the Mohorovičić (Moho) discontinuity zone. Below this zone, wave propagation accelerates. A more fundamental low seismic velocity zone is now recognized below the Moho and is used to define the lower boundary of the earth's lithosphere. Another seismic discontinuity 2,900 kilometers below the surface delineates where the solid mantle is separated from the fluid outer core. Later work revealed the existence of a solid inner core.

Studies of the magnetic nature of the earth have also given support to the zonal nature of the planet's interior. Hypotheses concerning the generation of the magnetic field within the earth take into account an iron-nickel-rich core (not unlike the iron-nickel meteorites) with a solid interior surrounded by a mobile liquid outer part. This combination of phenomena would provide for an electric dynamo that could sustain a magnetic field.

All these disciplines support the nebular hypothesis in its modified state. As more data are obtained, the picture of the earth's origin becomes clearer and more refined. The story of the planet's birth is not linked to one or two sources, but encompasses all the major scientific disciplines.

Context

An understanding of the genesis of ore bodies and the processes that produce them can be of invaluable use to prospecting and mining technology, perhaps even leading to the synthesis or recycling of important minerals or elements. Planetary engineering would also benefit from such information for the modification or preservation not only of planet Earth but also of the other planets. By knowing the products of various processes in the past, one is better able to predict technological impact on the present earth. Astronomers are now prepared to look for similar processes elsewhere in the galaxy, not only to find proof of the hypothesis but also to discover and refine existing ideas about the earth's origins.

Technological spinoffs from the investment in research projects relative to learning about the origins of the earth also play an important role in the daily routine of life. Instrumentation developed to analyze the various objects involved in solving this mystery has been applied to medical, forensic, and other disciplines requiring sensitive and accurate analyses. Meteorite size and shape studies, made long before the space age, were employed by space engineers in designing reentry vehicles and in studying their aerodynamic properties. Theories about material behavior in zero-gravity conditions similar to those in the solar nebula have led technologists to experiment in outer space in preparing perfect ball bearings and other substances impossible to produce on Earth.

Bibliography

Beatty, J. Kelly, and Brian O'Leary. *The New Solar System*. 2d ed. New York: Cambridge University Press, 1981. A general overview of the solar system and its components, this book has been organized around comparative planetology. A discussion of the various aspects of the genesis of the planets is scattered throughout. Draws heavily on the results of space exploration and on the interdisciplinary use of science to illustrate many concepts. Contains abundant illustrations, such as full-color photographs, artwork, graphs, and charts. An excellent source of material for the general reader.

Hutchison, Robert. *The Search for Our Beginning*. London: Oxford University Press, 1983. The author addresses the problem of determining the processes involved in the formation of the earth and other solar system bodies through the analyses of meteorites. Links astrophysics, geology, cosmochemistry, organic chemistry, and astronomy to one another using meteoritics as the common ground. Also, fact is resolved from theory using historical perspectives and recent space exploration results. Contains some fine illustrations, both in color and in black and white. Suggested for college-level readers.

McCall, G. J. *Meteorites and Their Origins*. New York: John Wiley & Sons, 1973. This book serves as a general text for college students as well as for amateur scientists. In addition to discussions on meteorite fall phenomena, petrology, mineralogy, and astronomical phenomena, special sections are included concerning their origins and planetological considerations. Also of interest are discussions of the possibility of life in meteorites and age-dating results. Contains many fine illustrations in the form of graphs, charts, and photographs.

Morrison, David, and Tobias Owen. *The Planetary System*. Reading, Mass.: Addison-Wesley, 1988. Designed as a text for a college course in planetology, this book contains many references to the origins of the solar system and its individual components. Comparative planetology based on space exploration results, meteoritics, and other sources are utilized throughout the text to illustrate some of the evolutionary phases in the development of the planets and other solar system objects. Extensively illustrated.

Ozima, Minoru. *The Earth: Its Birth and Growth*. Translated by J. F. Wakabayashi. New York: Cambridge University Press, 1981. This book traces the genesis of the earth and its growth while highlighting problems that are rapidly being solved through isotope geochemistry. The past 4.5 billion years are sketched out for the reader in terms that are easy to comprehend. Other hypotheses are introduced and are compared to, or used to amend, the general hypothesis. Suitable for advanced high school or college students.

Smart, William M. *The Origin of the Earth*. 2d ed. New York: Cambridge University Press, 1953. This older references is included for those interested in the earlier forms of the nebular hypothesis. Quite inclusive of the major concepts while omitting many of the details that would be of interest only to scientists. Divided into three major topics: a general description of the solar system's components, chronology and how it is derived, and a synopsis of other theories. Easily understood by high school and lower-division college students.

Wasson, John T. *Meteorites: Their Record of Early Solar-System History.* New York: W. H. Freeman, 1985. Written as a text for a course on solar-system genesis, this book includes topics on meteorite classification, properties, formation, and compositional evidence linking meteorite groups with individual planets. Several techniques are revealed that describe how researchers use meteorites to determine conditions in the formative periods of the earth and other planets. Includes many graphs, charts, and illustrations that are closely tied to the subject at hand. Suggested for upper-division college students.

Bruce D. Dod

Cross-References

Earth's Composition, 162; Earth's Oldest Rocks, 202; Earth's Structure, 224; The Geologic Time Scale, 272; Meteorite and Comet Impacts, 407.

Earth's Shape

It has been known for centuries that earth is not a perfect spheroid. The circumference of the planet is significantly greater at the equator than in the dimension of the meridians, the so-called polar circumference. This oblateness is the result of the substantial centrifugal force generated by earth's daily rotation around its axis.

Field of study: Geophysics

Principal terms

CENTRIFUGAL FORCE: an upward or outward force that results from mass in motion along a curved path

DEFORMATION: the warping of earth materials as a result of strain in response to stress

EARTH TIDE: the slight deformation of earth resulting from the same forces that cause ocean tides, those that are exerted by the moon and the sun

LITHOSPHERIC CRUST: the relatively thin outer portion of earth's "onion" structure, composed of solid rock

PERFECT SPHEROID: a three-dimensional body that has the same circumference regardless of the direction by which it is measured; that is, perfectly "round"

Summary

A view of earth from satellite distance in space would, to the naked eye, suggest that the planet is a perfect spheroid. Yet, finer measurements from space and on earth itself reveal that the planet is an oblate spheroid, meaning that it is deformed. In other words, earth is distended at its "waistline," the equator. Earth's flattening at the poles, or its oblateness, is about three one-thousandths in terms of its diameter, which, as measured at the equator, is 12,756 kilometers and, pole to pole, is 12,714 kilometers.

The discovery that earth is not a perfect spheroid dates to the seventeenth century, when measurements of the distance of 1 degree of latitude (one-ninetieth of the distance from the equator to a pole) demonstrated inconsistencies from one place to another. It eventually became evident that the distance from one latitude to another becomes less the farther from the equator the measurements were taken. Extreme precision in the measurements of earth's oblateness was not possible until the advent of twentieth century instrumentation.

Earth turns on its axis, making one complete rotation every twenty-three hours, fifty-six minutes, and four seconds, which means that on the equator at sea level, there is rotational velocity of about 1,670 kilometers per hour. At 45 degrees latitude (north or south), however, the surface velocity is approximately one-half that speed, and at the poles there is no rotational velocity. This differential in rotational velocities means that the centrifugal force at work in the low latitudes is great, while in the polar regions the centrifugal force is small or nonexistent. From this differential derives the equatorial deformation and thus earth's oblate spheroidal shape.

It is necessary at this point to consider why all planets are essentially spheroidal while the hundreds of objects that comprise the asteroid belt lying between Mars and Jupiter are not, which is known as the principle of gravitational equilibrium. Any object that has density also develops its own gravitational force. When the mass of an object is sufficiently large, it cannot sustain any shape other than spheroidal; the fragments of the asteroid belt are not sufficiently large to have met that requirement. On the other hand, Earth, all the other planets of the solar system, and most of the satellites of planets (the Moon, for example) are indeed spheroids. An object with large mass (density times volume) develops its own gravitational force in a direction downward from the surface of the object. This force results in unit weight or specific gravity; the greater the force of gravity, the greater an object's weight. Only a spheroidal shape will permit physical equilibrium in a large mass in which the forces of gravity exceed the strength of the material that makes up the mass.

One is easily misled as to the rigidity of planet Earth. Its solid lithospheric crust has a thickness that approximates an egg shell in relative dimensions. Therefore, it is understandable that the earth's "shell" has been broken, warped, and distorted through geologic time and into the present. It also becomes comprehensible that earth is extremely sensitive to distortion and deformation from forces imposed upon it, upsetting gravitational equilibrium, as in the case of centrifugal force created by rotational velocity, especially in the equatorial latitudes where the velocities are highest.

A comparison of the oblateness of earth with that of other planets in the solar system demonstrates that centrifugal force from rotational velocity is the principal cause of the earth's equatorial "bulge" (see table). Oblateness is not directly proportional to the surface velocity of rotation at the planet's equator; the strength or rigidity of a planet's material composition affects the degree of warping. Planets Mercury and Venus, however, each with very slow rotational velocities, have no discernible oblateness and are nearly perfect spheroids. Mars has three times more oblateness than Earth, yet with only about one-half of the rotational velocity. The average density of Mars is about 30 percent less than that of Earth, which suggests that its materials have less rigidity, perhaps accounting for Mars' excessive equatorial bulge.

Lesser forces imposed upon earth also affect the shape of the planet, a prime example being earth tides. The oceans of the world have two huge bulges, or regions where the ocean surface is relatively high. Sea level rises as the bulge is approached and falls as it is passed. Therefore, two high tides and two low tides are recorded each day. The ocean bulges are caused primarily by the gravitational attraction of the moon, which slightly counters earth's controlling force, its own gravitational attraction. The sun also imposes an attraction, although less than that of the moon. When both the sun and the moon are positioned in line with the earth, the oceans display the highest tides, as the negative attractions of both bodies are imposed collectively. The solid earth distorts very slightly from the same external force imposed by the moon and the sun, but the distortion is so slight as to render great difficulties in actual measurement. The cycle of these earth tides is much longer than that of ocean tides; they are sufficiently long to be called static tides.

PLANET SHAPES AND ROTATIONAL PERIODS AND VELOCITIES

Planet	Oblateness	Rotational Period	Surface Velocity of Rotation
Mercury	0.0	59 days	10.8
Venus	0.0	243 days	6.5
Earth	0.003	23 hours 56' 4"	1,670.0
Mars	0.009	24 hours 37' 23"	866.0
Jupiter	0.06	9 hours 50' 30"	45,087.0
Saturn	0.1	10 hours 14' 0"	36,887.0
Uranus	0.06	11 hours 00'	14,794.0
Neptune	0.02	16 hours 00'	9,794.0
Pluto	?	6 days 9 hours	128.0

Note: Surface Velocity of Rotation is expressed in kilometers per hour.

Methods of Study

The passing of geologic time has brought startling changes in the phenomena which control earth's shape. The distance from earth's surface to the moon was far shorter than at present, and earth's rotational velocity was much faster back in geologic time. Therefore, earth's shape does not remain constant. About 350 million years ago, earth appears to have had about 405 days in the year. Fossil coral of that age shows microscopic diurnal growth lines, and the year is measured by seasonal changes in patterns. On the basis of 405 days per year, the planet's rotational velocity must have been at least 10 percent faster. Such a rotational velocity would have generated significantly greater centrifugal force, and earth's oblateness would have been more severe than at present. One can speculate as to what effect that condition might have had on ancient dynamic processes; couple that thought with the moon's nearer proximity to earth, which would have resulted in far more severe ocean tides in the Paleozoic era (about 250-575 million years ago). The two phenomena are scientifically linked. It is the moon's relatively strong negative gravitational attraction on earth that is believed to be the principal force causing the slowing of earth's rate of rotation.

Until recently, it was generally assumed that, other than the two tidal bulges, the surface of the oceans represents the smooth curvature of the planet. Geodesists, however, have hypothesized for decades that the ocean surface should theoretically have highs and lows conforming to the peaks and deeps of the ocean floor. Actual measurements in the early 1980's proved those theories to be correct.

It has become possible to map the topography of an ocean surface to a precision of a few centimeters. The instrument used is the satellite-mounted radar altimeter, which makes continuous measurements of the distance from the satellite to the water surface. Assuming the satellite itself maintains a consistent orbital path, a perfectly curved ocean surface should reflect an equally consistent distance. The fact is, with wave crests and troughs averaged out, ocean surfaces show substantial deviations from a smooth curve. Major seamounts and suboceanic ridges are clearly represented by corresponding high places of ocean surface. Likewise, the major deep-sea troughs, as are common in the western Pacific, the Indian Ocean, and the Caribbean, reveal themselves with troughs in the ocean surface above. This phenomenon derives from geographic variations in the acceleration of gravity. In the case of a deep-sea trough, for example, the space within the trough is filled with seawater instead of with rock. The density or specific gravity of seawater is slightly greater than 1.0, whereas the density of suboceanic rock is typically about 2.85—which means that the acceleration of gravity at sea level directly over a deep-sea trough is less than normal. A compensating rise in the ocean surface maintains nature's equilibrium.

Context

Earth's inhabitants suffer little or no effect from the planet's distortion. It cannot be observed with the naked eye, and it does not appear to play a role in weather patterns and climate. The principal cause of deformation, however, affects the entire habitability of the planet. The daily rotation of earth around its polar axis and its 23.5-degree tilt, with respect to the plane of orbit around the sun, results in the seasons of the year. If the earth rotated today at the rate it rotated 350 million years ago, days and nights would be 10 percent shorter. An interesting meteorological challenge is to calculate exactly how shorter days and concurrently shorter nights would affect weather, climates, agriculture, and human health. Conversely, were rotation to slow appreciably, as surely it must in geologic time, the effect would be disastrous. The days would be far hotter and the nights far colder because of longer exposure to solar radiation in daylight and longer nocturnal radiation at night. It is doubtful that the agricultural systems of today's society could be sustained.

Earth's daily turn on its axis results in other familiar phenomena. The ocean currents of the Northern Hemisphere always flow clockwise, while those of the Southern Hemisphere flow counterclockwise. Witness the Gulf Stream of the north Atlantic and the Japanese (Alaskan) current of the north Pacific, always turning to the right, while the south Atlantic flow is to the left in rotation. This phenomenon is caused by earth's west-to-east rotation. If earth rotated in the opposite direction, the sun would rise in the west and set in the east, and the phenomenon would also be reversed. The planet's oblateness, however, would be unchanged, as the equatorial bulge results from centrifugal force. That force has no regard to compass direction, deriving as it does solely from the difference in surface velocity of the land surface: very high speed in the equatorial belt, diminishing to zero speed at the poles.

Bibliography

Dott, R. H., and R. L. Batten. *Evolution of Earth.* 3d ed. New York: McGraw-Hill, 1981. Describes the physical and paleontological evolution of earth from pre-Paleozoic to the present. Designed as an introduction to the natural history of the planet and to the fossil evidence of the past.

Heacock, John G., ed. *The Structure and Physical Properties of the Earth's Crust.* Geophysical Monograph 14. Washington, D.C.: American Geophysical Union, 1971. Contributors were drawn from pertinent disciplines of geophysics, physics, geochemistry, and geology. Serves primarily as a reference for advanced students of earth science.

Melchior, Paul. *The Earth Tides.* Oxford, England: Pergamon Press, 1966. A sophisticated treatment of the physical phenomenon of small distortions of earth resulting from gravitational forces imposed by the moon and the sun.

Munk, W. H., and G. J. F. MacDonald. *The Rotation of Earth: A Geophysical Discussion.* New York: Cambridge University Press, 1960. A detailed analytical treatment of the physics of earth's rotation. Designed for professional geophysicists. Includes discussion of the small fluctuations in rotation as a result of redistribution of angular momentum, thought to be caused by dynamics in the fluid outer core.

Plummer, Charles C., and David McGeary. *Physical Geology.* 4th ed. Dubuque, Iowa: Wm. C. Brown, 1988. An introductory textbook designed for the college student and the general reader. Contains excellent overviews of planet Earth and its interior.

Siever, Raymond. *The Solar System.* San Francisco: W. H. Freeman, 1975. A comprehensive and readable compendium of twelve parts by twelve authors, including part 6, "The Earth." This work is unique in that it could serve as a reference, as a textbook, or simply as reading for the inquiring mind.

Stacey, Frank D. *Physics of Earth.* 2d ed. New York: John Wiley & Sons, 1977. A reference volume on solid-earth geophysics, including radioactivity, rotation, gravity, seismicity, geothermics, magnetics, and tectonics. Carries detailed numerical tabulations on dimensions, properties, and unit conversions.

Terrall, Mary. "Representing the Earth's Shape." *Isis* 83, no. 2 (June, 1992): 218-238.

John W. Foster

Cross-References

Earth's Structure, 224.

Earth's Structure

Processes that are occurring in the interior of the earth have profound effects upon the surface of the earth and its human population. The results of processes operating in the interior include earthquakes, volcanic activity, and the shielding of life forms from solar radiation.

Field of study: Geophysics

Principal terms

ASTHENOSPHERE: a region of the upper mantle that has less rigid and probably plastic rock material that is near to but below its melting temperature

CORE: the central spherical region of the earth consisting of an outer liquid layer and a solid inner core

CRUST: the thinnest, outermost layer of the earth

DENSITY: the mass per unit of volume (grams per cubic centimeter) of a solid, liquid, or gas

GRANITE: a silica-rich igneous rock light in color composed primarily of the mineral compounds quartz and potassium- and sodium-rich feldspars

LITHOSPHERE: the outer layer of the earth, including the outer mantle and the crust

MANTLE: a layer of dense silicate rock that lies between the crust and core and comprises the majority of the earth's volume

P WAVE: the fastest elastic wave generated by an earthquake or artificial energy source; basically an acoustic or shock wave that compresses and stretches solid material in its path

PERIDOTITE: a silicate igneous rock consisting largely of the mineral compound olivine

PLATE TECTONICS: the theory that the crust and upper mantle of the earth are divided into a number of moving plates about 100 kilometers thick that meet at trench sites and separate at oceanic ridges

REFLECTION: the bounce of wave energy off of a boundary that marks a change in density of material

REFRACTION: the change in direction of a wave path upon crossing a boundary resulting from a change in density and thus seismic velocity of the materials on either side of the surface

Summary

Evidence that comes primarily from the study of earthquake waves reveals that the interior of the earth is not homogeneous. It is instead divided into a number of layers of varying thickness, some of which show a change in composition. The thinnest layer is the outermost one known as the crust. The crust of the earth varies in thickness from about 5 kilometers under parts of the ocean basins up to about 70 kilometers under the highest mountain ranges of the continents. The rock materials of the crust are composed of a number of different rock types, but if an average continental rock could be chosen, it would probably be best represented by a granite. Granite is an igneous rock, formed by crystallization from a hot liquid known as a magma. It characteristically is rich in the element silicon, which comprises about 68 percent of its composition. The ocean basin areas of the crust, on the other hand, are characterized by an igneous rock type known as basalt. Basalt is not as rich in silicon (48 percent of its composition) but does have a greater abundance of the elements magnesium and iron.

The base of the crust is marked by a boundary known as the Mohorovičić discontinuity, or Moho. In places, the Moho is quite sharp, such as under the Basin and Range province of the western United States. It is marked by a change in density of rock types on either side of the boundary. Density is the weight per unit of volume of materials. Thus, if a cubic centimeter of rock of a granitic composition under the continents and just above the Moho could be sampled, it would weigh 2.9 grams. Rocks below the Moho, however, would weigh 3.3 grams per cubic centimeter. This suggests a change in composition to a denser type of material. It is believed this material below the Moho is probably a rock type known as peridotite. Peridotite is similar to basalt in composition, but the former is richer in magnesium while having slightly less silicon than basalt.

Peridotite is believed to represent the basic composition of the layer of the earth underlying the crust known as the mantle. The mantle comprises the bulk of the earth, representing about 80 percent by volume. The mantle is also heterogeneous. In the upper mantle at depths beneath the surface ranging from 100 to 350 kilometers is a zone of less rigid and more plastic, perhaps even partially melted material. This zone has been termed the asthenosphere. The mantle and crust above it, acting as a more rigid unit or plate, are known collectively as the lithosphere. The change in physical properties in the asthenosphere occurs because at about 100 kilometers, temperatures in the upper mantle are close to the melting point of peridotite. Although temperature continues to increase below 350 kilometers, the tremendous pressures at those depths are high enough to keep melting of peridotite from occurring.

The asthenosphere has been suggested to play an important role in changes taking place in the lithosphere above. The theory of plate tectonics suggests that the lithosphere is divided into a number of plates about 100 kilometers thick that are in constant motion, driven by hot, convective currents of material moving slowly in the plastic asthenosphere. The heat rises along plate boundaries marked at the surface by volcanic mountain ranges in the ocean basins known as mid-ocean ridges. The slowly moving currents in the asthenosphere then move laterally away from the ocean ridges beneath the lithospheric plates, perhaps helping to carry the plates above away from the ridges. As they move laterally, these asthenospheric convection currents cool, eventually becoming denser and sinking back downward. The sites where the convection currents sink are also sites where lithospheric plates dive into the mantle, perhaps pulled by the sinking currents. At these sites, marked at the surface by gashes or trenches in the ocean-basin floor, crustal rocks may be carried into the upper mantle as deep as 670 kilometers.

Two other changes in properties occur within the mantle. At 400 and 670 kilometers below the surface, increases in density occur. Although one might suspect a change in composition to account for the jump in density, laboratory studies of rocks under pressure suggest a simpler explanation. The primary constituent of peridotite is olivine. At pressures that exist at 400 kilometers and again at 670 kilometers, there is a change that occurs that causes collapse of the crystalline structure and, as a result, produces a denser mineral compound with the same composition of iron and magnesium silicate. At pressures existing at 400 kilometers, olivine converts to the denser mineral compound spinel. At the even higher pressures at 670 kilometers, spinel will convert to yet a denser mineral compound with the same composition, known as perovskite.

Thus, the changes occurring in the mantle to produce the asthenosphere and the 400- and 670-kilometer boundaries or discontinuities are not related to changes in composition but to changes related to temperature and pressure. Recall that crustal materials may be carried downward into the mantle no deeper than 670 kilometers. Although some difference of opinion exists on this point, if it is true, as most believe, it may suggest that the rock below this level is simply too dense for the lithospheric plates to penetrate.

The next layer beneath the mantle is called the outer core. This layer begins at a depth of about 2,900 kilometers beneath the surface and continues to a depth of 5,100 kilometers. There is a large density increase across the core-mantle boundary. At the base of the mantle, density has increased to a value of 5.5 grams per cubic centimeter, compared to about 3.3 grams per cubic centimeter at the Moho. At the top of the outer core, the density is estimated to be 10 grams per cubic centimeter. Iron is the only abundant element that would have the required density at the tremendous pressure of millions of atmospheres at these depths. Thus, the core-mantle boundary represents a composition change from the silicate perovskites of the lower mantle. Pure iron would give too high a density, so an iron alloy has been suggested with silicon or possibly sulfur.

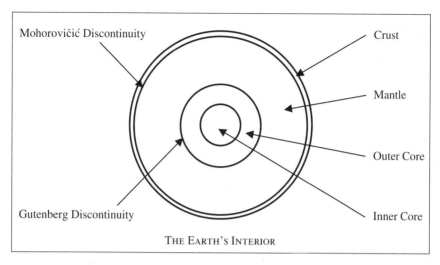

Mohorovičić Discontinuity

Crust

Mantle

Outer Core

Gutenberg Discontinuity

Inner Core

THE EARTH'S INTERIOR

At the pressures and temperatures that must exist at the depths in the outer core, iron compounds would be in a liquid state. Complex currents of metallic iron alloy, generated in the fluid outer core by the earth's rotation, give rise in some complex and, as yet, poorly understood way to the earth's main magnetic field. Some of the changes in the earth's magnetic field, such as a slow westward drift, are a direct consequence of this rotation-generated magnetic field.

The outer core mantle boundary is a sharp one, but whether it is a smooth, spherical shape or irregular with hills or peaks on its surface is not well known. There is also some evidence from seismology that the lower mantle within 100 kilometers of the core boundary represents a transition zone with a change of properties. It is not known whether this represents a mix of mantle and core material or some other composition that is less rigid than the mantle rocks above it.

The innermost layer of the earth's interior is the inner core. This region has a radius of about 1,200 kilometers and has a sharp boundary with the outer core. Increasing pressures at these depths within the earth require that the iron and perhaps nickel of the inner core exist in the solid state. There is some controversy as to whether the inner core has remained relatively constant in size throughout much of earth's history or has grown at the expense of the outer core.

Methods of Study

Much of what is known about the structure of the interior of the earth comes from the study of earthquake waves that pass through the body of the earth. These waves are of two varieties, the primary, or P wave, and the secondary, or S waves. The P wave is the same as an acoustic or sound wave. As the P wave travels through rock material, it causes the material to move back and forth in the line of wave travel, stretching and compressing it. Since the rock is not fractured by this response, or permanently deformed, this is called an elastic response, as with the stretch and release of a rubber band. It can be shown in the laboratory that there is a relationship between the amount of elastic response, or the specific physical property, and the velocity of the P wave. The ability to maintain a fixed shape, or the rigidity of the material, is one of these physical properties, while the resistance to squeezing or a change in volume is another. The S wave, however, which moves material in the path from side to side, is sensitive only to rigidity. Therefore, an S wave cannot travel across a substance that is not rigid, such as a gas or liquid, while a P wave can, but with reduced velocity. It has been found for the interior of the earth that both P waves and S waves cross the asthenosphere of the upper mantle, but with reduced velocity, suggesting lower rigidity but not a liquid state, since the S wave is propagated through the zone. Therefore, it seems that the asthenosphere may represent a plastic but still solid region.

Elsewhere in the earth's interior, only the earth's outer core shows a sharp drop in velocity of the P wave as it crosses the mantle outer core boundary. At this point, the S wave disappears, suggesting no rigidity of the material of the outer core. Since it is known that gases cannot exist at pressures at the depth of the outer core, the material of this region must consist of a liquid.

Still in other locations, the interior of the earth is marked by increases in velocity for both the P wave and S wave. There is a sharp increase in velocity at the base of the crust. Above the Moho, the P-wave velocities are around 7 kilometers per second, while below it they jump to 8 kilometers per second or more. Below the asthenosphere, the P-wave velocity increases gradually to a depth of about 400 kilometers. At this depth, there is a rather sharp increase in P-wave and S-wave velocities. Studies of wave velocities in rocks under pressure in the laboratory indicate that peridotite is an appropriate choice for upper-mantle rocks. At 400 kilometers, pressures are such that peridotite collapses to form spinel. This change in mineral compounds could account for the increased velocity in 400 kilometers.

At 670 kilometers depth in the mantle, a second discontinuity or increase in wave velocities occurs. Here, pressures may cause a second collapse to produce the yet denser mineral compound, perovskite. Once again, P and S waves passing through a mantle composed of perovskite would show an increase in velocity. Moreover, the velocities for this part of the mantle match those obtained in the laboratories from waves passing through perovskite samples placed under the kinds of pressures found at 670 kilometers. A final increase in velocity may be observed at the outer to inner core boundary. This could be explained by a phase transition from liquid to solid iron. Such a proposal is supported by the reappearance of the S wave in the inner core.

Another way in which the existence of structural boundaries within the earth can be shown from a study of seismic waves is to examine the behavior of the waves when they encounter the boundaries. Depending on the angle at which the waves approach the boundary, as well as on the properties of materials on both sides of the boundary, a seismic wave may bounce off or reflect from such a boundary, or refract or bend as it crosses the boundary.

The same bending occurs when light crosses from air to liquid in a glass. This can be shown by placing a straight straw in the liquid and looking down along it. The straw will appear bent when it is actually the light wave that has bent. P waves are reflected off the Moho, mantle-core, and outer-inner core boundaries, providing clear evidence that there are sharp boundaries between these layers. Waves have also been detected bouncing off of the 670-kilometer discontinuity.

The bending or refraction of waves yields further evidence. As P waves cross the mantle-core boundary, they are refracted away from the center of the earth because of the change in density. This lenslike focusing action for waves passing through the outer core leaves a gap on the other side of the earth where the earthquake waves emerge. This gap is known as the P-wave shadow zone because no P waves will reach the surface in this area. This shadow zone is also evidence of the existence of a liquid outer core.

Advances in computer science have allowed the identification of even subtler details about the earth's interior. Computerized tomography is a technique used in medicine, in which X rays from all directions are analyzed in a computer to give a three-dimensional picture of the human body. Seismic tomography is an analogous approach that uses seismic waves that travel from earthquakes to seismographs around the world to map the earth's interior. This includes both P and S waves in the interior as well as the results of the study of surface waves, which can also move rock material at great depths. By looking at the time of travel between two points, scientists are able to compare velocities along different paths. Such an approach has already resulted in maps of slow and fast regions of the mantle that probably represent warmer (less rigid) and colder (more rigid) regions.

Context

The interior of the earth has profound effects on humans and their environment. The interior acts as a complex great engine. The heat energy released affects the crust of the earth. This release of energy is the driving force behind plate tectonics that results in the formation and evolution of oceanic and continental crustal rocks. In the process, earthquakes and volcanic activity occur that create hazards for the human population on the earth's surface. Complete acceptance of the plate tectonic theory could not occur without the discovery of the asthenosphere, which makes the movement of the lithospheric plates more plausible.

The earth is fortunate to have a magnetic field. Without it, the age of discovery and exploration would not have been possible, for navigation by magnetic compasses allowed voyages across uncharted oceans. Modern-day navigation is equally dependent on the magnetic field. It is now known that the earth's magnetic field is generated from deep inside in the region of the outer core. Another important implication of the core-generated magnetic field is the changes it undergoes through time. In particular, at rather irregular intervals the magnetic poles switch places between north and south. The details of such a switch are uncertain; however, it is known that the magnetic field decreases in strength. Since the magnetic field shields life forms on the earth's surface from extremes of solar radiation, there is some concern for the effect on the human population. Some scientists suspect that genetic changes occur during polar reversal periods that aid the process of biologic evolution. Thus, the surface of the earth as well as the life forms on it depend upon and are strongly affected by changes occurring within the earth, in its mantle and core.

Bibliography

Bolt, Bruce A. "Fine Structure of the Earth's Interior." In *Planet Earth*. San Francisco: W. H. Freeman, 1974. An extremely well-illustrated review of how seismic waves have been used to discover and define the various layers of the earth's interior. This collection of articles from *Scientific American* is written at a general-interest college level. Little background or expertise in mathematics is required.

_____. *Inside the Earth: Evidence from Earthquakes*. San Francisco: W. H. Freeman, 1982. This book is written for undergraduate college students in physics and the earth sciences and for nonspecialists interested in a more detailed summary of knowledge of the earth's interior. The text is relatively free of mathematics and is clearly and well illustrated. It is a rather concise and readable treatment of the use of seismic waves to discover and interpret the earth's interior. A large list of useful references is included.

Cromie, W. J. "Windows to the Earth." *Mosaic* 15, no. 6 (1984): 28-37. The articles in this journal are written for the nonspecialist, providing a very readable review of the latest developments in research of the interior of the earth. A good summary of seismic tomography.

Heppenheimer, T. A. "Journey to the Center of the Earth." *Discover* 8 (November, 1987): 86-92. A very well-illustrated treatment of planet Earth with color illustrations and clear diagrams. Includes a short treatment of the latest advances in understanding of the earth's interior, as well as an excellent explanation of seismic tomography. This article describes the development of the relationships between the earth's interior and its processes and the dynamic changes occurring on the surface of the earth.

Jeanloz, Raymond. "The Earth's Core." *Scientific American* 249 (September, 1983): 46. This article is geared toward general science and undergraduate college audiences. It is well illustrated with excellent color photographs and diagrams. The references are restricted to a few key ones. Emphasis is on the earth's magnetic field and on the physical state and chemical composition of the inner and outer core.

McKenzie, D. P. "The Earth's Mantle." *Scientific American* 249 (September, 1983): 67. An excellent companion article to "The Earth's Core." Again, the color illustrations are excellent and helpful. There is an emphasis on the physical state and composition of the mantle and a strong development of the relationship between processes in the mantle and the dynamics of the crust.

"Modeling the Earth's Structure." *American Scientist,* September/October, 1984: 483-495.

David S. Brumbaugh

Cross-References

Continental Crust, 78; Earth's Core, 171; Earth's Crust, 179; Earth's Mantle, 195; Igneous Rocks, 315; The Lithosphere, 375; Plate Tectonics, 505.

Elemental Distribution

Ocean floors are composed of a dark, fine-grained rock called basalt that is more depleted in silicon and potassium and is richer in magnesium and iron than are the abundant light-colored granitic rocks on the continents. Igneous rocks that form where one oceanic plate is being thrust below another are generally intermediate in composition. Certain ore deposits occur only where certain plate tectonic processes take place, thereby enabling a geologist to focus the search for these deposits.

Field of study: Geochemistry

Principal terms

ANDESITE: a volcanic rock that is lighter in color than basalt, containing plagioclase feldspar and often hornblende or biotite

BASALT: a dark-colored, volcanic rock containing the minerals plagioclase feldspar, pyroxene, and olivine

GRANITIC ROCK: a light-colored, intrusive rock containing large grains of quartz, plagioclase feldspar, and alkali feldspar

LIMESTONE: a sedimentary rock composed mostly of calcium carbonate formed from organisms or by chemical precipitation in oceans

P WAVES: the first waves from earthquakes to arrive at a seismic station; because they travel at different speeds through different types of rock, they may be used to deduce the rock types below the surface

PERIDOTITE: a dark-colored rock composing much of the earth below the crust; it usually contains olivine, pyroxene, and garnet

PLATE TECTONICS: the theory that assumes that the earth's crust is divided into large, moving plates that are formed and shifted by volcanic activity

SANDSTONE: a sedimentary rock composed of larger mineral grains than those forming shales, thus deposited from faster-moving waters

SEDIMENTARY ROCK: a flat-lying, layered rock formed by the accumulation of minerals from air or water

SHALE: the most abundant sedimentary rock, composed of very tiny minerals that settled out of slowly moving water to form a mud

Summary

The surface of the earth may be broadly divided into the oceanic crust and the continental crust. The oceanic crust is on the average "heavier," or denser, than the continental crust. Both the continental and oceanic crusts are denser than the underlying rocks in the earth's mantle. The continental and oceanic crusts can thus be considered a lower-density "scum" floating on the denser mantle, somewhat analogous to an iceberg floating in water. Because the denser oceanic crust sinks lower into the mantle than the continental crust, much of the oceanic crust is covered by the oceans, but the less dense continental crust is mostly above the level of the oceans. Also, seismic waves from earthquakes indicate that the oceanic crust is much thinner (about 6 to 8 kilometers) than the continental crust (about 35 to 50 kilometers). The density difference between the oceanic and continental crusts is related to the kinds of minerals composing them. The oceanic crust contains more of the denser iron- and magnesium-rich minerals, olivine (iron and magnesium silicate) and pyroxene (calcium, iron, and magnesium silicate), than does the continental crust. The continental crust contains much more of the less dense minerals, quartz (silica) and alkali feldspar (potassium, sodium, and aluminum silicate), than does the oceanic crust. In addition, the oceanic crust contains much of the feldspar called calcium-rich plagioclase (calcium, sodium, and aluminum silicate) than does the continental crust.

This difference in mineralogy between the oceanic and continental crusts is reflected in their average elemental composition. The oceanic crust is enriched in elements concentrated in olivine, pyroxene, and calcium-rich plagioclase, and the continental crust is enriched in those elements concentrated in quartz and alkali feldspar. Thus, the continental crust contains larger concentrations of silicon dioxide (60 weight percent in the continental crust versus 49 weight percent in the oceanic crust) and potassium oxide (2.9 versus 0.4 weight percent) and lower concentrations of titanium dioxide (0.7 versus 1.4 weight percent), iron oxide (6.2 versus 8.5 weight percent), manganese oxide (0.1 versus 0.2 weight percent), magnesium oxide (3 versus 6.8 weight percent), and calcium oxide (5.5 versus 12.3 weight percent) than does the oceanic crust. The other major elements, aluminum and sodium, are fairly similar in concentration in both the oceanic and continental crusts. The mantle is even denser than the crust, since it contains the dense minerals olivine, pyroxene, and garnet (magnesium and aluminum silicate) in the rock called peridotite. It does not contain the less dense minerals, quartz and feldspar. Thus, the mantle is even more enriched in iron oxide and magnesium oxide and more depleted in potassium oxide, sodium oxide, and silicon dioxide than are the crustal rocks. (See the accompanying table for the typical elemental composition of rocks that form the mantle and continental crust.)

The above discussion summarizes the average characteristics of the oce-

TYPICAL COMPOSITION OF PRIMARY ROCKS OF THE EARTH'S MANTLE AND CRUST

Element Oxide	Unmelted Peridotie in the Mantle	Basalt Formed at Oceanic Ridges or Rises	Andesite Formed at Subduction Zones	Granitic Rock Along Continental Subduction Zones	Continental Rift Basalt	Shale	Sandstone Near the Source	Sandstone Far from the Source	Limestone
SiO_2 (silicon oxide)	45.0	49.0	59.0	65.0	50.0	58.0	67.0	95.0	5.0
TiO_2 (titanium oxide)	0.4	1.8	0.7	0.6	3.0	0.7	0.6	0.2	0.1
Al_2O_3 (aluminum oxide)	8.7	15.0	17.0	16.0	14.0	16.0	14.0	1.0	0.8
Fe_2O_3 (ferric iron oxide)	1.4	2.4	3.0	1.3	2.0	4.0	1.5	0.4	0.2
FeO (ferrous iron oxide)	7.5	8.0	3.3	3.0	11.0	2.5	3.5	0.2	0.3
MnO (manganese oxide)	0.15	0.15	0.13	0.1	0.2	0.1	0.1	—	0.05
MgO (magnesium oxide)	28.0	8.0	3.5	2.0	6.0	2.5	2.0	0.1	8.0
CaO (calcium oxide)	7.0	11.0	6.4	4.0	9.0	3.0	2.5	1.5	43.0
Na_2O (sodium oxide)	0.8	2.6	3.7	3.5	2.8	1.0	2.9	0.1	0.05
K_2O (potassium oxide)	0.04	0.2	1.9	2.3	1.0	3.5	2.0	0.2	0.3
Volatiles (water or carbon dioxide)	1.0	1.0	1.0	2.0	1.0	8.0	2.0	1.0	42.0

Note: Compositions are given as weight percentages of the element oxide in the entire rock.

anic and continental crusts, but they also vary substantially in composition. The continental crust is considerably more heterogeneous than is the oceanic crust. The oceanic crust consists of an upper sediment layer (about 0.3 kilometer thick), a middle basaltic layer (about 1.5 kilometers thick), and a lower gabbroic layer (about 4 to 6 kilometers thick). Basalts and gabbros both contain olivine, pyroxene, and calcium-rich plagioclase. They differ only in grain size; the basalts contain considerably finer minerals than do the gabbros. The basaltic and gabbroic layers are thus very similar in composition. They are also of fairly constant thickness across the oceanic floors. The gabbroic layers disappear over oceanic rises, or linear mountain chains on the oceanic floors. The basaltic rocks are believed to form at the rises by about 20 to 30 percent melting of the underlying peridotite in the upper mantle. The newly formed oceanic crust and part of the upper mantle are believed to be slowly transported across oceanic floors, at rates of about 5 to 10 centimeters per year, to where this material is eventually subducted or thrust underneath another plate.

The thickness of sediment on ocean floors varies considerably. It is nearly absent over the newly formed basalts at oceanic rises. It is thickest in basins adjacent to continents where weathering and transportation processes carry large amounts of weathered sediment into the basins. The composition of oceanic floor sediment varies considerably in composition. It contains varied amounts of calcite or aragonite (calcium carbonate minerals), silica (silicon dioxide), clay minerals (fine, aluminum silicate minerals derived from weathering), volcanic ash, volcanic rock fragments, and ferromagnesian nodules.

Finally, a few volcanoes composed of basalt form linear chains on the ocean floor, away from the rises or subduction zones such as the Hawaiian Islands. These ocean-floor basalts are similar in composition to those at oceanic rises, except that they contain greater amounts of potassium. The amount of basaltic rocks produced by these ocean-floor volcanoes is insignificant, however, compared to the vast amounts of basalt produced at oceanic rises.

In contrast to the oceanic crust, the continental crust is quite heterogeneous in mineralogy and chemical composition. About 75 percent of the surface of the continents is covered by great piles of layered rocks called sedimentary rocks. The average thickness of these sedimentary rocks on the continental crust is only about 1.8 kilometers, although they may locally range up to 20 kilometers in thickness. The main kinds of sedimentary rocks on the continents are the very fine-grained shales or mudrocks (about 60 percent of the total sedimentary rocks), the coarser-grained sandstones (about 20 percent of the total), and limestones or dolostones (about 20 percent of the total). The shales or mudrocks are composed of very small grains of mostly clay minerals and quartz. The resultant composition of the shales is often high in the immobile elements, aluminum and potassium, and low in the mobile elements, sodium and calcium. Sandstones vary in composition depending on which rocks weather to form the sandstone, the distance of the sandstone from the source, and the intensity of weathering. Sandstones formed close to a source of granitic rocks may have a composition similar to that of the granitic rock: high in silicon and potassium and low in magnesium, iron, and calcium compared to basaltic rocks. Sandstones formed a long distance from the source have more time to be weathered. Thus, these sandstones may have most of the unstable minerals weathered away to clays or soluble products in water (for example, sodium), and they may be enriched in silicone because of the abundance of the stable mineral quartz. Limestones typically form in warm, shallow seas by the action of organisms to produce most of the calcium carbonate in these rocks. Thus, limestones are enriched in calcium and depleted in most other elements. The dolostones are also enriched in magnesium as well as calcium. Some places, such as the Great Plains in the United States, consist mostly of alternating limestones and shales formed in ancient, shallow seas. (Thus, the average composition of the surface rocks in these areas may be considered an average of that of shale and limestone in whatever proportion they occur.) The average composition of sedimentary rocks on the continents is significantly different from that of the granitic rocks that weathered to form them. The average sedimentary rocks on continents are much more enriched in calcium (because of carbonate rocks), carbon dioxide (also because of carbonate rocks), and water (because of incorporation in clay minerals), and they are depleted in sodium (because of its solubility).

The thickness of these sedimentary rocks is still small compared to the 35- to 50-kilometer thickness of most of the continental crust. Only about 5 percent of the continental crust by volume is composed of sedimentary rocks. Most crustal rocks are igneous rocks or their metamorphic equivalents. Metamorphic rocks form in the solid state at high temperatures and pressures because of their deep burial in the earth. A substantial percentage of these igneous rocks of the upper continental crust are either granitic rocks (quartz and alkali feldspar rock) or andesitic rocks (plagioclase-rich rock). Basaltic rocks probably compose only about 15 percent of the upper continental crust.

Most of the granitic rocks and andesites originally formed along subduction zones, where oceanic crust is being thrust or subducted below either oceanic or continental crust. There also may be some basalts formed along these subducted plates. These basalts, andesites, and granitic rocks that formed along continental margins may eventually be plastered along the edges of the continents, resulting in the gradual growth of the continents. Other basalts are formed in portions of continents, called continental rifts, that are being stretched apart much like taffy. These basalts are considerably more enriched in potassium than basalts formed on ocean floors. For example, a large fraction of the states of Washington, Oregon, and Idaho is covered with these rift basalts extruded as lavas since about 20 million years ago. The total volume of about 180,000 cubic kilometers for these basalts is still comparatively insignificant; therefore, basalts make only a small contribution to the composition of the average upper continental crust.

The composition of the lower continental crust is much more difficult to determine than that of the upper continental crust because the rocks forming the lower crust are not exposed at the surface. Estimates of about 50 percent granitic and 50 percent gabbroic rocks in the lower crust have been reached. Thus, the lower continental crust is more enriched in the basaltic components, calcium, magnesium, iron, and titanium, and depleted in the granitic components, potassium and silicon, than is the upper continental crust.

Methods of Study

The average composition of the surface rocks of the continental crust may be easily estimated from the distribution of the different crustal rocks and their composition. This estimate of the surface composition of crustal rocks will be dominated by sedimentary rocks, as 75 percent of the exposed area are sedimentary rocks. The composition of the surface of the oceanic crust may be more difficult to estimate, because basalts on the ocean floor are often covered by a thin layer of sediment. Therefore, the areal extent and composition of the oceanic sediment is now known as precisely as is that of the continental crust. The average compositions of the middle and lower oceanic and continental crusts are difficult to determine because they cannot be directly sampled. Much of the information about the nature of the crust below the surface comes from the behavior of seismic waves given off by earthquakes, from heat-flow measurements, and from the composition of rock fragments brought up by magma passing through much of the crust. In addition, there are places in the crust where rocks from the lower crust have been uplifted to the surface, so their composition can be examined in detail.

The speed of the earthquake waves through the oceanic crust is consistent with the crust being composed of a thin upper layer of sediment (indicated by P-wave velocities of 2 kilometers per second), a thicker middle layer of basalt (P-wave velocities of 5 kilometers per second), and a thick lower layer of mostly gabbro (P-wave velocities of 6.7 kilometers per second). The thicker continental crust, however, has P-wave velocities (6.1 kilometers per second) consistent with mostly granitic rocks below the overlying sedimentary rock veneer (2 to 4 kilometers per second). The lower continental crust has P-wave velocities (6.7 kilometers per second) similar to those expected for lower-silica rocks like gabbro, so there is probably more gabbro mixed with granitic rocks in the lower crust.

How fast heat flows out of the earth may also be used to limit the composition of crustal rocks. Variation in heat flow at the surface depends on how much heat is flowing out of the earth below the crust; the distribution of radioactive elements in the crust, such as uranium, thorium, and potassium, that give off heat; and how close magmas are to the surface. Oceanic ridges and continental rift zones, for example, have high heat flow, suggesting that magmas are close to the surface. In contrast, the heat loss from much of the ocean floor and over much of the continents with old Precambrian rocks (older than about 600 million years) is considerably lower because of the lack of magma close to the surface. It is surprising, however, that the oceanic floor and continents with old Precambrian rocks have similar low heat flow, as the abundant granitic rocks in the continents ought to be more enriched in the heat-producing radioactive elements than is the oceanic crust. That suggests that many of the granitic rocks at depth in these parts of the continental crust are depleted in radioactive elements, perhaps because of melting processes carrying away the radioactive elements in the magmas during the Precambrian. Also, that is consistent with the presence of abundant basaltic rocks depleted in radioactive elements in the lower crust.

There are places on the earth, such as the island of Cyprus in the Mediterranean Sea, that appear to be ruptured portions of the entire oceanic crust and part of the upper mantle. In Cyprus, the lower zone is composed of peridotite, olivine-rich rocks, or pyroxene-rich rocks, as are predicted to occur in the mantle. These rocks correspond to the P-wave seismic velocities of 8 kilometers per second. There is a rather abrupt change to the next overlying layer of mostly gabbros that correspond to the sharp decrease in P-wave velocities to about 6.7 kilometers per second. These rocks grade upward into basalt corresponding to the upper igneous rock layers of the oceanic crust with P-wave velocities of about 5 kilometers per second. The basalt and gabbros are also penetrated by a multitude of tabular igneous dikes that were feeders of magma to the overlying basalt at the surface. Finally, there are overlying sedimentary rocks corresponding to the upper oceanic layers with P-wave velocities of about 2 kilometers per second.

Finally, foreign rock fragments and drill cores give up some information about the crust. A number of peridotite fragments are brought up with magmas derived from the mantle of the earth. There may be a variety of crustal rocks also brought up with them that may confuse the picture, as their depth of origin is unknown. Drill cores provide mostly information about the sedimentary rocks that might contain oil; they also give some information about the first igneous rocks just below the sedimentary rocks. Generally, wells are never drilled deep enough to obtain samples from the intermediate and lower continental crust.

Context

A knowledge of the overall distribution of rock types and the corresponding elemental compositions of these rocks over the earth can give geologists a guide to where to look for certain kinds of ore deposits, as certain ores occur in certain kinds of rocks. The most generalized pattern is the association of certain types of ores with certain tectonic environments. Both oceanic rises and subduction zones tend to heat waters and drive the resultant metal-rich waters toward the surface. Oceanic rises often contain sulfide-rich, copper and zinc hot-water deposits. These hot-water deposits at subduction zones are often enriched in copper, gold, silver, tin, lead, mercury, or molybdenum. One example of the subduction zones deposits are the copper porphyry deposits. These important ore deposits are formed in granitic rocks that crystallized at shallow depths below the surface in areas where an oceanic plate is being subducted, or thrust below, a second plate. They are especially abundant around the rim of the Pacific Ocean. The copper ores contain low copper concentrations (0.25 to 2 percent) and have some associated molybdenum and gold. These low-grade ores are often profitable to mine because of the large volume of ore (over a billion tons in some places) that can be rapidly extracted from the rock. A geologist looking for such ores designs an exploration campaign to search out only areas with active or inactive subduction zones. Also, the geologist looks for certain compositions of granitic rocks intruded at fairly shallow depths below the surface that have been exposed to erosion near the top of the intrusion, as these are the places where the copper porphyries form. Hundreds of these copper porphyry deposits have been discovered, accounting for about half of the copper ores of the world. Copper is used in wires to transmit electricity and in bronze and brass.

Bibliography

Ahrens, L. H. *Distribution of the Elements in Our Planet*. New York: McGraw-Hill, 1965. This book provides a clear summary of the composition of the solar system and the earth. The elements are grouped in a geochemical classification. Directed to the nonspecialist.

Craig, J. R., D. J. Vaughan, and B. J. Skinner. *Resources of the Earth*. Englewood Cliffs, N.J.: Prentice-Hall, 1988. This is an excellent book describing the distribution of ore deposits on the earth. Information is provided on the history and use of the elements, geologic occurrence, and reserves. For a nonscience major in college or interested layperson. There is a glossary of technical terms.

Skinner, B. J., and S. C. Porter. *The Dynamic Earth*. New York: John Wiley & Sons, 1989. This one of many introductory geology textbooks for college students that has a chapter on mineral and energy resources in the earth. The interested reader with some understanding of geology can find information here about the major ore deposits and their distribution within the earth.

Smith, D. G., ed. *The Cambridge Encyclopedia of the Earth Sciences*. New York: Crown, 1981. This reference is written for the reader with some background in science who needs to locate information on a specific earth science topic. Chapters 4 ("Chemistry of the Earth"), 5 ("Earth Materials"), and 10 ("Crust of the Earth") might be most appropriate for further reading related to elemental distribution. There are also chapters on plate tectonics.

Utgard, R. O., and G. D. McKenzie. *Man's Finite Earth.* Minneapolis, Minn.: Burgess, 1974. This book is written as supplementary reading for college geology courses. A section on earth resources that gives some insight on ore distribution and how it relates to public policy is suitable for a layperson.

Wedepohl, Karl Hans. *Geochemistry.* Translated by Egon Althaus. New York: Holt, Rinehart and Winston, 1971. This book gives nontechnical descriptions of the elemental distributions within the solar system and the earth. Some knowledge of chemistry and geology is necessary for full use of the book. Chapter 7 gives specific information on the distribution of elements in the earth's crust.

Robert L. Cullers

Cross-References

Continental Crust, 78; Igneous Rocks, 315; Plate Tectonics, 505; Sedimentary Rocks, 568; Siliciclastic Rocks, 584; Subduction and Orogeny, 615.

Floodplains

Floodplains historically have been the locations of dense human population, providing fertile land and water that enabled early human settlements to attain some degree of permanence. They have also been the locations for recurring damage from floods.

Field of study: Sedimentology

Principal terms

ALLUVIUM: sediment that is deposited by a stream

BLUFF: the edge of the remnant higher-elevated land that marks the margin of the floodplain

DISCHARGE: the volume of water that is transported by a stream, normally stated as cubic meters per second or cubic feet per second

FLUVIAL: pertaining to running water; for example, fluvial processes are those in which running water is the dominant agent

GEOMORPHOLOGY: the study of the origins of landforms and the processes of landform development

LOCAL BASE LEVEL: that elevation below which a stream of equilibrium will not degrade

MEANDER: a large sinuous curve or bend in a stream of equilibrium on a floodplain

100-YEAR-FLOOD: a hypothetical flood whose severity is such that it would occur on an average of only once in a period of one hundred years; equates to a 1 percent probability each year

OXBOW: a lake that is a remnant of a cutoff stream meander

STREAM OF EQUILIBRIUM: a stream that is carrying its maximum load of sediment; will not erode its channel any deeper but instead will establish a floodplain

Summary

Floodplains are well named. They are some of the most level surfaces on earth, and they are known for frequent floods. They also are probably the most agriculturally productive lands on earth. The earliest concentrations of civilization are associated with the floodplains of historical rivers whose names are well known: the Nile, Ganges, Euphrates, Yangtze, and Huang. These floodplains are still among the most populous areas on earth.

A floodplain is a landform, a physical feature that is studied in the discipline of geomorphology. Its origin is linked with frequently recurring floods. One of the most studied floodplains is that of the Mississippi River. Floods on the lower Mississippi Valley today are to a large degree under human control; at least the floodwaters are restricted to certain areas where they will do the least damage. A large flood, the so-called superflood or 100-year-flood, can still do tremendous damage. A large river cannot be prevented from flooding; it is a part of the natural fluvial process. It can only be altered to some extent so that its floodwaters will do a minimum of damage.

The variables of climate and topographic relief will result in floodplain variation around the world, but all will possess certain identifiable characteristics (see figure). The foremost distinguishing characteristic is the very low relief, or its almost "flat" appearance. Of course, the land is not in fact level, or the river would not flow. The floodplain possesses a very slight gradient to the sea. It is an expanse of sediment that was deposited by the river itself. Thousands of years of flooding have spread layer upon layer of river-borne sands, silts, and clays, referred to as alluvium. As the waters receded after each flood, the river returned to its normal channel.

In order to understand the floodplain, it is necessary to understand the processes that brought it into being. The originating river did not always possess the floodplain, and many rivers today do not have floodplains. Imagine the stream in the figure as it might have been perhaps millions of years ago. The stream was possibly flowing in a narrow channel with higher land along either side. There was no level land adjacent to the stream—no floodplain; its formation would require eons of time of erosion and deposition. As the stream eroded, its channel deepened and also attempted to erode laterally, but the vertical downcutting was dominant and the lateral cutting ineffective. At some point in the fluvial process of downcutting, the stream reached an elevation, referred to as local base level, beyond which the stream could not downcut. The local base level is not some hard, resistant rock that stops downcutting, but rather is a particular elevation in respect to that stream, which in relation to the stream's mouth at sea level determines the gradient of the stream. When the stream downcuts its valley, the stream's gradient, or slope of flow, is lowered. As the gradient is lowered, so is the stream's velocity, and consequently the stream's ability to carry sediment load is lowered. At some gradient, the stream's ability to carry its sediment load equals the sediment load that has been brought to the master stream by its tributaries. When that happens, the stream is said to be in equilibrium. Its sediment-load carrying capacity is then equal to the sediment load. This "stream of equilibrium" condition is a prerequisite for the development of a floodplain. The Mississippi River is a classic example of such a stream. When the stream attains equilibrium and stops downcutting, its lateral cutting becomes the dominant stream activity. Thus, the processes for floodplain development are set in motion. The lateral cutting by the stream results in a widening of the valley, but without any deepening. A broader-level valley floor then begins to develop, as the stream swings side to side, without downcutting. Over many years, the floor of the valley becomes wider than the channel of the stream itself. This level land on the valley floor adjacent to the stream, small at first, is the floodplain. It will be enlarged over time. The stream's volume of water, or discharge, will naturally vary with seasonal precipitation. During a flood, the discharge exceeds the channel capacity, and the excess overflows onto the level adjacent land. The term "floodplain" is now appropriate.

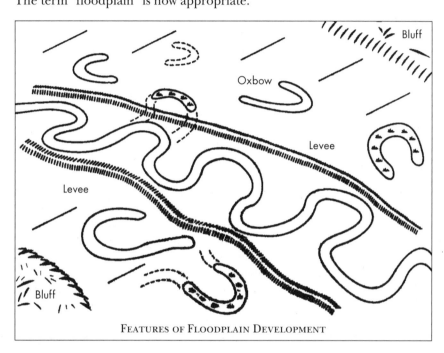

FEATURES OF FLOODPLAIN DEVELOPMENT

The floodplain is continually widened as the stream cuts laterally. The "bluffs" that mark the edge of the floodplain are driven farther back from the river as more of the adjacent higher land is eroded. When wide enough, the stream channel develops great curving, sinuous "meanders," as depicted in the diagram. Meanders are perhaps the most prominent identifying characteristics of a stream of equilibrium, and true meanders are always on a floodplain. Meanders may subsequently be cut off from the main channel by its persistent migrations or wanderings over the floodplain. Such cutoff meanders may persist for hundreds of years as lakes, called oxbows. The oxbows will gradually be silted in and become oxbow swamps for a period of time. Even after thousands of years, the oxbow lakes and oxbow swamps may survive as meander "scars" on the floodplain, showing up as vegetation and soil differences. These curving scars are even visible in fields of cropland. The river continues to flood periodically, frequently on an annual basis.

Each time the floodwaters spread over the floodplain, an additional layer of silt is deposited. The entire floodplain is a compilation of layer upon layer of alluvial deposits from flood after flood, over thousands of years. The silts that are deposited on the plain have been eroded earlier from soils upstream. They tend to be fertile topsoils from upstream drainage basins, and thus the floodplains tends to receive almost annual increments of fertile topsoil. The recurring flooding and silting shortens the lifespan of the oxbows. Older oxbows are slowly obliterated, while new ones are cut off by the meandering stream. Floodplains of large rivers can attain impressive dimensions. The Mississippi floodplain is 60 to 100 kilometers wide and some 800 kilometers long, entending from the confluence of the Ohio River in southern Illinois to the Gulf of Mexico.

The Mississippi floodplain is superb agricultural land. Great floods in 1927 and 1937, however, wrought such havoc that massive flood-control measures were initiated to prevent further occurrences. The controls are not complete and probably never will be, as it is a never-ending battle to stay abreast with the river. The major protective measures are in place, however, and the great Mississippi flood of 1973 was a success story, in which floodwaters were largely confined and a flood of catastrophic proportions was avoided.

A major aspect of all floodplains is the relationship between the rivers' natural tendencies to flood and the human occupants' efforts to protect themselves. There are a number of flood-control measures and devices. Principal among them are levees. Levees are human-made earthern ridges constructed to parallel the channel. The diagram shows these levees, which generally are placed about one-half kilometer back from the riverbanks. The function of the levees is to confine the overbank floodwaters to a narrow flood zone along the river and thus protect the greater portion of the floodplain. Without the levees the entire floodplain could be flooded in a major flood all the way to the bluffs. A levee failure would be disastrous. Another major flood-control measure is the construction of reservoirs on tributary streams, where water can be held back when the master stream is in flood. These reservoirs are not on the floodplain itself and may be located many miles upstream on the tributaries. An areawide, integrated plan must be adopted to ensure efficient flood control.

Methods of Study

Floodplains have been studied for almost as long as human civilization has existed. Most of the earliest permanent human settlements were in fact associated with floodplains, where fertile land and water were available. Early concerns were focused on two issues: distribution of irrigation water for agriculture and protection from excess water during floods. Modern concerns are similar. In the Mississippi floodplain, the concern is primarily flood control. The area's climate is humid, and river water has never been important for irrigation in the region. Research and study, therefore, have been directed toward developing effective measures and devices for control of overbank waters. An additional area for study, and for considerable expenditure as well, has been channel management for navigation purposes. The maintenance of an efficient channel is necessary for both navigation and flood control. Flood control, however, is concerned with maximum flow. Channel management for navigation purposes is primarily concerned with minimum flow, that is, maintaining minimum shipping channel depths and widths during dry seasons and droughts.

Studies and data collection are directed toward the mechanics of stream flow. Stream-gauging stations are established in which measurements are taken to record discharge, velocity, and flow turbulence. Many of the gauging stations are automatic, producing data for long-term analysis and also alerting warning monitors to flash-flood conditions. Gauging networks are established for tributary streams as well as for the master stream on the floodplain. The morphology of the channel is studied. The width, depth, and cross-sectional areas of the streams are analyzed for their capabilities to pass the flood flows. Great effort and expense are applied in river-engineering works to improve channel flow and to increase capabilities for high water discharge. Some of the adjustments are channel straightening, bank clearing, dredging and sandbar removal, and the construction of levees and dikes. Models of the stream channels, floodplain, and levees are constructed to precise scale. Water is passed through the models so as to determine the best configurations for efficient flow. Dikes are placed in critical locations in the model and the flow is studied to observe the effect of the dikes on bank erosion and sandbar formation. While the river conditions cannot be perfectly replicated in a scale model, its use does aid in the selection of sites for river-engineering works.

Context

Floodplains will continue to play an important role in human activities. They have always been areas of concentrated population and will be even more so in the future as the world population increases. Floodplains will acquire even more significance for agriculture. More people will live on the wide level surfaces that were once the undisputed domains of the rivers, yet the rivers can no longer be allowed to flood the plains. It is natural for a stream to flood the plain periodically. It is not natural to confine the flood to a narrow strip along the channel as has been done with the Mississippi. More than 3,000 kilometers of levees have been constructed along the lower Mississippi River. The levees have been a success in preventing disastrous floods like those that occurred in 1927 and 1937. They have brought protection to the valuable farmlands and to the towns and settlements on the plains. Nevertheless, there are problems. The protection offered by the levees has served as an invitation for continued settlement and development, which proceed today on the world's floodplains as if there were no threats of flooding. A levee failure today can be more catastrophic than one in the past, even with the same amount of flooding, because so much more lies in the path of the flood. It is probably correct that the greatest flood is yet to come.

Adding to the threat of flooding on the floodplain is the fact that human habitation of floodplains increases the likelihood of floods. The floodplain can be thought of as the bottom of a funnel. Runoff from all basins upstream merges on the floodplain. Many flood-control measures that are applied to tributary streams, such as channel straightening, actually increase the likelihood of flooding downstream by speeding up the runoff. Additionally, changes in land use today excerbate flooding. Deforestation and the drainage of wetlands increase runoff and intensify flooding. The increase in urbanization, with its growing expanse of asphalt paving and rooftops, accelerates runoff onto areas downstream. As the drainage basins undergo change, so does the potential for flooding on the floodplain. The floodplain is a dynamic and changing environment, and it will require continued study and adjustment. Humans are now settled densely on the world's floodplains. The plains have become more significant as food-producing areas in a world of growing demands. Great engineering works have been constructed and have doubtless prevented many floods, yet the potential for disastrous floods has only increased. Continued study and monitoring are thus essential.

Bibliography

Bayley, Peter B. "Understanding Large River-Floodplain Ecosystems." *Bioscience* 45, no. 3 (March, 1995): 153-159.

Bloom, Arthur L. *Geomorphology: A Systematic Analysis of Late Cenozoic Landforms.* Englewood Cliffs, N.J.: Prentice-Hall, 1978. A thorough text on geomorphology. Four chapters are devoted to the processes and landforms of streams. Diagrams illustrate the stages of floodplain development. The book assumes some knowledge on the part of the reader but is not difficult. An extensive list of references is provided with each chapter.

Chorley, Richard J., ed. *Introduction to Fluvial Processes.* London: Methuen, 1975. A paperback that is a technical treatment of the various aspects of fluvial processes. Includes diagrams, graphs, and formulas explaining stream hydraulics and channel morphometry.

Lobeck, A. K. *Geomorphology: An Introduction to the Study of Landscapes.* New York: McGraw-Hill, 1939. A well-known text on geomorphology. The diagrams are particularly helpful in understanding floodplains. The text is easy to read and is supplemental to the illustrations. Although dated and out of print, it is considered a classic in the field.

Morisawa, Marie. *Streams: Their Dynamics and Morphology.* New York: McGraw-Hill, 1968. A paperback that is a more detailed and technical treatment of stream processes than are the other listed references. Includes data on graphs, as well as maps and diagrams. The hydraulics of stream flow are explained.

Smith, Keith, and Graham Tobin. *Human Adjustment to the Flood Hazard.* New York: Longman, 1979. A paperback that deals specifically with flood hazards. Provides information on the nature of floods, planning strategies, and techniques of control. This specialized book is for the more advanced student.

Strahler, Arthur N., and Alan H. Strahler. *Modern Physical Geography.* 3d ed. New York: John Wiley & Sons, 1987. A general introductory text on physical geography. It gives a good explanation of fluvial processes and the development of floodplains. Well illustrated with maps and photographs. Recommended as a first source of information.

Tarbuck, Edward J., and Frederick K. Lutgens. *Earth Science.* 5th ed. Westerville, Ohio: Charles E. Merrill, 1988. A general beginning text for college earth science, easily readable by the layperson. The book employs quality color diagrams to explain the development of a floodplain, plus excellent illustrations throughout. Recommended as an initial source.

Thornbury, William D. *Principals of Geomorphology.* 2d ed. New York: John Wiley & Sons, 1969. A geomorphology textbook for the more advanced student. Three chapters deal with fluvial processes. Includes diagrams of floodplain features and the Mississippi Delta. References are listed.

John H. Corbet

Cross-References

Alluvial Systems, 1; Drainage Basins, 141; Lakes, 357; River Flow, 526; River Valleys, 534; Weathering and Erosion, 701.

Folds

Folds are the warping or bending of strata, foliation, or rock cleavage from an original horizontal or undeformed position into high and low areas. A fold is generally considered to be a product of deformation. Much of the folding of the earth's crust takes place near or along lithospheric plate boundaries and is considered to be a result of compressional stress.

Field of study: Structural geology

Principal terms

ANTICLINE: an arched upward fold of stratified rocks from whose central axis the strata slope downward; at the center it contains stratigraphically older rocks

AXIAL PLANE: a surface connecting all hinges of a fold; it may or may not be planar; that is, the axial plane of a fold may vary from a flat plane to a complexly folded plane

AXIS: a line parallel to the hinges of a fold, also called fold axis or hinge line

CRESTAL PLANE: a plane or surface that goes through the highest points of all beds in a fold; it is coincident with the axial plane when the axial plane is vertical

FLANKS: the sides of a fold, also called limbs, legs, shanks, branches, or slopes; anticlines share syncline flanks and synclines share anticline flanks

HINGE: the line of maximum curvature or bending of a fold

PLUNGE: the inclination and direction of inclination of the fold axis, measured in degrees from the horizontal

STRUCTURAL GEOLOGY: the study of the architecture of rocks insofar as it is the result of deformation

SYNCLINE: a bent downward fold of stratified rock from whose central axis the strata slope upward; at the center it contains stratigraphically younger rocks

TECTONICS: the study of the form, pattern, and evolution of large-scale units of the earth's crust, such as basins, geosynclines, and mountain chains

TROUGH: a line occupying the lowest points of a bed in a syncline; the trough plane connects the lowest points on all beds

Summary

Folding is a very common type of deformation seen in crustal rocks. It is obvious in the dipping (sloping) beds of mountains, where fold axes most often extend parallel to the length of the mountain chains. The folded beds may be observed in cliffs, roadcuts, and quarries, where the observer can see not only the steeply dipping beds but also some of the smaller anticlines and synclines. When viewed from the air, mountain chains are often found to be composed of sinuous ridges of sedimentary strata, such as limestone and sandstone, which are more resistant to erosion, while valleys are underlain by rocks such as shale, which are more susceptible to erosion. This phenomenon can easily be seen from the highway west of Denver, Colorado, in the front ranges of the Rocky Mountains, or near Harrisburg, Pennsylvania, where Interstate Highway 81 follows a ridge north for miles toward Scranton. In the North American midcontinent region, folds are less obvious, and the beds dip with inclinations of as little as 0.5 degree and upward to about 5 degrees; the folds are usually only discernible by surveying techniques. Occasionally there are very large folds in this region, hundreds of miles across and with very low dips. These folds are more circular than elongate in shape and are referred to as basins (downfolds) and arches, or domes (upfolds). Examples are the Michigan Basin, the Cincinnati Arch, and the Nashville Dome.

Rock mechanics is the study of the mechanical behavior of rocks; one branch of this study is concerned with the response of rocks to force fields in their geologic environments. Associated with this field is the study of tectonics, which relates the large-scale structures of the earth to these forces. For example, in the crust of the earth there are differential forces caused by the interaction of the lithospheric plates associated with the processes of plate tectonics along with hydrostatic (fluid) and lithostatic (gravity) pressures. Each may cause folding. The processes of plate tectonics form folded mountain chains at subduction zones. Differential lithostatic pressures associated with differences in specific gravity of different rocks and the effects of gravity form folds as in salt domes; lithostatic pressure, however, may act only as a confining pressure.

Rocks under stress behave in an elastic, plastic, or brittle manner, as do all solid materials. These three properties are related by the sequence of occurrence during deformation. Solid materials, when first subjected to a force, behave elastically; that is, they change shape or volume, but if the force is removed they return to their original shape. If, however, the force, instead of being removed, is increased, permanent deformation occurs: First the material changes shape plastically, and then it becomes brittle, breaking apart. Factors that affect the manner in which a rock behaves are time, temperature, confining pressure, and the pore pressure of water. Higher temperatures, water in pore space, lower confining pressures, or a longer length of time during which the rocks are subjected to a constant force tends to weaken rocks. Conversely, lower temperatures, lack of water, higher confining pressure, or a short period of time during which the force is applied tends to strengthen rocks.

Classification of fold types is a means by which structural geologists group the various kinds of folds in order to understand them and their origins. There are five major divisions in fold classification: descriptive or geometric, morphologic, by mechanics of origin and internal kinematics, by external kinematics and tectonic forces, and by position in the tectonic framework.

Descriptive or geometric classifications are based primarily on shape, that is, the attitude of the limbs, axial plane, and hinge line of the fold. The major use is for geologic mapping. The classification most commonly used is attitude of the axial plane, or the appearance of a fold in cross section or vertical section normal to its axis. In this classification there are symmetrical folds, in which the axial plane bisects the fold; asymmetrical folds, in which the limbs have different dips and are therefore asymmetrically disposed about the axial plane; overturned folds, in which one limb has been rotated or tilted through the vertical so that the original bed is upside down; and recumbent folds, which have been rotated so that their axial planes are nearly horizontal. Plunging folds are those in which the axis of the fold is not horizontal. Upright folds have vertical axial planes, and inclined folds have inclined axial planes.

Another descriptive classification is based upon fold symmetry: Ortho-rhombic folds have the axial plane and the plane normal to the axis of the fold as symmetry planes (mirror planes); monoclinic folds have either the axial plane or the plane normal to the axis of the fold as a symmetry plane but not both; and triclinic folds are folds with no symmetry planes. A third descriptive classification is based on the orientation of the axes of a fold. It includes cylindrical folds, which are described by the rotation of a line parallel to itself at a fixed distance from a central point and with parallel hinges (noncylindrical folds are folds with randomly oriented axes), and conical folds, which have axes divergent from an apex. A descriptive classification based upon flanks of folds includes isoclinal folds, with both flanks essentially parallel; open folds, which may not be folded more tightly without rock flowage. Monoclines are folds in which the strata dip or flex from the horizontal position in one direction only and are not a part of an anticline or syncline.

Morphologic classifications are based upon the shape of the fold with depth, map view, and spatial relationships with adjacent folds. The changes in shapes and patterns formed by folds are not always apparent when an individual fold is viewed in the field; therefore, the distinction is made between descriptive and morphological classifications. Concentric or parallel folds maintain a constant thickness of beds, which means an anticline will decrease in size downward, whereas a syncline will decrease in size upward. Similar folds are bent into similar curves and do not increase or decrease in size downward but maintain curves by thinning of flanks and thickening of crests and troughs. Disharmonic folds are different beds in a sequence of strata that have different amplitudes and wavelengths. Supratenuous folds die out downward. Nappes are large recumbent anticlines generally isolated by thrust faults.

Folds may also be considered in groups. Structural salients are observed when a sequence of folds is viewed in map view, and the curved fold axes are oriented convex toward the outer edge of the fold belt. Embayments are observed when a sequence of folds is seen in map view, and the curved fold axes are concave toward the outer edge of the fold belt. Homomorphic folding is the condition where folds cover an entire area, whereas idiomorphic folding is the condition where there is an intermittence of fold locality and often the folds are not linear in habit. Anticlinoriums are a series of anticlines and synclines forming a large arch, whereas synclinoriums are a series of anticlines and synclines forming a large trough. Both are generally tens of kilometers across. *En echelon* folds are a series of folds whose lengths are not extreme but which have axes that overlap in an oblique manner.

Fold classifications based upon mechanics of origin and internal kinematics describe the processes of folding in rocks. The mechanisms of folding reflect the elastic, plastic, or brittle manner of deformation dominant at different times during folding. Most folds appear, superficially, to be the result of plastic deformation. If, however, one takes a plastic material and

A fold at a California magnesium mine. *(U.S. Geological Survey)*

pushes it from both sides, only the edges are deformed. In the case of mountain chains there are often wide fold belts, which suggest that stress was transmitted equally across the width of the fold belt—the result of elastic deformation. The study of folds in thin sections (a thin slice of rock 0.03 millimeter thick prepared for viewing under the microscope), however, may show folding developing along minute shear planes and therefore must be the result of brittle deformation.

Flexure folding or flexural-slip folding, at times called true folding, is a form of elastic deformation. It can be demonstrated by taking a sheet of paper and folding it into one-inch accordion pleats. When stretched out a little, then compressed slowly, the folds become tighter; when released, the folds will spread out again—elastic rebound. In folding the paper, permanent folds were induced, which is plastic deformation, but the entire "fold belt" behaves elastically. What happens vertically can be demonstrated by taking a book and opening it to the middle, flat on the tabletop. The pages will be bent or folded on each side of the binding into anticlines. Folding of the book has proceeded by each page sliding upward relative to the page below, resulting in a narrow edge of all pages being visible at the side. Here again, there is only elastic deformation, because the pages of the book are not folded permanently. Shear folding—an example of brittle deformation—results from minute displacements along closely spaced fractures, which can be demonstrated with a stack of playing cards. Draw a vertical line in the middle of one side when the cards are neatly stacked and another line on the opposite side, then hold the deck of cards on end so that the line is horizontal. Next, indent the middle of the deck on the upper side forming a U-shaped trough. The original line now is bent into a synclinal form by each card slipping relative to the adjacent card.

Fold classifications based upon external kinematics and tectonic forces focus on external causes of folding. Block folding results from block uplift. Injection folding results from pluton emplacement. Folds caused by general crumpling can be attributed to two basic hypotheses: the tangential-compression hypothesis, or bench vise concept, and the vertical-tectonic concept, with material movement resulting from specific gravity differences. The first hypothesis assumes crustal shortening; the second assumes little or no difference in crustal length. Gravity gliding results when uplift occurs and rock masses move downslope along faults, producing folds and nappes. Rotational folding results from the rotation of blocks of the earth's crust, which are generally bounded by major faults. Regional coupling results from drag on blocks of the earth's crust sliding past one another. Differences in specific gravity produce folds in conjunction with the force of gravity. Two examples are salt anticlines and domes. Differential compaction produces folds where beds are draped over areas of less compaction. Ice shove may produce folds when a glacier encounters frozen, loosely consolidated rocks and sediments and shoves them into folds. Geosynclines are mobile down-warpings of the earth, generally elongate but also basinlike, measured in hundreds of kilometers, which subside as sedimentary and volcanic rocks accumulate to thicknesses of thousands of meters. This part of the tectonic cycle is often followed by orogeny, or mountain-building processes. Geosynclines are related to subduction zones of the plate tectonics theory. Geanticlines are mobile upwarpings of crustal material of regional extent; the term is particularly applied to anticlinal structures developed in a geosyncline.

Classification of folds based upon position in the tectonic framework is a way of describing components of a mountain chain. There have been few attempts made to devise a formal classification, but there are fold types characteristic of particular tectonic regimes. A belt of crustal folds and metamorphism, generally centrally located within a mountain chain, is a region of shear folding vertical uplift. Examples include the Pennine Nappes of the Alps and the Piedmont and New England upland provinces of the Appalachian Mountains. The outer belt of shallow folding and thrusting of mountain chains is characterized by sedimentary rocks folded into anticlines and synclines with associated thrust faults and is often underlain by a plane of décollement. Examples of this portion of a mountain chain include the Valley and Ridge province of the Appalachian Mountains, the Front Ranges of the Rocky Mountains, and the Jura Mountains of Europe.

Methods of Study

Classically, the study of folds is divided into field and laboratory techniques, but, currently, scientists also employ geophysical techniques as well as those of remote sensing, which often bridge field and laboratory studies. Field studies of folds involve direct measurements of the attitudes and positions of rock. These measurements are generally recorded on a base map—often a topographic quadrangle map—which, when completed, becomes what is called a geologic map. The processes of geologic mapping in the field may include various surveying techniques that identify the positions of the various rock types and their attitudes. Sedimentary sequences of strata, if folded, are recorded with their strikes and dips. Strike is the direction the line extends where a sedimentary bedding plane intersects the horizontal plane; it may be considered the direction the bed extends on a geologic map. The dip is the angle measured downward to the bedding plane from the horizontal plane normal to the strike of the bed (fold); it is a measurement of how fast an inclined bed is descending and in which direction. These measurements are most often made with a Brunton compass and its clinometer (an instrument for measuring angles of inclination or elevation).

In the laboratory, the geologic map data concerning the attitude of beds, foliation, rock cleavage, or fractures may be plotted on pi-diagrams and beta-diagrams to ascertain the stress field which produced the deformation and other information needed in the structural analysis of the region being mapped. Pi-diagrams are the plot of the poles of the bedding planes, whereas the beta-diagrams are the plot of the actual planes of the beds.

These techniques are extensively used in analyzing folds. Currently, as in other forms of statistical analysis, the computer is employed to do most of the work.

Geophysical techniques for the study of folds have been extensively developed to aid in the prospecting for petroleum. This application greatly extends the data available by direct observation of folds and is invaluable where bedrock geology is covered by surficial materials such as glacial drift or river deposits. Geophysical data are most successfully used when combined with direct geologic information observed by the geologist. There are six principal geophysical methods: granitational, magnetic, seismic, electric, radioactive, and thermal. Although all of them have been employed to study folds, the seismic methods are the most extensively used. The seismic methods involve the detailed observation of the elastic earthquake waves generated naturally or artificially. (Artificial waves are often induced by explosives, but other devices are used, including the electric spark.) The energy released produces elastic waves. The velocity of the wave indicates the nature of the rock through which it is propagated. The waves are also reflected and refracted wherever they encounter a rock interface, that is, the surface between two different beds or types of rock. From the detailed analysis of data recorded at receiving stations, a fold or other structure may be deduced. Again, computers perform nearly all the calculations.

Remote sensing, which at one time was limited to aerial photography, has been expanded to film sensitive not only to visible light but also to infrared and ultraviolet light. In addition, a wealth of information is available from satellite images in several wavelengths of the electromagnetic spectrum. Although these data may be useful, the most valuable information for studying folds is derived from high- and low-level aerial photography in the visible spectrum (black-and-white photography), and SLAR (side-looking airborne radar). Aerial photographs may be used to map geology. SLAR is a valuable tool for geologic analysis because it essentially penetrates clouds and murky atmosphere and "sees" the ground surface. It also has an active sensor; the system provides its own source of illumination in the form of microwave energy and therefore can be used day or night as well as during variable weather conditions. The product is much like a black-and-white aerial photograph that looks unusually clear.

Context

Folds are related to the occurrence of petroleum and ore bodies. One of the earliest references to this relationship occurred in 1842, when Sir William Logan noted seepage of oil from anticlines near Gaspé, Canada, at the mouth of the St. Lawrence River. Petroleum deposits form if five conditions are met. There must be a source, which is generally considered to be fine-grained organic-rich sediment. The source rocks must be buried deep enough to have temperatures and pressures high enough to cause chemical transformation of the organic material. The petroleum must then migrate into rocks that are permeable enough to allow production of the oil by wells; these rocks are usually limestone, dolomite, or sandstone and are referred to as the reservoir rock. Because petroleum floats on water, and there is water in the sedimentary rocks, there must be an impermeable cap rock covering the reservoir rock so that the petroleum does not escape to the surface. Finally, there must be some type of trap to collect the migrating petroleum, such as an anticline, that acts like a giant inverted bowl under which the petroleum collects until it is produced by wells. Other traps possible are fault planes, facies changes (often called stratigraphic traps), and differences in fluid pressures. Similarly, ore bodies may be related to anticlines. They may be formed by ascending ore fluids that need traps such as anticlines, faults, or facies changes where the metals may be emplaced. By identifying the occurrence of these phenomena, geologists are able to assist in the retrieval of such resources for general use.

Bibliography

Badgley, Peter C. *Structural and Tectonic Principles*. New York: Harper & Row, 1965. Perhaps the finest text in structural geology at the time it was published. The tectonics are outdated; nevertheless, it has a very fine section on folds, folding, and rock mechanics. Written at the level of an advanced college student.

Billings, Marland P. *Structural Geology*. 3d ed. Englewood Cliffs, N.J.: Prentice-Hall, 1972. Perhaps the easiest-to-read introductory text to the field of structural geology for the college student. Although an older book, it is essentially as modern as any text, except for the final chapter on geophysics. Provides perhaps the most comprehensive introductory descriptions of folds available.

Dennis, John G. *Structural Geology*. Dubuque, Iowa: Wm. C. Brown, 1987. A college-level introductory text to the field of structural geology. Includes a review of modern geotectonics and a good section on folding.

Hamblin, W. K., and J. D. Howard. *Exercises in Physical Geology*. Edina, Minn.: Burgess, 1986. An excellent introductory laboratory manual for beginning college courses in geology. Offers one of the better treatments of folds, with fine colored diagrams; however, it contains no information beyond the very elemental descriptions of folds.

Suppe, John. *Principles of Structural Geology*. Englewood Cliffs, N.J.: Prentice-Hall, 1985. Designed as a concise introduction to the deformation of the earth's crust and written for the advanced college student. Besides the usual material in a structural geology text, this book has a section on regional structural geology of the Appalachian Mountains, with good illustrations of folds, as well as a similar chapter devoted to the cordilleran ranges of the western United States.

Tarbuck, Edward S., and Frederick K. Lutgens. *The Earth*. Westerville, Ohio: Merrill, 1987. An example of many college freshman books in physical geology, with a minimal treatment of folds and an excellent section devoted to plate tectonics.

Charles I. Frye

Cross-References

Oil and Gas: Distribution, 481; Oil and Gas: Origins, 489; Plate Tectonics, 505; Stress and Strain, 607; Subduction and Orogeny, 615.

The Fossil Record

The fossil record provides evidence that addresses fundamental questions about the origin and history of life on earth: When did life evolve? How do new groups of organisms originate? How are major groups of organisms related? This record is neither complete nor without biases, but as scientists' understanding of the limits and potential of the fossil record grows, the interpretations drawn from it are strengthened.

Field of study: Geochronology and paleontology

Principal terms

BODY FOSSIL: the petrified remains of a plant or animal
FOSSIL: evidence of organic activity, usually preserved in sedimentary rock strata
GRADUALISM: an explanation of how evolution works involving slow, constant change through time
LAGERSTATTE: an assemblage of exceptionally well-preserved fossils
MACROFOSSIL: a fossil that is large enough to study with the unaided eye, as opposed to a microfossil, which requires a microscope for examination
MORPHOLOGY: the appearance (shape and form) of an organism
PALEONTOLOGIST: a scientist who studies ancient life; invertebrate paleontologists study fossil invertebrate animals, vertebrate paleontologists study fossil vertebrates, and micropaleontologists study microfossils
PUNCTUATED EQUILIBRIA: an alternative model of how evolution works, by rapid speciation events that involve major changes in morphology
SPECIATION: the evolutionary process of species formation, the process through which new species arise
SPECIES: a group of similar, closely related organisms
TAPHONOMY: the study of the sequence of events that lead to the burial and preservation of fossils
TRACE FOSSIL: indirect evidence of an organism's presence through tracks, trails, and burrows

Summary

The term "fossil" originally referred to any object dug up from the earth and included minerals as well as the petrified remains of once-living organ-

257

isms. The terms is now used in a restricted sense to describe the preserved remains of organic life, both plant and animal. Probably the most familiar kinds of fossils are body fossils, which are the fossilized remains of the actual organisms, but there is also indirect fossil evidence of organic activity, such as footprints as evidence of walking. Collectively, these tracks, trails, and burrows are termed "trace fossils."

The term "fossil record" refers to the sum total of fossils preserved in geological strata on earth. The fossil record extends back in time to rocks 3 billion years old. The first entries in the fossil record are single-celled, plantlike organisms. There are about 250,000 known fossil species of plants and animals. That seems to be a large number until one compares it to the approximately 4.5 million species of plants and animals that are alive today. The entire fossil record of ancient life amounts to 5 percent of the total number of modern species. What is the reason for the paucity of fossils compared to the abundance of modern species? Does this difference in number of fossil and recent species reflect true differences in diversity, or does it reflect limitations in the compilation of the fossil record? In other words, were there fewer species in the geological past than in the recent past, or is the fossil record very incomplete?

A fossil quarry at Utah's Dinosaur National Monument. *(Utah Travel Council)*

Charles Darwin, famous for his contributions to evolutionary theory, came to the conclusion that the fossil record is incomplete. Darwin was one of the earliest naturalists to publish on what he termed the "imperfection" of the fossil record. In his book, *On the Origin of Species* (1859), Darwin compared the preserved record of life on earth to a set of books in which several volumes are missing, and from the remaining volumes chapters are missing, and from the remaining chapters pages are missing, and from the surviving pages words are missing. Darwin and others concluded that not every kind of organism that once lived is fossilized and that the fossil record is in fact biased toward preservation of some forms over others.

The biases inherent in the fossil record stem from the fact that fossilization of organic material is the exception, not the rule, and very specific and relatively rare conditions must be met for an organism to become fossilized. Fossilization favors organisms with hard parts, for example an exterior shell (exoskeleton) or internal skeleton (endoskeleton). Fossilization also favors organisms living in certain environments. Two particular environmental conditions favor fossilization: rapid burial and anoxia (lack of oxygen). Rapid burial protects organic remains from predators or scavengers and physical reworking by tides and waves. Oxygen supports bacteria and decomposition of organic material. Burial in an oxygen-free (reducing) environment insulates organic material from decay and thus favors fossilization.

The most exceptional fossils known are from environments in which one or both of these two environmental conditions were met. German paleontologists call exceptionally preserved fossil assemblages fossil "lagerstatten," or mother lodes. Famous fossil lagerstatten include the Mazon Creek fauna, from Pennsylvanian-period strata (300 million years old) in Illinois, in which insects, crustaceans, and previously unknown, problematic soft-bodied organisms are preserved in iron-stone concretions; the Burgess Shale fauna, from Middle Cambrian (500 million years old) strata in British Columbia, famous for the discovery of a great variety of unusual arthropod and annelid-like animals; the Solnhofen Limestone, from Jurassic strata (200 million years old) in Bavaria, in which *Archaeopteryx* (a reptile with feathers) was discovered; insects preserved in amber (fossilized tree sap), of the Oligocene epoch (40 million years old) in Germany; and the La Brea Tar Pits, from the Pleistocene epoch (10,000 years old) in California, in which a variety of animals, including saber-toothed tigers and mastodons, were ensnared and preserved in natural asphalt springs. From this brief survey, it is clear that fossil lagerstatten occur in a variety of geological and geographical settings and through a wide range of geologic time. These lagerstatten share the characteristic of rapid burial or burial in a biologically inert environment. They provide information on the morphology of previously unknown groups or a better record of known groups, but do such assemblages reveal anything about how the organisms lived? Some of these lagerstatten represent environments in which the animals died rather than the environments in which they lived; the La Brea animals, for example, certainly did not live in the tar pits. Lagerstatten, however, can be used to reconstruct paleocommunities, to create a picture of organisms that lived contemporaneously, even though that picture of paleocommunity structure or paleoecological relationships might not be complete. Each different fossil lagerstatten provides an exceptional view of a geologically unique situation.

Because of the preservational biases inherent in the fossil record, most fossilized species represent only a few major groups—the numerically abundant and well-skeletonized organisms that lived in or near an anoxic environment or an environment that was subjected to rapid, episodic influxes of

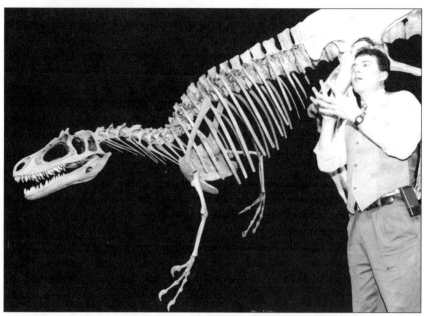

A paleontologist poses with a fossil reconstruction of *Deltadromeus,* a dinosaur that lived in Africa during the Cretaceous era some 65 million to 100 million years ago. *(Reuters/ Luc Novovitch/Archive Photos)*

sediment. The environment with the best fossil preservation potential on earth is the shallow marine shelf: Most marine life lives in the shallow shelf; the shelf is subject to rapid influxes of sediment (via storms and rivers, for example), and many marine invertebrates have exoskeletons. Thus, hard-shelled invertebrates from shallow marine environments constitute the bulk of the fossil record. The major marine invertebrate groups that dominate the fossil record include corals, bryozoans, brachiopods, mollusks (clams, cephalopods, and snails), arthropods (especially trilobites), and echino-derms (starfish and their relations).

The fossil record of other groups, including marine and terrestrial verte-brates and plants, is neither as abundant nor as complete as the marine invertebrate record. Leaves and flowers are rare as fossils, often preserved as imprints in sediments deposited in ancient inland lakes. Woody tissue of trees is often preserved by a process called permineralization, in which mineral-rich water percolates through the porous wood and minerals (espe-cially quartz) precipitate from the water, filling the voids in the plant structure. The logs of the Petrified Forest in Arizona are examples of permineralized wood. Pollen spores and seed pods are important constitu-ents of the paleobotanical record, because these structures are abundant and often have tough outer coverings. Paleobotanists face special problems in identification and classification of plant fossils, because entire plants are rarely found as fossils. Stems, roots, leaves, flowers, and seeds are described separately as they are found. Consequently, a single plant species is likely to have separate species names given to each of these structures.

Quality and quantity of fossil material are restricted for the vertebrate paleontologist, compared to the wealth of fossil material available to the invertebrate paleontologist, because vertebrates have an endoskeleton that is more easily broken apart (disarticulated) after death of the organism than the well-calcified exoskeletons of marine invertebrates. Terrestrial vertebrates also live and die in environments in which their remains are subject to destruction by predators, scavengers, and bacterial decomposition. The vertebrate fossil record is dominated by teeth, which contain the mineral apatite, which is hard and resistant to weathering. Often, entire fossil vertebrate species are defined on the basis of solitary bits of jaws, teeth, and bone.

The biases associated with fossilization do not diminish the importance of the fossil record for scientists' understanding of the origin and evolution of life. The fossil record not only enables scientists to reconstruct the morphology of long-extinct individuals but also provides evidence from which ecological relationships—for example, interactions between different organisms and the structure of ancient communities of organisms, or paleo-communities—can be inferred. Sometimes, the evidence for ancient ecological relationships is striking, as in the find of an ammonite (related to the modern genus *Nautilus*) with circular holes in the shell associated with circular teeth of a mosasaur (marine reptile)—evidence of a predator-prey relationship.

Trace fossils contribute their own unique information to scientists' understanding of ancient life. They are often found in rocks devoid of body fossils and provide the only evidence of biological activity in these ancient sediments. Unlike body fossils, trace fossils cannot be transported by wind or currents. Traces were made where they are found, and there is little fear of erroneous interpretations based on mixing trace fossils from different environments. Trace fossils also provide important evolutionary information: The earliest evidence of multicellular organisms is small vertical burrows in rocks 2 billion years old. There are no body fossils in these rocks; the trace-makers were undoubtedly soft-bodied organisms that were poor candidates for fossilization. The presence of the burrows in these ancient rocks suggests that the evolutionary transformation from unicellular to multicellular life took place in a shallow marine, nearshore environment. The diversity of trace-fossil forms reveals a variety of animal behaviors: burrowing, crawling, walking, feeding, and grazing. For example, the gait of some extinct arthropods and details of their appendages can be determined from the pattern of their fossilized footprints, even though body fossils of these animals are rarely found with intact appendages. The spacing of dinosaur footprints has led to the idea that many dinosaurs were able to move rapidly and were not the slow-gaited, lumbering giants formerly imagined.

The fossil record documents the fact of evolution and provides data from which scientists can infer the processes by which evolution works. Darwin believed that evolution was a slow and gradual process of small changes, accumulating over time to transform one fossil species into another. This gradualistic view of evolution, however, is not strongly supported by the fossil record. Gradualism predicts that numerous intermediate forms (transitional in appearance between the old species and the new species) should exist. In fact, many fossil groups have left no record of these intermediate forms. Darwin attributed the absence of transitional forms to the incompleteness of the fossil record. This view of the fossil record as very incomplete provided a convenient explanation for gradualists, and gradualism was widely accepted as an explanation of how evolution works for many years. The gradualist view was challenged by paleontologists who saw a different pattern in the geological record of some groups of organisms. Many fossil groups appear suddenly in the fossil record and persist largely unchanged in appearance through most of their geologic history. This pattern suggests that evolution (as measured by change in appearance of an organism) happens very quickly and involves a major change in appearance, as the old form "jumps" to the new form. According to this alternative view, the sudden appearance of new species in the fossil record and the absence of intermediate fossils is real and not an artifact of an imperfect fossil record. This alternative model of evolution, termed "punctuated equilibrium," holds that evolution is not the gradual, constant process envisioned by Darwin but rather a rapid process involving major changes in morphology. Speciation in this punctuated model of evolution does not require the hypothetical (and problematic) transitional intermediates proposed by Darwin. Both gradualistic and punctuated modes of evolution have been documented for different groups of organisms. The evidence suggests that different groups of organisms evolved through different evolutionary pathways, some gradual, some punctuated.

Methods of Study

Study of the fossil record begins in the field, where paleontologists collect fossils and record observations on the orientation of the fossils and the character of the sediment in which the fossils are found. The specific technique used in collecting fossils depends on the kinds of fossils sought, the nature of the research, and time constraints. A major problem faced by paleontologists in collecting fossils in the field is separating the fossils from the surrounding sediment. Certain microfossils that are commonly found in limestone are collected by dissolving many kilograms of the rock in hydrochloric acid. The microfossils themselves are insoluble in weak acid and are easily recovered from the acid bath. Other microfossils may be picked from deep-sea cores that consist of unconsolidated mud. Macrofossils can be removed from the outcrop with the aid of hammers and chisels, although vertebrate paleontologists sometimes use brushes in excavating fragile vertebrate material. Unless the fossils have weathered naturally from the outcrop, they probably need to be cleaned of adhering sediment. Macrofossils can be cleaned by boiling in solvents, by sandblasting with a miniature air-abrasion unit, or by meticulous, time-consuming hand cleaning with small probes (old dental instruments make good fossil-cleaning tools).

Strategies for collecting fossils from outcrop exposures include collecting fossils exposed at the surface from weathering, and "mining," digging down to or along targeted fossiliferous bedding surfaces. First-time collectors usually begin by collecting everything in sight. Once the novelty has worn off, they select only the best, or most perfect, specimens and toss aside broken or deformed fossils. Perfect specimens make good museum exhibits, but experienced paleontologists realize that important information is often revealed by the broken or imperfect specimen. The pattern of shell wear or breakage may reveal important information about the transport and burial history of the fossil after its death. The study of these postmortem processes is called taphonomy. Taphonomic studies require detailed observation about the preservation of the fossils and the original orientation of the fossils in the outcrop.

In the laboratory, fossils are examined with a binocular microscope, or they may be ground down to translucent slices and examined under a petrographic microscope. This thin-section technique permits microscopic examination of shell structure. The scanning electron microscope is used to examine surficial features of the shell, at magnifications on the order of 30 to 5,000 times. The composition of fossil shells might be determined by microprobe or isotopic analysis.

Much paleontological study concerns description of fossil species. Paleontologists seek not only to differentiate one species from another but also to discover the relationship between different species. This is accomplished by describing and comparing the key morphologic characters of different species. Computers and digital imaging equipment enable paleontologists to measure many morphologic features quickly and process large sets of numerical data rapidly. Relationships between species can be represented numerically through a variety of mathematical techniques and the results presented in graphical form in computer printouts.

Not all paleontological discoveries are made in the field. There is a wealth of fossil material residing in museum collections, awaiting future study. For example, a new dinosaur species was discovered when crates of fossils collected a century earlier were unpacked and examined in detail for the first time.

Context

The fossil record documents patterns in the history of life. The fundamental pattern revealed by this record is that life forms have changed through time. This pattern of change is seen most clearly in mapping the distribution of familiar organisms through time. The oldest fossiliferous rocks contain very few fossils of organisms that are alive today but contain a high proportion of unfamiliar organisms that are now extinct. Younger strata contain progressively higher proportions of fossils that have living relations. The very youngest sedimentary rocks contain the highest proportion of fossils that are represented today by modern living forms. This pattern of organic change through time is evolution.

Fossils change through time in a definite and determinable order. Once the pattern of change is known for any one group of organisms, the members of the group can be used to determine the relative order of deposition of the beds in which they are found. In other words, different fossils are characteristic of certain periods of geologic time, and once their order of occurrence is known, the rocks in which the fossils are found can be arranged in relative chronologic order. This is the principle of faunal succession, first enunciated by British surveyor William Smith. Smith supervised digging of canals in Great Britain during the 1800's and noticed that the sequence of fossils uncovered in the digging was the same throughout Great Britain; certain fossils appeared first in the lower strata, others appeared in a consistent and predictable order in overlying strata. Using this order of fossil occurrence, Smith was able to correlate strata (that is, to demonstrate equivalence of strata exposed in separate outcrops) and produce a geologic map of England.

Fossils are a fundamental tool of correlation, but not all fossils are useful in correlation. Those that are, called index or guide fossils, have the following characteristics: They are abundant, geographically widespread, distinctive in appearance, and geologically short-lived (in other words, they are characteristic of a single slice of geologic time). The divisions of the geologic time scale were originally defined on the basis of faunal succession.

Another type of pattern that is revealed through the fossil record is the interrelationship between different groups of organisms. For example, *Archaeopteryx*, a fossil that is reptilian in many aspects but has feathers, is evidence that reptiles and birds are closely related, that is, reptiles and birds have a common ancestor from which both groups are descended.

The fossil record also reveals patterns of biological crises through time. The record documents several episodes in the history of life when many different groups of animals became extinct over a relatively short time period. The record most likely holds the information needed to understand the cause or causes of these mass extinctions. Answers will come from studying the groups of organisms that were affected during each extinction and recording the time of the extinction of each group as precisely as possible.

There is an economic aspect to the fossil record as well. Fossil fuels, on which humans rely for heat, electricity, and transportation, come from accumulations of organic material. The hydrocarbons (oil and natural gas) are derived from accumulations of microscopic animals that inhabited ancient seas. Coal is formed from the burial and compression of large amounts of plant material that accumulated in ancient coastal swamps.

Bibliography

Darwin, Charles. *On the Origin of Species: A Facsimile of the First Edition.* Cambridge, Mass.: Harvard University Press, 1975. Darwin's seminal 1859 work is worth reading to appreciate the depth and breadth of evidence that he marshaled for his theory of evolution by natural selection. Of special interest is Darwin's struggle to reconcile the evidence from the fossil record (the absence of transitional fossils) with his gradualist theory of evolution and his description of the imperfection of the fossil record. Although the Victorian-era writing style requires patience on the part of the reader, Darwin's book is accessible to a general audience.

Gould, Stephen J. *Hen's Teeth and Horse's Toes: Further Reflections in Natural History.* New York: W. W. Norton, 1983. This is one of several volumes of essays in which Gould interprets paleontological principles for a lay audience (other volumes are *The Panda's Thumb: More Reflections in Natural History, Ever Since Darwin: Reflections in Natural History,* and *The Flamingo's Smile: Reflections in Natural History*). A recurring theme in many essays concerns the mechanism of evolution; Gould is a leading proponent of punctuated equilibria. These essays are eminently readable, often humorous, and always enlightening.

Hamblin, W. Kenneth. *The Earth's Dynamic Systems.* Minneapolis, Minn.: Burgess, 1975. This textbook gives an overview of physical geology. Chapter 5 contains a clear discussion of the principle of faunal succession, with particular emphasis given to William Smith's work. Useful illustrations and a glossary are included.

Laporte, Leo F., ed. *The Fossil Record and Evolution.* San Francisco: W. H. Freeman, 1982. This volume comprises sixteen articles reprinted from *Scientific American,* united by the theme of the fossil record. Included are articles on the origin and earliest fossil record of life, the Burgess Shale fossil lagerstatte, and a reinterpretation of dinosaur behavior from fossil evidence. These articles are suitable for a high school audience and are characterized by numerous well-constructed illustrations. An index and a bibliography for further reading are included.

Rudwick, Martin J. S. *The Meaning of Fossils: Episodes in the History of Palaeontology.* London: Macdonald, 1972. Rudwick traces the history of humankind's understanding of the fossil record, from the time of the ancient Greeks, when "fossil" referred to any object dug up, through the Dark Ages, when fossils were viewed as tricks of the Devil, to the modern understanding of fossils as evidence of past life. Accessible to a general audience, the book includes a glossary, an index, and a bibliography.

Spears, Iain R. "Biomechanical Behaviour of Modern Human Molars: Implications for Interpreting the Fossil Record." *American Journal of Physical Anthropology* 106, no. 4 (August, 1998): 467-483.

Thompson, Ida. *The Audubon Society Field Guide to North American Fossils.* New York: Alfred A. Knopf, 1982. One of the most comprehensive and useful books of its kind, the Audubon guide is published in a slim, softcover format that is meant to accompany the reader into the field. Primarily a guide to the fossil record of marine invertebrates of North America; vertebrates and plants receive coverage proportional to their fossilization potential. More than five hundred full-color plates of the most common North American fossils are accompanied by detailed descriptions of each. Also included is a chapter on where to find fossiliferous rocks in North America.

Danita Brandt

Cross-References

Catastrophism, 53; The Cretaceous-Tertiary Boundary, 118; The Geologic Time Scale, 272; Uniformitarianism, 665.

Geologic and Topographic Maps

Topographic and geologic maps are basic tools for wise management and development of the earth's resources. Topographic maps represent on paper the earth's surface and its various landforms. Geologic maps show the distribution of the rocks that underlie the landforms and provide a view of the earth's surface to a depth of several thousand feet.

Field of study: Geochemistry

Principal terms

CONTOUR LINES: on a topographic map, lines of equal elevation that portray the shape and elevation of the terrain

GEOLOGIC MAP: a representation of the distribution of mappable units (formations)

MAP SCALE: the scale that defines the relationship between measurements of features shown on the map compared with those on the earth's surface

TOPOGRAPHIC MAP: a line-and-symbol representation of natural and selected human-made features of a part of the earth's surface, plotted to a definite scale

Summary

A topographic map is a line-and-symbol representation of natural and selected human-made features of a part of the earth's surface, plotted to a definite scale. Topographic maps portray the shape and elevation of the terrain by contour lines, or lines of equal elevation. The physical and cultural characteristics of the terrain are recorded on the map. Topographic maps thus show the location and shape of mountains, valleys, prairies, rivers, and the principal works of humans.

In the past, topographic maps were constructed by labor-intensive field methods, which involved detailed field measurements made with telescopic-type instruments. These data were translated in the field to actual distances and plotted by hand on field sheets for eventual office compilation and printing. More recently, however, most maps have been prepared using

adjacent pairs of aerial photographs. Highly accurate, these photograpahs
are further checked for accuracy by reference to global positioning satellites.
In some instances, laser-beam surveys provide extremely precise control in
areas that are subject to earthquakes, such as California, to monitor stress
buildup. Complex stereoscopic plotting instruments are used by a trained
observer to delineate contour lines and various features on a base map. Field
verification of place-names and features is required before the map is
printed. The maps are then compiled on a stable base (a type of plastic that
does not change dimensions during temperature and humidity fluctuations). Such a base helps to ensure the map's accuracy by preventing
distortions. Modern photographic and photochemical techniques are used
to prepare the map for printing.

All maps must meet accuracy standards established by the government.
Special standardized symbols, each with its own meaning, are used to convey
a wide variety of information. Also, colors are used frequently to show the
more common features. Generally, blue indicates water bodies, brown indicates contour lines, red indicates map features with special emphasis (chiefly
land boundaries), pink indicates built-up urban areas, and purple indicates
revisions based on new photographic information since the original map
was made.

Geologic maps are a representation of the distribution of mappable units
(formations). These maps provide the data for an accurate compilation of
the rock units at the surface or in the subsurface. A geologist makes
hundreds of observations each day in the field. Many of these observations
are recorded in a notebook or on the field sheet that eventually becomes a
geologic map. Some geologic maps are prepared from aerial photographs
or from remotely sensed images created through satellites. New detailed
geologic mapping frequently reveals information that may require reevalu-
ation of previously mapped areas. Large-scale (1:24,000) geologic maps
require detailed examination of the area being mapped. Field investigation
describes outcrops as close as a spacing of several hundred feet using a
topographic map as a base. Outcrop descriptions include determinations of
fossils and rock type and mineralogy, along with descriptions of rock properties, such as color, thickness, and type of bedding units, and attitudes of
the rocks. Some studies are supplemented by geophysical surveys. Drilling
(cores or cuttings) is integral to many studies, including oil and gas exploration, mining, and engineering. Samples are examined later in the laboratory. Some of the laboratory work includes microscopic study of thin slices
of rock to establish mineral relations, binocular study and identification of
fossils, or detailed chemical analyses of the whole rock or of separated
minerals, and radiometric age dating.

Geologic maps are compiled in the office. This compilation requires
review of field notes and observations, laboratory data, and information
from the scientific literature. Data are transferred to a topographic map base
made of plastic to ensure stability during temperature or humidity vari-

ations. Finalized contacts are drawn that divide rocks of one unit from those of another. The degree of certainty of the contacts is shown by a standard set of line symbols. The orientation of the various rock units is indicated by uniform symbols. When a geologic map is complete, it is prepared for publication by conventional drafting methods or, more frequently, by digitization and computer plotting methods. More recently, geologic and topographic maps are added to large computer databases known as Geographic Information Systems (GIS's). Through computer manipulation of layers of information such as geology and topography, informed decisions can be reached by combining these with other data layers. Individual geologic maps are issued in a numbered series by state and federal geological surveys. Other geologic maps accompany formal, numbered geologic reports issued by the same agencies. A nominal fee of a few dollars is charged to recover printing costs. These maps and reports may be purchased over the counter in major cities or by mail from the respective agencies.

Applications of the Method

Map scale defines the relationship between measurements of features shown on the map and measurements of features on the earth's surface. These comparisons are numerically expressed as a ratio, for example, 1:24,000, 1:125,000, 1:250,000, 1:500,000, and 1:1,000,000 scale. Large-scale maps (1:24,000 or larger) are used when highly detailed information is required. Examples include proposed projects (roads, large construction projects, and so on) in highly developed or populated areas. Intermediate-scale maps are quite useful in land- and water-management planning projects and in resource management. The 1:100,000 scale (metric) maps have become popular for a growing number of applications, particularly environmental protection and planning. Small-scale maps (1:250,000 to 1:1,000,000) cover very large areas. They are useful mainly in regional planning.

The topographic map series of the National Mapping Program includes quadrangle and other map series published by the U.S. Geological Survey. A map series is a family of maps conforming to the same specification or having common characteristics, such as scale. Adjacent maps of the sample quadrangle series can be combined to form a single large map manually, photographically, or by computer methods. Geologic maps are prepared using existing topographic and/or planimetric base maps (maps showing boundaries but no indications of relief). Thus, the scales of geologic maps generally correspond to those of the common topographic and/or planimetric maps. In special cases, such as a major engineering project (for example, a dam or nuclear power plant), preparation of a site-specific large-scale topographic map may include detailed geologic mapping.

Geologic maps are used for metallic, nonmetallic, and energy resource exploration assessment and development. They are also used in land-use and planning studies to determine technically suitable and environmentally safe locations for subsurface solid, hazardous, or low-level and high-level nuclear waste repositories and excavations, waste disposal, water resources investigations, and military applications. In addition, geologic maps are used to identify potential hazard areas of faults, volcanoes, and ground failure.

Context

Topographic and geologic maps help to reveal the structure and re-sources of the surrounding environment. These maps are basic tools for resource management and planning and for major construction projects. They are used in the planning of roads, railroads, airports, dams, pipelines, industrial and nuclear plants, and basic construction. Both types of maps are also used in environmental protection and management, water quality and quantity studies, flood control, soil conservation, and reforestation plan-ning. In addition, topographic maps receive wide use in recreational activi-ties such as hunting, fishing, boating, rock climbing, camping, and orien-teering.

Geologic maps provide baseline data for the identification and orderly development of the earth materials required for modern civilization. Exam-ples include sand and gravel, crushed stone and aggregate, clay, metal deposits and hydrocarbon fields. Geologic maps are also used extensively in environmental monitoring and protection, in local regional planning, and in scientific studies. They help to identify areas prone to landslides or earthquakes, groundwater recharge areas, and potential sand and gravel resources.

Bibliography

Compton, Robert R. *Manual of Field Geology*. New York: John Wiley & Sons, 1962. This somewhat dated publication provides an excellent idea of the practical aspects of conducting fieldwork and the steps involved in making a geologic map.

Lindholm, Roy Charles. "Soil Maps as an Aid to Making Geologic Maps, with an Example from the Culpeper Basin, Virginia." *Journal of Geologi-cal Education* 41, no. 4 (September, 1993): 352-358.

National Research Council. *Geologic Mapping: Future Needs*. Washington, D.C.: National Academy Press, 1988. This publication presents the results of a national survey on geologic maps. The survey was designed to identify current usage of geologic maps as well as future needs. Most important, the survey identified the relative needs for geologic maps by map scale, style of presentation, and type of user (for example, exploration, basic research, engineering, and hazard assessment).

Steger, T. D. *Topographic Maps*. Denver, Colo.: U.S. Geological Survey, n.d. This free brochure provides a concise overview of topographic maps, their production, and their use.

U.S. Geological Survey. *COGEOMAP: A New Era in Cooperative Geological Mapping*. Circular No. 1003. Denver, Colo.: Author, 1987. Single copies of this circular are free upon application to the U.S. Geological Survey. Provides an overview of how cooperative geologic mapping between state and federal geological surveys is attempting to meet the need for large- and intermediate-scale geologic maps and other types of earth science maps.

_____. *Digital Line Graphics from 1:24,000-Scale Maps: Data Users Guide.* Denver, Colo.: Author, 1986. This free publication is a key reference to understanding how digital data are used to make topographic maps. This publication will be of value to students with an interest in computer applications.

_____. *Finding Your Way with Map and Compass.* Denver, Colo.: Author, n.d. This free brochure shows the hiker how to use a topographic map and describes the various map scales used on maps in the national topographic map series.

_____. *Large-Scale Mapping Guidelines.* Denver, Colo.: Author, 1986. This free publication provides basic information and aids in preparing specifications and acquiring large-scale maps for a variety of uses. Contains a large number of practical maps and an extensive applied glossary.

_____. *National Geographic Mapping Program: Goals, Objectives, and Long-range Plans.* Denver, Colo.: Author, 1987. This free publication provides a nontechnical overview of the use and importance of geologic maps in the United States.

_____. *Topographic Map Symbols.* Denver, Colo.: Author, n.d. This free brochure summarizes all the symbols used on large-scale maps that are in the National Mapping Program. It indicates where and how to order topographic maps.

Wang, Fan. "Odorant Receptors Govern the Formation of a Precise Topographic Map." *Cell* 93, no. 1 (April 3, 1998): 47-61.

Jeffrey C. Reid

Cross-References

Elemental Distribution, 232; Oil and Gas: Distribution, 481.

The Geologic Time Scale

Geologic science has contributed to modern thought the realization of the immense time involved in the earth's history. So vast is this time span that the term "geologic time" is used to distinguish it from other kinds of time.

Field of study: Geochronology and paleontology

Principal terms

BRACHIOPOD: a bivalved filter feeder; clams and oysters are the modern equivalents

BRYOZOAN: a colonial marine animal very much like modern sponges

FAUNAL SUCCESSION: the sequence of life forms, as represented by the fossils within a stratigraphic sequence

GEOLOGIC MAP: a map illustrating the age, structure, and aerial distribution of rock units

HOLOTYPE: the definitive example of a specimen, used to compare all others

LITHOLOGY: the mineral composition and texture of a rock

PALEONTOLOGY: the science of ancient life forms and their evolution as studied through the analysis of fossils

PLUTONISTS: a school of thought that attributed the formation of all rock strata to volcanic action

STRATIGRAPHIC SEQUENCE: a set of rock units that reflect the geologic history of a region

TONGUE STONES: an ancient colloquial term used to describe what is now recognized as fossil sharks' teeth; if viewed from the convex side, a large shark's tooth might resemble a tongue turned to stone

TRILOBITE: a many-legged arthropod named for its three symmetrical lobes; the principal index fossil for the Cambrian period

UNCONFORMITY: a surface that separates two strata; represents a gap in time in which no geologic records remain

UNIFORMITARIANISM: the general principle that the earth's past history can be interpreted in terms of what is known about present natural laws, as these processes differ neither in degree nor in kind

Summary

The ancient civilizations were very indefinite regarding time periods. For Strato (d. c. 270 B.C.) and Eratosthenes (c. 276-c. 194 B.C.), invertebrate fossils were evidence only of ancient seas having existed. Herodotus (c. 485-c. 425 B.C.) associated vertebrate fossils with Greek mythology, concluding that large fossilized bones were remnants of battles between giants and their gods. During the mid-1600's, the science of historical geology began to branch from the trunk of natural philosophy and develop its own identity. The first step toward that development was a new understanding about fossils. In the fall of 1666, fishermen fishing off the west coast of Italy caught a great white shark. Word of this unusual fish spread to the Medici court in Florence, where the Grand Duke Ferdinand II ordered the head cut off and brought for examination to Neils Stensen (known as Nicolaus Steno, 1638-1686), a young Danish doctor serving the court. Steno recognized the strong resemblance of the shark's teeth to tongue stones. He made the intuitive leap that, contrary to common belief, tongue stones did not grow in the ground but had their origin in the heads of sharks. The problem was to account for the transposition of the teeth from the shark's head into the solid rock that enclosed them. Through a series of critical observations and deductions, Steno arrived at three basic tenets of modern geology: Layered rocks result from sediments settling out of water; the oldest strata are on the bottom; and the strata originally are deposited in an essentially horizontal position. Steno published these conclusions in *Prodromus*, his great work of 1669.

Robert Hooke (1635-1703) gave Steno's ideas a wide hearing in lectures he presented before the Royal Society of London in 1667-1668. Hooke's main contribution to the budding science of historical geology was his support of the fossils' organic origin, the extinction of species, the change within species over time, and the theory that subterranean forces have caused the continents to rise and fall with respect to the sea. Thomas Burnet's very controversial *Sacred Theory of the Earth* (1681-1689) sparked further debate about the origin and changes in the earth's surface features. Burnet (1635-1715) called his theory sacred because it would justify by reason the biblical doctrines, specifically the Fall and the Universal Deluge. Burnet held that the earth was around 4,000 years old and that the Universal Deluge occurred about 1,600 years later. Benoit de Maillet (1656-1738), a French diplomat, proposed in *Telliamed* (1748) a theory that put the age of the earth at more than 2 billion years. Maillet based his theory on the sun's life expectancy and on observations of the fall of sea level. In this work, he supported two fundamental ideas: Thales' belief that terrestrial life originated in the sea and Aristotle's "infinite age of the earth," which became Maillet's "vast amount of time" to build mountains layer by layer from strata once submerged.

Another description of the earth's age appeared in the efforts of a second

Frenchman, Georges-Louis Leclerc, the Comte de Buffon (1707-1788). Beginning in 1749, Buffon published a comprehensive multivolume work with the modest title *Natural History*, in which he divided the history of the earth into seven epochs. Buffon's contribution to the question of age was an empirical study of the cooling rates of iron. In his own foundry, he heated to incandescence and then cooled iron balls of different diameters. He recorded the time for each ball to cool, extrapolating from his results the time it would take for a ball the size of the earth to cool to the current level: 96,670 years and 132 days. Buffon generated a second timetable based on sedimentation rates observed in oceans. The variable deposition rates reported revealed a considerable range of time. The longest estimate placed the earth's age at nearly 3 million years and had life appearing between 700,000 and 1 million years. These ideas circulated widely because of Buffon's strong influence on the intelligentsia of his time, which included correspondence with Benjamin Franklin and Thomas Jefferson. Buffon contributed two important ideas that helped to build a sense of time. First, he expanded the age of the earth to millions of years. Second, he showed that we could understand past geologic events by observing the causes of change that are in operation today.

In 1785, James Hutton (1726-1797) published "Theory of the Earth" in the *Transactions of the Royal Society of Edinburgh*. The ideas expressed in this essay and his later elaborations of them are the beginning of modern geology. He concluded that if one could measure the rate at which erosion destroys lands, that rate could be used to calculate the time needed to form the strata observed in the field. The rates Hutton observed were so small, however, that he questioned the possibility of measuring them in a year or even a lifetime. The vastness of the time required to describe the earth's history came to Hutton as he observed what modern geologists call an unconformity (a gap of time in the geologic record). Hutton observed this sequence of strata at Siccar Point near St. Abbs, Scotland, where a sequence of horizontal strata rested on a sequence of vertical strata. He concluded that if all strata began as horizontal, a vast amount of time must have elapsed to produce the present configuration. According to Hutton, first the sediments had to be deposited in a marine environment, where they solidified. Next, as a result of the earth's internal heat, the horizontal sediments rose above sea level to a vertical position and eroded. Then, they were submerged in their vertical position and a new deposition placed horizontal strata on top of the vertical strata. Finally, the whole structure again rose and eroded to expose the structure. Accounting for this history required a time factor of enormous scale. In Hutton's closing essay he suggested an infinite time frame: "The result, therefore, of this present inquiry is, that we find no vestige of a beginning, no prospect of an end."

In the late eighteenth and early nineteenth centuries, a new principle for determining the geologic ages of fossiliferous strata emerged from field studies in England and France. This was the principle of faunal succession.

In a simplified form it stated that within sequences of strata, different kinds of fossils succeed one another in a definite order. The Englishman William Smith (1769-1839) most graphically demonstrated this principle. Smith was a self-taught surveyor and civil engineer working on the construction of canals in England in 1794. In the course of these excavations, he discovered that each of the formations revealed a distinctive fossil species. Extrapolating this knowledge to other regions allowed him to identify and predict the stratigraphic sequence. In 1814, Smith consolidated his findings in what geologists often consider the first geologic map. The next year he published his major work, *Delineation of the Strata of England and Wales, with Part of Scotland*. Smith was recognized in his own time as the father of English geology. His contribution to dating was to utilize the fossils in determining the relative ages of strata.

Already at the turn of the nineteenth century, European geologists had begun a systematic classification of fossiliferous strata into coherent units or periods. The first of the geologic time periods appeared in 1799 with the work of Alexander von Humboldt (1769-1859). Humboldt applied the name "Jurassic" to a coherent sequence of fossiliferous limestone strata found in the Jura Mountains of Switzerland and France. Just to the west of these mountains, the limestones and their associated fossils dip under a dominantly chalky sequence studied by the Belgian geologist Omalius d'Halloy (1783-1875). He gave it the name "Cretaceous" in 1822. The pattern of naming rock units, based on the lithology and associated fossils, for the geographic region in which they were first described continued through the nineteenth century.

The original boundaries of these periods were neither distinct nor easily translated outside the holotype regions. Methods for defining the boundaries resulted from a dispute in England over a sequence of strata that Adam Sedgwick (1785-1873) and Sir Roderick Impey Murchison (1792-1871) described in 1835. Murchison based his chronological sequence on the fossil order that he observed and named it the Silurian. Concurrently, Sedgwick had relied on lithology to establish a sequence of strata that lay below the Silurian and was therefore older. He called it the Cambrian. Initially, the two periods seemed to create order for these strata; however, under closer examination the periods overlapped, and dispute developed between Sedgwick and Murchison over the commonly held strata. The answer appeared in 1879 after their deaths, when Charles Lapworth (1842-1920) separated the systems based on fossils. Lapworth collected evidence that illustrated that the Cambrian-Silurian sequence actually contained three distinct fossil assemblages and resolved the dispute by removing the lower Silurian from its previous classification and renaming it the Ordovician. The discrimination of fossil assemblages provided the key to distinguishing other increments of geologic time, such as the Devonian and Permian periods.

By the middle of the nineteenth century, geologists recognized that the rock units they studied in one location were not universal geographically.

Their systems were highly variable in lithology and thickness from region to region. The distinctiveness of the fossils within each group enabled them to translate from one geographic area to another. The power of this approach was that the order of the sequence was not random but predictable; for example, bryozoans and corals characterize the Ordovician period the world over. Each has its own distinctive suite of fossils. Recognition of the sequencing of life forms through time had a profound effect on Charles Darwin (1809-1882). Indeed, Darwin's paleontological investigations led him to focus on the idea of a species changing through time. Later, in his *On the Origin of Species by Means of Natural Selection* (1859), Darwin pointed out that natural selection was a viable concept if and only if enough time had elapsed for its operation.

In 1862, Lord Kelvin (1824-1907) made the first attempt to calibrate the geologic time scale. Working with the thermodynamic laws of heat production and radiation, he calculated the cooling rates of the earth and the sun. Kelvin's calculations set a "natural" upper limit to their ages and indicated that the earth had solidified from the original molten state between 20 million and 400 million years ago. This was much less than the time scale advocated by the uniformitarians. Geologists thus found themselves no longer in a position to assume unlimited time; their theories had to fit the time interval established by Kelvin's thermodynamic studies. Resolution of the time interval came in 1896 with the discovery of radioactivity by Antoine-Henri Becquerel (1852-1908) and Pierre (1859-1906) and Marie Curie (1867-1934). This was the beginning of a chain of events that revolutionized science and expanded the geologic time range into thousands of millions of years.

In 1902, Pierre Curie announced that radioactive minerals constantly radiate heat. Two years later, Ernest Rutherford (1871-1937) established that the amount of heat they radiate is proportional to the number of alpha particles they emit. John Joly (1857-1933) provided additional support for Rutherford's discovery in his 1909 publication, *Radioactivity and Geology.* Joly demonstrated that the heat from radioactive decay within the earth could alter the earth's actual cooling rate to make it appear younger than it really was. Kelvin's calculations of 20 to 400 million years did not include the masking effect of internal radioactive heat, and thermodynamics alone therefore no longer established the boundaries. Geologists and biologists legitimately could claim a longer time interval for the evolutionary process. Continued investigations revealed that radioactive minerals might serve not only as sources of heat but also as clocks to date the rocks that contained them.

In 1905, Lord Rayleigh (1842-1919) and Sir William Ramsay (1852-1916) calculated an age of 2,000 million years for a specimen containing uranium. The American physical chemist Bertram Boltwood (1870-1927), noting that uranium ores always contain lead, speculated that lead might be the end product of a uranium decay series. By comparing the ratio of lead to

THE GEOLOGIC TIME SCALE

Eon	Era	Period	Epoch	Age
			Recent	
		Quaternary	Pliestocene	1.6
	Cenozoic		Pliocene	5.3
			Miocene	23.7
		Tertiary	Oligocene	36.6
			Eocene	57.8
			Paleocene	66.4
Phanerozoic	Mesozoic	Cretaceous		144
		Jurassic		208
		Triassic		245
	Paleozoic	Permian		286
		Carboniferous	Pennsylvanian	320
			Mississippian	360
		Devonian		408
		Silurian		438
		Ordovician		505
		Cambrian		570
Proterozoic	Late			900
	Middle			1,600
	Early			2,500
Archean	Late			3,000
	Middle			3,400
	Early			3,800?

Note: Ages are measured in millions of years.

uranium in forty-three minerals, he calculated their ages and obtained results ranging from 400 to 2,200 million years. Boltwood's results were the first quantitative proof of the earth's age. Then, in 1913, Frederick Soddy (1877-1956) and, independently, Kasimir Fajans (1887-1975) demonstrated that radioactive elements can have the same chemical properties but slightly different atomic masses. Their discovery of isotopes was the next step in establishing intervals on the time scale. This development enabled researchers to measure the decay rate or half-life of a radioactive isotope. By the late 1920's, Francis William Aston (1877-1945) had begun to use mass spectrometry to make significant improvements in isotopic analysis. Alfred Otto Carl Nier continued to improve the mass spectrometric technique in the 1930's and 1940's, providing the most accurately determined values of uranium-lead ratios found in naturally occurring uranium minerals. Earlier, in 1913, the English geologist Arthur Holmes (1890-1965) had made the first attempt at a quantified time scale using Boltwood's calculations of radioactive decay. Later, in a 1948 paper, he developed an expanded time scale based on Nier's more recent values. By the 1960's, J. L. Kulp was continuing Holmes's investigations, introducing the rubidium-strontium decay series to date rock samples.

Mass spectrometric investigations continued throughout the twentieth century, and by the mid-1970's, the discovery and dating of Precambrian rocks in Greenland, South Africa, Australia, and Canada yielded ages of about 3.7 billion years. The earth appears to be even older. Its age seems bracketed somewhere between these ancient terrestrial rocks and the lunar rocks, which date from 4.6 billion years. Isotope analysis of meteorites supports this same age range.

The geologic time scale has expanded with the maturation of the geologic and physical sciences. The age of the earth has not changed; only the paradigm changes. Undoubtedly, science and technology will continue to define the age of the earth more accurately and without the widely ranging extremes of the past.

Methods of Study

Scientists have used two methods to study the earth's time scale. The first method depends on the succession of fossils through time. Its premise is that life developed in an orderly fashion over time from simple to more complex forms. The fossil record records the succession by showing the less advanced life forms dominating the lower, or earlier, strata and the more advanced forms appearing in the later, or more recent, strata. It illustrates clearly the sequencing of the events in the earth's development.

The second method uses the half-life of an atom to decode the radioactive clock hidden in a type of rock broadly described as igneous. The word igneous means "fire formed," indicating the formation of these rocks deep within the earth's crust or from volcanic action. Analyzing the ratio of radioactive isotopes and their daughter products gives absolute ages. Because this method is effective only for igneous rocks, scientists infer absolute ages for other rock types by bracketing them between igneous intrusions and then "splitting the difference" of the ages.

Context

The concept of geologic time has contributed significantly to the development of human thought. It has expanded from a few thousand years to billions of years and has become the common denominator for all the historical sciences. Nevertheless, the growth of the geologic time scale has not been a smooth development yielding a gradual expansion of knowledge. In the seventeenth century, the age of the earth appeared to be a few thousand years; estimates of millions and even thousands of millions followed in the eighteenth century, yet during the late nineteenth century, the large numbers shrank to as low as a few million years. Then, the "new physics" of the twentieth century replaced a relative scale with an absolute one and described the age of thousands of million years. As capricious as these fluctuations seem, they reflect the growth and development of the competing paradigms of their times.

The unfolding of a sense of prehistoric time parallels geology's maturation as a science. That the earth must be ancient beyond comprehension was as radical an idea in the nineteenth century as replacing the earth with the sun at the center of the universe was in the sixteenth century. Once again, empirical knowledge redefined the place of humans in the cosmos and challenged what had become a faith-based paradigm. Not until "the vastness of time" became the dominant paradigm were many of the sciences able to advance.

Bibliography

Adams, Frank D. *The Birth and Development of the Geological Sciences.* Mineola, N.Y.: Dover, 1938. An excellent resource for the advanced reader and interested student. It begins with ancient times and continues to the birth of the new sciences. Many fine illustrations and woodcuts from primary sources.

Albritton, Claude C., Jr. *The Abyss of Time: Unravelling the Mystery of the Earth's Age.* San Francisco: Freeman, Cooper, 1980. A very readable history emphasizing the people who shaped the concept of geologic time. Suitable for college freshmen and the interested reader.

Berry, William B. N. *Growth of a Prehistoric Time Scale, Based on Organic Evolution.* New York: W. H. Freeman, 1968. A good introduction to the history of the geologic time scale. Well written and illustrated. Suitable for college-level students.

Holmes, Arthur. *The Age of the Earth: An Introduction to Geological Ideas.* New York: Harper & Row, 1960. A complete synopsis of the question of the age of the earth. Well grounded in the paradigm of historical research. Suitable for the informed reader and the college student.

Smith, Patrick E. "Ages of Glauconies: Implications for the Geologic Time Scale and Global Sea Level Variations." *Science* 279 no. 5356 (March 6, 1998): 1517-1520.

Woodward, Horace B. *History of Geology*. New York: Putnam, 1911. Reprint. New York: Arno Press, 1978. An appropriate background for the reader interested in this subject, suitable for the college-level reader.

Anthony N. Stranges

Richard C. Jones

Cross-References

Glacial Landforms

Glacial landforms are common in many parts of Canada, the Northern United States, northern Europe, Asia, and many mountain ranges of the world. The material composing these landforms, as well as the shapes of some of the landforms, is essential to many human activities.

Field of study: Glacial geology

Principal terms

ARÊTE: an extremely narrow mountain ridge created between adjacent U-shaped valleys
CIRQUE: a bowl-shaped depression near mountaintops where glaciers originate
DRUMLIN: a streamlined hill formed under actively moving ice
ESKER: a sinuous ridge of stratified drift formed in a tunnel under the ice
KAME: a conical hill of stratified drift
KETTLE: a depression created by the melting of a chunk of ice
MORAINE: landscapes of till, varying from fairly flat terrain to gently rolling hills to long ridges
STRATIFIED DRIFT: material deposited by glacial meltwaters; the water separates the material according to size, creating layers
TILL: a mixture of varisized materials deposited by glacier ice
U-SHAPED VALLEY: the classic shape of glaciated mountain valleys

Summary

Glacial landforms are distinguished primarily on the basis of shape and composition, both of which relate directly to the mode of origin of the glacial features. Some landforms originate by erosion or deposition directly by the ice, while other features are created by erosion or deposition of the meltwaters from the ice. It is important to distinguish whether the ice or its meltwaters produced the landform. Another factor in recognizing glacial landforms is the realization that many glacial features are commonly associated with a particular kind of glacier. Two types of glacier are commonly recognized, continental glaciers and valley glaciers. Although some landform features are common to both types of glacier, others are relatively unique to each type.

An understanding of glacial ice movement is also necessary in order to understand how glacial landforms develop. Whether the ice is moving because of pressures from great thicknesses of continental ice or is responding to gravity down a valley slope, the ice always moves forward. Even when the glacier is melting back, and it looks like it is getting shorter or smaller, internally the ice is still moving forward.

Glaciers can erode the material over which they move, and if they scrape down to solid bedrock, several glacial features can result. If the glacier is carrying small, hard particles along its base, these particles may scratch the bedrock, creating many roughly parallel striations. If large rocks are being transported at the base of the ice, they can gouge out varisized grooves, which can range from a few centimeters to several meters deep. On the other hand, if the ice has ground its load into flour-size particles, called rock flour, the flour can actually polish the bedrock. The orientation of striations and grooves can be used to determine ice-flow direction.

Of all the glacial landforms, perhaps the most spectacular are the erosional products of mountain glaciers. Mountains that have experienced glaciation exhibit some the world's most breathtaking scenery. Glaciers tend to sharpen peaks and ridges and to steepen valley walls, producing avenues for numerous spectacular waterfalls. As snow accumulates near the tops of mountains, the forming ice tends to carve out a basin-shaped depression on the side of the mountain. This basin is called a cirque and is the home base of the glacier; it is in this basin that more snow will accumulate, thickening the ice, until the ice begins to move out of the cirque and down the valley. If three or more glaciers form around the same mountain peak, the crowding of their cirques around the peak will erode it into a sharper than usual pyramidal shape. This type of mountain peak is called a horn. Many of the peaks in the Alps have the word "horn" incorporated into their names; the Swiss Matterhorn is probably the most famous glacial horn. If two glaciers are moving down a mountain in adjacent valleys, the ridge between them may be eroded into a narrower, sharper ridge called an arête.

Mountain glaciers tend to widen the base and steepen the sides of the valleys through which they move, creating obviously U-shaped valleys. Large glaciers have more power to erode more deeply, resulting in deeper U-shaped valleys. Small tributary glaciers produce valleys that are not carved as deeply; therefore, when the glaciers retreat from such an area, the floors of the tributary valleys are left high above the floor of the main valley. Such valleys are appropriately called hanging valleys. If streams are present, they become waterfalls that cascade over the edges of the hanging valleys. U-shaped valleys usually contain strings of lakes that are impounded by irregularities of the valley floor. These lakes are quite picturesque in that the rock flour they contain from glacial meltwaters gives the water a distinctive turquoise color.

Perhaps the most spectacular of the glacial valleys are fjords, most common in Norway, British Columbia, Alaska, and Chile. A fjord is a coastal glacial valley that was carved out when sea level was lower during times of glaciations. Upon retreat of the ice and subsequent rise in sea level, seawater completely inundated the valley. The water in many fjords is quite deep—often more than 1,000 meters deep. In some areas where fjords extend inland for considerable distances and intersect with other fjords, it may be difficult to distinguish a fjord from a large lake in a U-shaped valley. The fjord, being inundated with marine water, will contain seaweed and experience tidal changes; the lakes will be freshwater and relatively static.

Glacial deposits are widespread and create interesting landform features. Glaciers deposit till, an unconsolidated mixture of clay, sand, silt, pebbles, cobbles, and boulders. Depending on the area over which the glacier moved, the relative abundance of these components will vary; some till can be very sandy, while other till can be clayey. Abrasion during ice transport causes the particles in till to be usually angular in shape. Because it is the glacier ice itself that deposits this material, the till is unsorted. This lack of sorting results because the materials in till are let down to the surface as the ice melts in the same relative positions they had in the ice. Till can be formed into various types of morainal landforms, depending on how the glacial ice moved. Till can be deposited as ground moraine as the ice retreats. It is laid down under the ice as the ice moves along, resulting in a landscape called a till plain, which usually takes on the appearance of gently rolling, hummocky hills.

When a glacier front becomes stationary, material will continue to be moved forward through the internally forward-moving ice and be deposited along the front of the stationary glacier. It will build up into a ridge called an end moraine. If the glacier then advances, the end moraine will be destroyed by the moving ice. If the glacier retreats, the end moraine will become part of the landscape. The end moraine that is created at the maximum forward position of the ice is called the terminal moraine. Any end moraines that form as the ice periodically retreats are called recessional moraines and usually form concentric bands of ridges in the landscape. If a glacier advances and then retreats without spending any significant length of time in a stationary position, end moraines cannot form; the till will be formed into ground moraine.

Two types of moraines are unique to mountain glaciers. Because these glaciers are restricted to valleys, accumulations of debris tend to gather along the valley walls at the sides of the glaciers. These are called lateral moraines and are left behind as long ridges along the valley walls as the ice melts away. A medial moraine forms when two glaciers join and their lateral moraines are consequently trapped between the two ice rivers, which are now flowing along as one glacier. Several medial moraines can form as more and more glaciers from tributary valleys join, giving the glacier an interesting striped appearance.

Glacial meltwater is an important agent in creating several landform features. The power of glacial meltwater can be extraordinary, often carving deep gorges. The Channeled Scablands of Washington attest the effects of glacial meltwaters. When glaciers melt, they not only release water but also the huge amounts of debris they carry. The meltwater transports the debris, eventually depositing it. The material deposited by glacial meltwater is called stratified drift and differs from till in that it is sorted and layered. The heavier particles drop out first, followed by progressively smaller particles as the water travels farther from the ice, gradually slowing down. Fine clays and silts are usually transported the farthest, sometimes completely out of the area, giving the stratified drift its distinctive sand and gravel characteristics. As the glaciers eventually release less and less water, smaller and smaller particles are deposited over the coarser material deposited previously. Unlike the ice-deposited till, transport in water tends to round the particles of stratified drift.

Outwash plains are huge flat areas of stratified drift located adjacent to where glaciers existed. If outwash is confined to a river valley, it is called a valley train. Large outwash areas are usually devoid of vegetation cover at first and are subjected to strong winds. Often silt-sized particles in the outwash are then transported by wind and deposited in very thick layers downwind from the deposit. Windblown silt is called loess and is an important ingredient in soils. Soils developed on loess-capped till plains are particularly good for agriculture.

Meltwater sands and gravels can form some rather interesting landscape features if conditions of the melting glacier are just right. If the area near the snout of the glacier becomes too thin, the ice will cease to move internally and become stagnant. When that happens, the stagnant ice may be eroded by its own meltwater as well as that coming off the active ice. Meltwater streams may form on the ice and drill out tunnels in and under the ice. The movement of the streams determines what type of landform feature will evolve from the deposition of the sands and gravels transported by the stream. For example, if streams in or on the ice cascade down fairly vertical shafts (called moulins) in the ice or tumble down a fairly steep face of the ice edge, the sand and gravel will spill into a conical-shaped pile. This pile, called a kame, can be more than 60 meters high. Usually several kames are formed in the same area, dotting the landscape as isolated hills.

Another example of meltwater stream deposition is an esker. Eskers also form in zones of stagnant ice where sand and gravel are deposited within stream tunnels that are in and under the stagnant ice. When the ice melts away, long, sinuous ridges are left in the landscape. New England has some very impressive eskers that are more than 40 meters high in places and wind for miles through the countryside.

Other features commonly associated with outwash and stagnant ice areas are kettles. Kettles form when chunks of ice break off from the glacier and become buried in outwash or in till as the glacier melts. Gradually the ice chunks melt, leaving depressions in the ground, sometimes in the shape of pioneers' kettles, which is how the name originated. Kettles that intersect the groundwater table or collect precipitation contain water and are called kettle lakes, some of which can cover several acres. An outwash area that contains numerous kettles is called pitted outwash.

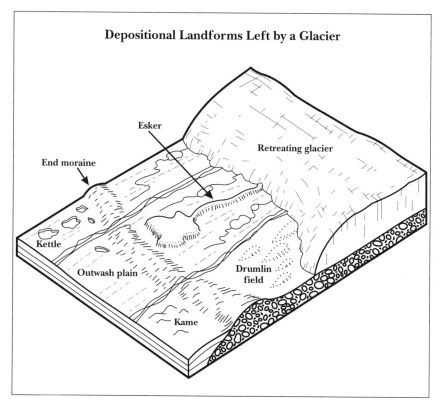

Depositional Landforms Left by a Glacier

Actively moving ice sometimes has the ability to form streamlined shapes through either erosion or deposition. A roche moutonnée is an eroded bedrock form with a gentle up-ice side and a steep, rough down-ice side. Hence, ice-flow direction can be determined by the orientation of roches moutonnées. These features vary in size from a meter to large hills in the landscape.

Drumlins are streamlined hills that look like half-buried whales or inverted spoons. Drumlins can be composed of a variety of materials. The cores can be composed of bedrock or, more commonly, stratified drift and are covered with a veneer of till. The formation process of drumlins is unclear, but agreement on some aspects does exist. They appear to have been formed only by actively moving ice near, but not at, the edges of the ice. Drumlins always appear in groups, called drumlin fields, composed of literally thousands of drumlins. The most famous drumlin fields are in New York, Wisconsin, and New England.

The Great Lakes exist because of the action of continental ice sheets in the Midwest. Tongues of ice gouged out old river valleys, creating the lake basins. Meltwater and precipitation filled the basins. The history of the glacial Great Lakes from the time of the ice sheets to the present day is extremely complicated. As the ice retreated and made minor advances in

285

various parts of the upper Midwest, different sizes and shapes of lakes emerged in each basin. The lakes had different outlets at different times, creating different water levels. Most of the old lake levels can be traced today because old beach and dune deposits exist at various elevations along the present-day Great Lakes shorelines. The old clay-rich lake beds that cover large parts of Ohio, Indiana, and Illinois have some of the richest soils in the country.

Methods of Study

Glacial landforms are studied using a variety of methods. The general forms of the features are analyzed using topographic maps and various types of aerial photographs. Such tools not only illustrate the forms of each glacial feature but also show its relationship to other nearby features. Patterns in the landscape can be observed quite easily, and ice-direction movement may be interpreted. Topographic maps are very detailed, showing all natural and human-made features in a small area. These maps also include contour lines that represent elevations above sea level, and with some practice, hills and valleys can be identified. All landforms that are large enough to be incorporated by the given contour interval can be shown on the map. All kinds of glacial features can be identified using topographic maps, because their shapes are distinctively outlined by the contours.

Aerial photographs can be used in the same way as the topographic maps, except that with photographs, the actual feature in the landscape can be seen. There are several types of aerial photographs that can be used for analysis, depending on what features are being studied. Standard black-and-white, color, and a variety of other wavelength bands (from satellite imagery) each emphasize distinct characteristics of the landscape.

Recall that a diagnostic characteristic for many glacial landforms is their composition. The composition of any particular feature is determined by digging into it. Sometimes quarries or roadcuts have already exposed the interior of the structure, making the job of the geologist easier. At other locations, core samples are obtained by drilling. Sometimes well records from already existing water wells may be utilized to determine what kind of material has been deposited in an area. Another clue to the composition of a feature may be obtained by the type of soil that has developed on the material. County soil surveys are useful for this kind of analysis.

On a more detailed scale, geologists may want to understand the mechanics of ice flow when determining the creation of the landform, the direction of ice flow, the source area of the material (including clues to ore deposits), or the period in which the ice existed in a particular area. These pieces of information can be obtained by analyzing samples taken from the landscape features. Till fabric, pebble lithology and roundness, particle size, radiocarbon dating of organic material, and the amount and type of weathering are all commonly used observations.

Till fabric is the orientation of the pebbles and cobbles in the till.

Measurements of these orientations give clues about ice-flow directions. Pebble counts of the lithologies of rocks found in the deposits indicate possible source regions and general flow paths of the ice. Pebble roundness and particle size give evidence as to the transportation of the materials, whether by meltwaters or by ice, and if by ice, where in the ice the particles were most likely carried. Radiocarbon dating is indispensable in developing the time frames of glacial deposition, as is the more locally used weathering profile, but with less dependability.

Many areas rely on glaciers as a major water resource for either general consumption or power generation. Mountain glaciers are carefully monitored in order to determine the amount of water that can be expected from them each year. The monitoring is done by either digging into the ice at various points and measuring ice layers and rates of melting or, more commonly, by taking precipitation and meltwater stream gauging measurements throughout the entire glacial basin.

Context

Continental glaciers have covered large portions of North America and lesser amounts of Europe and Asia. In their wake, they have left a variety of landforms that have benefited not only people in the glaciated regions but also those in many other parts of the world. Most of the till plains and lake plains have become major agricultural areas because of the thick deposits in which rich soils have developed since the retreat of the ice. Crops grown in the glaciated regions are shipped to many other areas of the world.

Stratified deposits are often utilized for construction purposes. An area of such deposits can be recognized because of the number of gravel pits excavating the volumes of sand and gravel from outwash plains, valley trains, kames, eskers, and drumlins. In many places, the glacial landscape containing these features has been completely changed by having totally eliminated these features through quarrying practices. Most of the quarried material remains in the local area for use because it is too expensive to ship.

The recreation industry is important in many glaciated regions. In continental areas where numerous kettle lakes exist, people build vacation homes in aesthetically pleasing places, and other tourists flock to the lake areas to fish, swim, boat, and relax. In mountain regions, the scenery resulting from glacial activity attracts thousands of people every year. Skiing abounds in many such areas, and many mountain-climbing adventures take place in areas previously carved by glaciers.

Of particular importance is the role of water in the glaciated regions. Besides adding to the aesthetics of areas, water resources are quite precious. In some gravel areas, wells can be made cheaply, because they do not have to be made particularly deep. The groundwater flows easily through such material and often in great abundance, but it can also be easily polluted.

The Great Lakes, carved out by the huge continental ice sheets, are very important sources of potable water for some surrounding cities. Their greatest advantage, however, is their commercial value, benefiting not only those working on the Great Lakes but also those receiving goods shipped away from the Great Lakes, including food products from the farms of the glaciated region, various mineral resources (iron ore, copper, and coal), and finished products.

Some glaciated regions are hazardous to humans. Oversteepened mountain topography is often the site of snow avalanches or landslides. Numerous places in Norway have experienced landslides that have wiped out entire villages. In many cases, the destruction was not directly a result of the landslide but was caused by its debris falling into one of the deep lakes or fjords, creating huge waves that rushed the narrow, confined shores along the water. Less exciting hazards include the siting of landfills. Landfills placed in coarse glacial deposits often contribute pollutants to groundwater.

Bibliography

Adams, George F., and Jerome Wyckoff. *Landforms*. New York: Golden Press, 1971. This Science Golden Guide covers all aspects of landforms. The section on glaciers and glaciated land is brief but contains excellent diagrams and photographs to illustrate the formation of glacial landscapes. The section on useful publications not only lists general references but also indicates where to obtain regional, state, and local information about geologic features.

Brunsden, Denys, and John C. Doornkamp, eds. *The Unquiet Landscape*. Bloomington: Indiana University Press, 1974. The introduction to this book contains a good general introduction to geology, maps, and general methods of geological study. A brief bibliography and an index are included. The two chapters explaining glacial erosion and deposition and related landforms are illustrated with excellent photographs. The chapters in the book are interrelated; for example, information about glacial effects on landscape is also explained in the chapters on lakes and changing sea level.

Colby, William E., ed. *John Muir's Studies in the Sierra*. San Francisco: Sierra Club Books, 1960. John Muir was the first to recognize the dramatic impact of glaciation in the Sierra Nevada. His observations and methods of reasoning showed geologists the folly of their previously incorrect interpretations of the development of the Sierra landscape. This book contains Muir's observations and explorations written in his easy-to-understand style. It includes his own drawings as well as several photographs. Because glacial geology is related to other geologic effects, Muir considers all aspects of the Sierra geology.

Forrester, Glenn C. *Niagara Falls and the Glacier*. Hicksville, N.Y.: Exposition Press, 1976. This small book explores all the geology of Niagara Falls, especially the influence of the continental glacier and the development of the Great Lakes. It is well illustrated with maps, photographs, and diagrams and includes a walking tour. Although there is no index, there is a brief bibliography.

John, Brian S. *The Ice Age: Past and Present.* London: Collins, 1977. This book is very thorough in its examination of glacial and meltwater features. Many photographs, maps, and diagrams are used, and both a good index and a bibliography are included.

Larter, R. D., and L. E. Vanneste. "Relict Subglacial Deltas on the Antarctic Peninsula Outer Shelf." *Geology* 23, no. 1 (January, 1995): 33-37.

Matsch, Charles L. *North America and the Great Ice Age.* New York: McGraw-Hill, 1976. Numerous photographs, maps, and diagrams illustrate the chapter on glacial erosional and depositional features. There is also a chapter on lakes and sea-level changes. Brief selected readings and an index are included.

Moore, W. G. *Glaciers.* London: Hutchinson Educational, 1972. This is a small book of excellent black-and-white photographs with commentaries on glaciers and glacial landscapes from around the world. Although there is no bibliography or index, there is a list of terms as well as questions accompanying each photo.

Price, R. J. *Glacial and Fluvioglacial Landforms.* New York: Hafner Press, 1973. Most of this book treats glacial and meltwater landforms in a more technical, yet nonmathematical, manner, with excellent diagrams, photographs, and maps. The index is good, and the bibliography is very thorough.

Schultz, Gwen. *Glaciers and the Ice Age.* New York: Holt, Rinehart and Winston, 1963. The chapters on glacial landforms are intriguingly named "The Great Detective Story" and "The Artistry of Ice." Both describe erosional and depositional features and discuss how they can be interpreted in the landscape. Illustrated with photographs, maps, and diagrams. Although both the bibliography and the index are brief, there is a glossary.

Diann S. Kiesel

Cross-References

Alpine Glaciers, 8; Continental Glaciers, 93; Lakes, 357.

Granitic Rocks

Granitic rocks are coarse-grained igneous rocks consisting mainly of quartz, sodic plagioclase, and alkali feldspar, with various accessory minerals. These rock types occur primarily as large intrusive bodies that have solidified from magma at great depths. Granitic rocks can also occur to a lesser degree as a result of metamorphism, a process referred to as granitization.

Field of study: Petrology

Principal terms

APHANITIC: a textural term that applies to an igneous rock composed of crystals that are microscopic in size

CRYSTAL: a solid made up of a regular periodic arrangement of atoms

CRYSTALLIZATION: the formation and growth of a crystalline solid from a liquid or gas

GRANITIZATION: the process of converting rock into granite; it is thought to occur when hot, ion-rich fluids migrate through a rock and chemically alter its composition

ISOTOPES: atoms of the same element with identical numbers of protons but different numbers of neutrons, thus giving them a different mass

MAGMA: a body of molten rock typically found at depth, including any dissolved gases and crystals

MIGMATITE: a rock exhibiting both igneous and metamorphic characteristics, which forms when light-colored silicate minerals melt and then crystallize, while the dark silicate minerals remain solid

PHANERITIC: a textural term that applies to an igneous rock composed of crystals that are macroscopic in size, ranging from about 1 millimeter to more than 5 millimeters in diameter

PLUTON: a structure that results from the emplacement and crystallization of magma beneath the surface of the earth

PORPHYRITIC: a texture characteristic of an igneous rock in which macroscopic crystals are embedded in a fine phaneritic or even aphanitic matrix

Summary

The term "granitic rocks" generally refers to the whole range of plutonic rocks that contain at least 10 percent quartz. They are the main component of continental shields and also occur as great compound batholiths in folded geosynclinal belts. Granitic rocks are so widespread, and their occurrence and relation to the tectonic environment are so varied, that generalizations often obscure their complexity. Basically, major granitic complexes are a continental phenomenon occurring in the form of batholiths and migmatite complexes.

When large masses of magma solidify deep below the ground surface, they form igneous rocks that exhibit a coarse-grained texture described as phaneritic. These rocks have the appearance of being composed of a mass of intergrown crystals large enough to be identified with the unaided eye. A large mass of magma situated at depth may require tens of thousands, or even millions, of years to solidify. Because phaneritic rocks form deep within the crust, their exposure at ground surface reflects regional uplift and erosion, which has removed the overlying rocks that once surrounded the now-solidified magma chamber.

As with other rock types, granitic rocks are classified on the basis of both mineral composition and fabric or texture. The mineral makeup of an igneous rock is ultimately determined by the chemical composition of the magma from which it crystallized. Feldspar-bearing phaneritic rocks containing conspicuous quartz (greater than 10 percent in total volume) in addition to large amounts of feldspar can be designated as granitic rocks. This nonspecific term is useful where the type of feldspar is not recognizable because of alteration or weathering, for purposes of quick reconnaissance field studies, or for general discussion.

Granitic rocks consist of two general groups of minerals: essential minerals and accessory minerals. Essential minerals are those required to be present for the rock to be assigned a specific name based on a classification scheme. Essential minerals in most granitic rocks are quartz, sodic plagioclase, and potassium-rich alkali plagioclase (either orthoclase or microcline). Accessory minerals include biotite, muscovite, hornblende, and pyroxene.

When an initial phase of slow cooling and crystallization at great depths is followed by more rapid cooling at shallower depths or at the surface, porphyritic texture develops, as is evident in the presence of large crystals enveloped in a finer-grained matrix or groundmass. The presence of porphyritic texture is evidence that crystallization occurs over a range of temperatures, and magmas are commonly emplaced or erupted as mixtures of liquid and early-formed crystals.

Classification of granitic rocks can be based either on the bulk chemical composition or on the mineral composition. Chemical analysis units are in the weight percent of oxides, whereas mineral composition units are in approximate percent in total volume. The mineral composition of granitic rocks, unlike that of volcanic rocks, provides a reliable basis for classification. Because the two primary feldspars may be difficult to distinguish as a result of extensive solid solution and unmixing, the chemical composition of the rock in terms of the normalized proportions of quartz, plagioclase, and

ORE MINERALS ASSOCIATED WITH GRANITIC PLUTONS

Mineral	Metal	Comments
Chalcopyrite [$CuFeS_2$]	Copper	"Porphyry copper," finely disseminated grains
Native gold [Au]	Gold	Hydrothermal quartz veins, placer deposits
Molybdenite [MoS_2]	Molybdenum	Used in steel alloys
Cassiterite [SnO_2]	Tin	Mostly mined as placer deposits
Urananinite [UO_2]	Uranium	Mostly mined in granite pegmatites
Rutile [TiO_2]	Titanium	Granite pegmatites, quartz veins
Scheelite [$CaWO_4$]	Tungsten	Granite pegmatites
Wolframite [$(Fe,Mn)WO_4$]	Tungsten	Granite pegmatites
Beryl [$Be_3Al_2Si_6O_{18}$]	Beryllium	Pegmatites; gem crystals are emerald and aquamarine
Lepidolite mica [$K(Li,Al)_{2\text{-}3}(Al,Si_3O_{10})(OH)_2$]	Lithium	Pegmatites; used in lubricants and antipsychotic drugs
Spodumene [$LiAlSi_2O_6$]	Lithium	Pegmatites, with lepidolite
Columbite-tantalite	Niobium, tantalum	Used in heat-resistant alloys, stainless steel

alkali feldspar is recast. Thus, specific rock types are defined on the basis of their ratios. Accessory minerals may or may not be present in a rock of a given type, but the presence of certain accessory minerals may be indicated in the form of a modifier (such as biotite granite).

Granitic rocks include granodiorite, quartz monzonite granite, soda granite, and vein rocks of pegmatite and aplite. The mineralogy of these vary, and the distinction between different granitic rocks can be gradational. Granodiorite is composed predominantly of andesine-oligoclase feldspar, with subordinate potassium feldspar and biotite, hornblende, or both as accessory minerals. Quartz monzonite is composed of subequal amounts of potassium and oligoclase-andesine feldspars, with biotite, hornblende, or both as accessory minerals. Granite is composed predominantly of potassium feldspar with subordinate oligoclase feldspar and biotite alone or with hornblende or muscovite as accessory minerals. Soda granite is composed predominantly of albite or albite-oligoclase feldspar, with small amounts of algerine pyroxene or sodic amphibole.

Pegmatite is a very coarse-grained and mineralogically complex rock. Structurally, pegmatites occur as dikes associated with large plutonic rock masses. Dikes are tabular-shaped intrusive features that cut through the surrounding rock. The large crystals are inferred to reflect crystallization in a water-rich environment. Aplite is a very fine-grained, light-colored granitic rock that also occurs as a dike and consists of quartz, albite, potassium feldspar, and muscovite, with almandine garnet as an occasional accessory mineral. Most pegmatites can be mineralogically simple, consisting primarily of quartz and alkali feldspar, with lesser amounts of muscovite, tourmaline, and garnet; they are referred to as simple pegmatites. Other pegmatites can be very complex and contain other elements that slowly crystallize from residual, deeply seated magma bodies. High concentrations of these residual elements can result in the formation of minerals such as topaz, beryl, and rare earth elements, in addition to quartz and feldspar.

Other rocks that are occasionally grouped with granitic rocks are migmatites. Migmatites, meaning mixed rocks, are heterogeneous granitic rocks which, on a large scale, occur within regions of high-grade metamorphism or as broad migmatitic zones bordering major plutons. Migmatites appear as alternating light and dark bands. The light-colored bands are broadly granitic in mineralogy and chemistry, while the darker bands are clearly metamorphic.

Geochemically, granitic rocks vary in several ways, including isotopic composition; proportion of low-melting constituents, such as quartz and alkali feldspars, to high-melting constituents, such as biotite, hornblende, and calcic plagioclase; relative proportion of the low-melting constituents; alumina saturation; and accessory mineral content. Granitic rocks can further be divided into two groups, S-types and I-types, according to whether they were derived from predominantly sedimentary or igneous sources, respectively. This distinction is based on strontium isotope ratios. For example, a sedimentary source is characterized by a high initial ratio, whereas an igneous source is characterized by a low initial ratio. Most isotope applications are conditional on the magma not having been subsequently contaminated by crystal material. Thus, isotopic composition in granitic rocks must be used with caution when attempting to identify source rocks or locate source regions at specific levels within the mantle or deep crust. S-type granites occur in some regionally metamorphic and migmatitic complexes. Enormous volumes of I-type granites, which constitute most plutons, occur along continental margins overriding subducting oceanic material.

A common feature of granitic rocks, notably in granodiorite to dioritic plutons, is the presence of inclusions or rock fragments that differ in fabric and/or composition from the main pluton itself. The term "inclusions" indicates that they originated in different ways. Foreign rock inclusions, called xenoliths, include blocks of wall rocks that have been mechanically incorporated into the magma body. This process is referred to as stopping. Some mafic inclusions are early-formed crystals precipitated from the magma itself after segregation along the margin of the pluton, which cools first. These inclusions are called antoliths. Other inclusions may reflect clots of solidified mantle-derived magma that ascended into the granitic source region or residual material that accompanied the magma during its ascent.

Along continental margins, belts of granitic rocks developed as batholiths composed of hundreds of individual plutons. Formation of batholithic volumes of granitic magmas generally appears to require continental settings. Some of the more prominent batholiths in North America are the Coast Range, Boulder-Idaho, Sierra Nevada, and Baja California. The largest are more than 1,500 kilometers long and 200 kilometers wide, and have a composite structure. The Sierra Nevada, for example, is composed of about 200 plutons separated by many smaller plutons, some only a few kilometers wide.

Methods of Study

Granitic rocks are widespread, but the greatest volume is in areas underlain by continental crust in orogenic regions (that is, where mountain building has occurred). Granitic magmas can be derived from a number of sources, notably the melting of continental crust, the melting of subducted oceanic crust or mantle, and differentiation. The problem facing the geologist is to decide on the importance of these sources in relation to their tectonic environment. To accomplish this objective, both petrographic and geochemical information is used.

The standard method of mineral identification and study of textural features and crystal relationships is by the use of rock thin-sections. A thin-section is an oriented wafer-thin portion of rock 0.03 millimeter in thickness that is mounted on a glass slide. The rock thin-section shows mineral content, abundance and association, grain size, structure, and texture. The thin-section also provides a permanent record of a given rock that may be filed for future reference.

Thin-sections are studied with the use of a petrographic microscope, which is a modification of the conventional compound microscope commonly used in laboratories. The modifications that render the petrographic microscope suitable to study the optical behavior of transparent crystalline substances are a rotating stage, an upper polarizer or lower polarizer, and a Bertrand lens, used to observe light patterns formed on the upper surface of the objective lens. With a magnification that ranges from about 30 to 500 times, the petrographic microscope allows one to examine the optical behavior of transparent crystalline substances or, in this case, crystals that make up granitic rocks.

The study of granitic rocks may be greatly facilitated by various staining techniques of both hand specimens and thin-sections. Staining is employed occasionally to distinguish potassium feldspar from plagioclase and quartz, the three main mineral constituents of granitic rocks. A flat surface on the rock is produced by sawing and then polishing. The rock surface is etched using hydrofluoric acid. This step is followed by a water rinse and immersion in a solution of sodium cobaltinitrate. The potassium feldspars will then turn bright yellow. After rinsing with water and covering the surface with rhodizonate reagent, the plagioclase becomes brick red in color. Staining techniques are available for other minerals, including certain accessory minerals such as cordierite, anorthoclase, and feldspathoids.

Measurement of the relative amounts of various mineral components of a rock is called modal analysis. The relative area occupied by the individual minerals is estimated or measured on a flat surface (on a flat-sawed surface, on a flat outcrop surface, or in thin-section) and then related to the relative volume. Caution must be used, because the relative area occupied by any mineral species on a particular planal surface is not always equal to the modal (volume) percentage of that mineral on the rock mass.

When a rock specimen is crushed to a homogeneous powder and chemically analyzed, the bulk chemical composition of the rock is derived. Chemical analyses are normally expressed as oxides of the respective elements, which reflects the overwhelming abundance of oxygen. Analysis of granitic rocks shows them to be typically rich in silica potassium and sodium, with lesser amounts of basic oxides such as magnesium, iron, and calcium oxides. Magnesium, iron, and calcium oxides are present in higher abundance in basalts, which contain plagioclase feldspar, pyroxene, and olivine.

Isotopes such as strontium, oxygen, and lead can also be used as tools in evaluating granitic magma sources such as a mantle origin and crustal melting of certain rock types. Some accessory minerals reflect the trace element content of the magma and thus the possible nature of their source. Some, such as garnet or topaz, are products of contamination, while others, such as andalusite, magnetite, or limenite, are products of late hydrothermal alteration. Much research is focused on chemical tracers. Tracers help distinguish the source region of a granitic magma, such as lower-crustal igneous or sedimentary rock or mantle material.

Context

Granitic rocks have been used as dimension stones for many years. Dimension stones are blocks of rock with roughly even surfaces of specified shape and size used for the foundation and facing of expensive buildings. When crushed, granitic rocks can be used as aggregate in the cement and lime industry. In addition to these uses, granitic rocks are valued because of their geographic association with gold. Gold ores are found in close proximity to the contacts of the granitic bodies within both the granitic rocks and surrounding rocks.

Pegmatites can also be very valuable. Simple pegmatites are exploited for large volumes of quartz and feldspar, used in the glass and ceramic industries. Complex pegmatites can also be a source of gem minerals, including tourmaline, beryl, topaz, and chrysoberyl. In spite of varied mineral composition, relatively small size, and unpredictable occurrence, pegmatites constitute the world's main source of high-grade feldspar, electrical-grade mica, certain metals (including beryllium, lithium, niobium, and tantalum), and some piezoelectric quartz.

Bibliography

Carmichael, Ian S., Francis J. Turner, and John Verhoogen. *Igneous Petrology*. New York: McGraw-Hill, 1974. A well-known reference presenting both the mineralogical and geochemical diversity of granitic rocks and their respective geologic settings. Chapter 2, "Classification and Variety of Igneous Rocks," and chapter 12, "Rocks of Continental Plutonic Provinces," are particularly recommended.

Hutchison, Charles S. *Laboratory Handbook of Petrographic Techniques*. New York: John Wiley & Sons, 1974. Stresses the practical aspects of laboratory and petrographic methods and techniques.

McKee, Edwin H., and James E. Conrad. "A Tale of Ten Plutons—Revisited: Age of Granitic Rocks in the White Mountains, California." *Geological Society of America Bulletin* 108, no. 12 (December, 1996): 1515-1518.

Phillips, William Revell. *Mineral Optics: Principles and Techniques*. San Francisco: W. H. Freeman, 1971. A standard textbook discussing mineral optic theory and the petrographic microscope and its use in the study of minerals and rocks.

Smith, David G., ed. *The Cambridge Encyclopedia of Earth Sciences*. New York: Crown, 1981. Chapter 5, "Earth Materials: Minerals and Rocks," gives a good discussion of the processes involved in the formation of granitic rocks and description of the mineral and chemical composition of granitic rocks. A well-illustrated and carefully indexed reference volume.

Stephen M. Testa

Cross-References

Continental Crust, 78; Igneous Rock Bodies, 307; Igneous Rocks, 315; Magmas, 383; Minerals: Physical Properties, 415.

Hot Spots and Volcanic Island Chains

Crustal hot spots and volcanic island chains are geologic features that result from the mechanisms associated with plate tectonics. A few hot spots are found on the continents, but most are associated with oceanic plates. Both types of hot spots are related to rising magma plumes. Hot spots are generally isolated near plate centers and remain relatively stationary. Volcanic island chains are formed along plate margins and result from the sinking and melting of oceanic plates.

Field of study: Tectonics

Principal terms

ASTHENOSPHERE: a layer of the earth's mantle (about 50-200 kilometers beneath the surface) where the shock waves of earthquakes travel at reduced speeds, probably because of its low rigidity

CONVERGENT PLATE BOUNDARY: the boundary between two plates that are moving toward each other, which may result in island arc development or volcanic arcs on land

DEEP-OCEAN TRENCH: an elongate depression in the sea floor produced by bending oceanic crust during subduction; examples include the Mariana, Puerto Rico, and Aleutian trenches

DIVERGENT PLATE BOUNDARY: the boundary where two plates move apart, resulting in the upwelling of magma from the mantle; the Mid-Atlantic Ridge is a good example

HOT SPOT: a concentration of heat in the mantle, capable of producing magma that, under certain circumstances, can extrude onto the earth's surface

ISLAND ARC: a curved chain of volcanic islands generally located a few hundred kilometers from a trench where active subduction of one oceanic plate under another is occurring

MAGMA: a body of molten rock material found at depth and capable of intrusion and extrusion; igneous rocks are derived from this material through the process of solidification

MANTLE: the 2,885-kilometer-thick portion of the earth's interior located beneath the crust; it is believed to consist of ultramafic material

PLATE TECTONICS: the theory which proposes that the earth's outer layer consists of a series of individual plates that interact and thereby produce earthquakes, volcanoes, and mountain building and recycle the crust

PLUME: a generally isolated convection cell believed to carry heat and mantle material from lower levels up to the crust, resulting in hot spots at the surface

SUBDUCTION ZONE: an area where oceanic crustal plates plunge into the mantle along a convergent zone; examples include Japan and the Aleutian Islands

Summary

In the late 1960's, a theory emerged which suggested that all continents were once joined as a supercontinent that later broke apart. This theory had been debated for decades but finally achieved acceptance on the basis of information by the Deep Sea Drilling Project (begun in 1969). The theory of plate tectonics envisioned that the earth's crust consists of continental and oceanic components that are mobile and "float" upon a semiliquid mantle. The element of mobility offers various explanations for mountain building, volcanism, and earthquake activity. More significant, the theory enabled scientists to see how most geological processes fit into a single unifying mechanism that shapes the earth.

The big question of what caused plate movement, however, had yet to be answered. One theory proposed that convection cells existed in the upper portion of the mantle. In this process, hot, less dense, semisolid material rises up toward the crust, where lower temperature and pressure conditions exist. Contact with the solid crust results in both an upward stress and lateral movement (deflection) of this material. Occasionally, the crust fractures and volcanism results. The lateral movement of the convection cell drags the crust away from the volcanic centers, thus giving the appearance of continental drift if continents are present. The Mid-Atlantic Ridge is an excellent example of a spreading ridge pushing plates apart, and it typifies the convection cell process.

One of the more interesting aspects of plate tectonics involves the occurrence of hot spots. A hot spot is a relatively small region (a few hundred kilometers in diameter at most) of higher than average temperatures that produces melting in the mantle and volcanic activity above. Hot spots occur under both oceanic and continental plates, but they are more common under oceanic plates. Most hot spots are found along spreading ridges, such as those found in Iceland and the Azores. A few do exist at the centers of plates, including the hot spot of the Hawaiian Islands. Although quite rare, hot spots do occur on land, including the hot spot at Yellowstone National Park.

The origin and nature of hot spots are still something of a mystery to scientists. Hot spots undoubtedly have their origin within the earth's mantle and extrude magma up through the crust. It is believed that a hot spot originates beneath the convective flow zone of the mantle and is localized at a much deeper position. This situation could cause the rise of less dense material upward along cylindrical channels called plumes. Plumes can be relatively isolated and produce a single volcanic structure when melting takes place at shallower depths and magma extrudes to the surface. This process is in contrast to the more elongate upwelling associated with spreading axes such as the Mid-Atlantic Ridge. Hot spots and plumes may be the mechanism that initiates plate spreading through development of new plate boundaries.

Over the last several years, more than one hundred hot spots have been identified as having been active during the past 10 million years. This suggests that hot mantle material rises and spreads out laterally in the asthenosphere from perhaps twenty major thermal plumes. These plumes in turn give rise to hot spots with associated volcanic activity. Such hot spots appear to be stable because they occupy the same positions relative to each other as the overlying mobile crust passes over. This stability became quite apparent when sea-floor mapping of the Pacific Ocean basin revealed a linear chain of volcanic structures stretching from the Hawaiian Islands to the Aleutian trench southwest of Alaska. Most of these are submerged seamounts, while others such as Midway Island break the surface. Collectively, these islands and seamounts form the Hawaiian-Emperor chain.

When rocks from the Hawaiian-Emperor chain were dated by the potassium-argon method, an unusual sequence of dates emerged. The Suike seamount, positioned near the Aleutian trench, proved to be 65 million years old, while the island of Hawaii at the opposite end of the chain was less than 1 million years old. The implication was clear: During its continual movement, the overlying Pacific plate was passing over a stationary hot spot that was extruding magma to the surface and thereby building up large volcanic structures. As the plate carried an older lava pile away, a new one would arise to take its place over the hot spot. In this way, the entire chain was built up over time.

As the Hawaiian Islands typify an oceanic hot spot setting, Yellowstone National Park is an example of a continental setting. The basic mechanism for each type of hot spot is similar. The generation of a basaltic magma through partial melting of mantle material is responsible for the volcanism both at Yellowstone and at Hawaii. The mantle at both places can be found at relatively shallow depths, which permits the partial melting of materials with low melting points. Unlike the crust at Hawaii, where there is no significant variation in crustal composition, the crust under Yellowstone is composed of material higher in silica content, easily subjected to partial melting. Magma thus formed is of a much lower density and is capable of rapid upward movement. The eventual volcanism was of a greater explosive nature than that of Hawaii, and in fact did create the giant explosion caldera that is Yellowstone today.

Although the "rising magma plume" explanation for hot spot development is an attractive theory, there are notable objections. One problem is that hot spots remain stationary for extremely long periods of time (more than 50 million years). Recent studies have shown some movement (a few millimeters per year) that resembles the motion of a swaying palm tree. This motion, however, seems rather insignificant in relation to the larger plate tectonics picture. Seismic studies have not strongly supported the existence of mantle plumes but merely offer mild evidence suggestive of their presence.

Presently, the best evidence for the existence of deep mantle plumes comes from chemical comparisons of the lavas extruded. The volcanism of most oceanic hot spots like Hawaii tends to be of an alkali basaltic nature (rich in sodium-potassium), while that of the oceanic spreading ridges is mostly tholeiitic (with little or no olivine present). This fact supports widely separated sources for the respective magmas and tends to isolate the hot spot phenomenon from the more common volcanism characteristic of plate boundaries. Scientists presently cannot offer a conclusive explanation for the existence of hot spots or how they fit into the overall mechanism of plate tectonics.

In contrast to the hot spot phenomenon, much is known about volcanic island chains and their associated plate boundaries. One plate descends beneath the other; as it penetrates into the asthenosphere, partial melting occurs, with low-melting-temperature materials forming an upwelling magma. Quite often, these magmas break through the crust and form a chain of small volcanic islands arranged in an arc pattern relative to the nearby trench. These islands are usually located a few hundred kilometers behind the trench and are collectively called an island arc system. The Aleutian, Mariana, and Tonga islands are excellent examples.

The geological situation that produces a volcanic island chain always occurs at plate boundaries where two oceanic plates are converging. This area is also referred to as a subduction zone, as one of the plates is pushed under the margin of the other. The immediate result is the formation of a trench, and the movement of the descending plate generates an inclined zone of seismic activity. This movement can be seen on seismic reflection profiles because the solid material must reach a considerable depth (more than 50 kilometers) before it melts and remains intact for a long period of time. The descending plate itself is composed of three basic layers: unconsolidated sediment, lithified sediment, and basalt. The top layer of unconsolidated sediment is scraped off by the overriding plate and piles up to form an accretionary prism. Both the lithified sediment and the basaltic layers are eventually remelted and give rise to the volcanism that will create the island arc.

This satellite view of Java shows a chain of cloud-capped volcanoes marking the island's spine. *(National Aeronautics and Space Administration)*

The material that eventually is remelted in the asthenosphere consists of relatively cold, wet oceanic crust. Water, which was originally absorbed during the long journey from the spreading ridge to the trench, is driven off during various stages of increasing temperature initially derived from metamorphic reactions. As the water-rich melt rises into the surrounding mantle, it acts as a flux that lowers the melting temperature sufficiently to produce magma distinctive to subduction zones. Andesite is the principal rock formed, with various other forms of silica-rich rock also present. Even in the case of remelting of the basaltic layer of the plate, differentiation (separation by density) and remelting of digested crustal material would also produce these silica-enriched magmas.

Given sufficient time and continued volcanic activity, a rather large and geologically mature landmass may result from an island arc beginning. Volcanism, coupled with the buoyancy of the intrusive rock being emplaced within the crust below, will gradually increase the surface area and elevation of the island arc. As the island arc grows, greater erosion will increase the amount of sediment being accumulated offshore. Through various transportation means (such as currents and mudslides), some of these sediments will reach the trench and will be subjected to metamorphism and even be melted by the compressional forces of the converging plates. In this way, a geologically mature island arc will result that has all the familiar forms of volcanic and intrusive igneous, sedimentary, and folded metamorphic rocks present. Good examples of a mature island arc system are the Alaska Peninsula, the Philippines, and Japan.

A continental equivalent of an island arc system exists. Here, an oceanic plate is being subducted beneath continental crust, resulting in a trench lying in front of the margin and a series of andesitic volcanoes present a few hundred kilometers inland. This situation would also produce deformation of the continental margin into folded mountain belts. The resulting metamorphism and partial melting of the descending plate would produce high-grade metamorphic rocks, along with granitic intrusions on the overlying plate. A fine example is the Andean mountain chain and is referred to as an Andean-type arc. In some cases, such arcs may be covered by seas and may somewhat resemble island arcs. Examples can be seen in some of the eastern-end Aleutian islands and in Sumatra, Indonesia.

Methods of Study

The methods employed to study crustal hot spots and volcanic island chains involve both direct and indirect approaches. Both geologic features are part of the plate tectonics theory and must be explained in the light of that concept. Hot spots and volcanic island chains therefore must be examined from the perspective of a localized occurrence as well as a global situation. Only when the two are blended will a detailed picture emerge showing their close relationship in both origin and formation processes.

The search for a better understanding of hot spots and volcanic island chains began shortly after World War II, when sonar, a radio pulse used to detect submarines, was used to measure the topography of the sea floor. The time it takes for a sonar signal to strike an underwater object and reflect back to its source is measured, and the rate of return plots the highs and lows of the sea floor. By using this technique, scientists were able to measure accurately ocean depths and discover the geological features that constitute the sea floor. Long mountain chains, deep trenches, and flat abyssal plains were revealed, giving the sea floor a geological characteristic as complex and interesting as the continental surfaces. It was in this manner that the Hawaiian Islands were revealed to be a related series of volcanic seamounts that stretch several thousand kilometers across the Pacific Ocean basin. As a result of such discoveries, the theory of continental drift was given new consideration.

Once the nature of the sea floor was characterized, other techniques were employed to investigate the nature of the crust that underlies the deep ocean sediment. The earliest studies employed a combination of sediment penetrators to collect samples and limited drilling along the continental shelves. These first studies—partly a result of offshore oil exploration—began to reveal a somewhat different view of the sea floor. Later studies made by the research vessel *Glomar Challenger* provided long core sections of both the sediment accumulation on the sea floor and the underlying rock. Samples were taken from numerous worldwide locations, and comparisons were made that revealed a much clearer picture of the global implications of plate tectonics.

The deep-ocean core samples provided scientists with direct evidence of the underlying oceanic crust. From this evidence, age determinations were made by using radioisotope dating techniques. These studies revealed that rocks found near the oceanic volcanic ridges were much younger than those found in front of the deep oceanic trenches. A progression of younger to older rocks was clearly evident and provided direct evidence for plate movement. It became apparent that new crust originated at the spreading centers along the submerged ridges as active volcanoes, with the older crust being pushed aside toward the distant trenches. Collision with a thicker plate would cause the older, thinner crust to be diverted downward, thus producing the trench. As the older rock plunges deeper into the mantle, it melts and separates into denser (sinking) and less dense (rising) material. It is this less dense material that rises and later forms the volcanic island chains behind the trenches.

An additional research technique that provides evidence to support this scenario is seismic reflection profiling. In this method, shock waves are generated from a controlled explosion. Because different shock waves travel at different speeds and behave differently as they pass through rocks of varying densities, these aspects can be used to determine the nature of the underlying rock. Primary shock waves, or P waves, travel faster than secondary shock waves, or S waves: P waves can penetrate liquids that S waves cannot. Each has its specific velocity as it passes through rocks of varying densities, and this difference is used to recognize specific rock types. In some situations, the shock waves can bounce off a particular rock material and thus give an image of its shape and thickness. This image is seen clearly where a trench and island arc are present. The seismic-reflection images reveal a relatively thin oceanic plate being diverted underneath the thicker plate's edge, upon which the island arc chain forms from the volcanic activity generated by the melting plate.

Perhaps the final piece of evidence needed to understand hot spots and volcanic island chains comes from the study of the chemistry and petrology of the rocks. Their chemistry and mineral compositions can reveal much about the depth, temperature, and pressure conditions that existed when the original magma formed. Experimental studies dealing with the melting and the recrystallization of the rocks can provide much evidence for the rocks' origin and formative processes. When rocks from widely separated locations are studied, the resulting data reveal the rather complex yet uniform mechanisms governing plate tectonics.

Context

The understanding of the mechanisms of crustal hot spots and volcanic island chains not only is important for scientific purposes but also has a bearing on everyday life. Associated with both geological occurrences are earthquakes and volcanic activity. Hundreds of millions of people live in areas threatened by intense earthquake activity. Every year, several earthquakes of large magnitude rock the earth and cause the loss of life and inflict great property damage. Although scientists have collected much information on seismic activity, there is no reliable means for predicting earthquakes.

In contrast, hot spots and volcanic island chains occur in areas of sustained earthquake activity, and much can be learned from their constant monitoring. Where the frequency of earthquake activity is high, patterns can develop over relatively short periods of time, and theories can be tested. The large number of earthquakes that would normally occur over millions of years happen relatively quickly in these active regions. The greater the amount of available data, the better the chances for making more accurate predictions. If earthquake predictions can eventually be made with a high degree of accuracy for these localized areas, then perhaps the process could be extended to include all earthquake-prone regions. Loss of life could then be greatly reduced. Perhaps even a method of earthquake prevention could be developed from a better understanding of what occurs at hot spots and volcanic island chains.

Volcanoes do not present the same danger that an earthquake does, as their impending eruptions are usually quite predictable. Loss of life caused by volcanic activity therefore is generally much less. What is gained from volcanic studies is a greater awareness of the physical processes responsible for continuous crustal renewal and the development of new landmasses. The study of volcanoes has also provided evidence to show how volcanic activity can influence the weather and global climatic patterns. Most of the volcanic activity that occurs over hot spots such as Hawaii is rather gentle and quite predictable. It slowly continues to increase the surface area of the islands and replenish the soil with new mineral nutrients. Volcanism that occurs along plate boundaries and forms island arc chains can be very explosive and has produced some of the world's greatest and most destructive eruptions. The hot spot that currently lies beneath Yellowstone National Park once erupted with a tremendous explosive force and covered most of North America with ash. It represented a plate boundary at that time.

Perhaps the greatest benefit from the study of hot spots and volcanic island chains is the realization of how the earth works over relatively short periods of time. A volcanic eruption or an earthquake demonstrates the enormous energy contained within the earth and how that energy is released. Over the great span of time, this energy has shifted plates and led to the development and breakup of a supercontinent. From scientific observations, the earth can be seen as a living organism as it recycles its surface materials. In the eruption of a Hawaiian volcano, the past and the future come together in a demonstration of earth processes. Such events are dramatic reminders that the earth is a dynamic, evolving planet.

Bibliography

Burke, Kevin C., and J. Tuzo Wilson. "Hot Spots on the Earth's Surface." *Scientific American* 235 (February, 1976): 46-57. An excellent discussion on the worldwide occurrences of crustal hot spots. Emphasizes the mechanisms of hot spot generation, along with worldwide geographical distribution. The relationship between hot spots and plate tectonics is dealt with quite well. Excellent maps and diagrams. Best suited for advanced high school and undergraduate college students.

Condie, K. C. *Plate Tectonics and Crustal Evolution.* Elmsford, N.Y.: Pergamon Press, 1976. This book presents a good basic summary of the development of continental and oceanic crust, based on the principles of plate tectonics and continental drift. Good background information that is necessary to understand hot spot development and its various occurrences. Well illustrated. Suitable in part for high school through graduate levels.

Cox, A., and R. Hart. *Plate Tectonics: How It Works.* Palo Alto, Calif.: Blackwell Scientific, 1986. Incorporates most of the latest developments in the theory of plate tectonics and provides a good overview as well as an in-depth look into the mechanisms behind plate tectonics. Included are discussions on hot spots and related volcanic island arcs. Best suited for advanced high school and undergraduate college students.

Heezen, Bruce C., and Ian D. MacGregor. "The Evolution of the Pacific." In *Readings from Scientific American Ocean Science.* San Francisco: W. H. Freeman, 1977. This article presents results from the Deep Sea Drilling Project, which deals with the evolutionary history of the Pacific sea floor as a result of plate tectonic processes. Good information is presented on the techniques of data collection and the acquired results. Discussions of trench and seamount development, along with the varying ages of deep-sea sediment, give a good picture of the true nature of the Pacific Ocean basin. Suitable for advanced high school and undergraduate college students.

Macdonald, G. A., A. T. Abbott, and F. L. Peterson. *Volcanoes in the Sea: The Geology of the Hawaiian Islands.* 2d ed. Honolulu: University of Hawaii Press, 1983. This book deals specifically with the geology of the Hawaiian Islands, which represents one of the earth's most active hot spots. Among the many topics covered are past and present volcanism on Hawaii and how it fits into the overall picture of global plate tectonics. A detailed reference to the many geological activities involved in the formation and continued evolution of the Hawaiian Islands as a stationary hot spot center. Best suited for undergraduate college students.

Van Andel, T. H. *New Views on an Old Planet: Continental Drift and the History of the Earth.* New York: Cambridge University Press, 1985. This book provides a general and popularized history of the earth and the evolution of scientific ideas. Provides a good discussion of continental drift and plate tectonics to aid the reader in better understanding those concepts. Presents a rather different approach not found in standard references on these topics. Recommended for college students.

Wilson, J. Tuzo, ed. *Continents Adrift and Continents Aground.* San Francisco: W. H. Freeman, 1976. This work is a collection of articles on plate tectonics which originally appeared in the journal *Scientific American.* The articles provide both an overview and specifics of plate tectonics in a way that is understandable for most readers with a basic interest in earth sciences. Well illustrated through its use of maps, charts, and diagrams. Best suited for high school and undergraduate college levels.

Wyllie, Peter J. *The Way the Earth Works.* New York: John Wiley & Sons, 1976. An often-cited work that is highly readable and well illustrated, dealing with various tectonic subjects. Coverage of continental drift and plate tectonics is extensive and should provide the reader with the background knowledge needed to understand crustal hot spots. Best suited for advanced high school and undergraduate college students.

Paul P. Sipiera

Cross-References

Island Arcs, 332; Magmas, 383; Ocean Basins, 456; The Ocean Ridge System, 464; Plate Tectonics, 505.

Igneous Rock Bodies

The geometry of the bodies in which igneous rocks are found can provide useful information about the physical characteristics of the magmas or lavas that produced them and about the stress field that was active in the region at the time of their formation.

Field of study: Petrology

Principal terms

ASH: solid particles from an erupting volcano, usually formed as ejected molten material cools during its flight through the atmosphere

CONCORDANT: having sides (contacts) that are nearly parallel to the layering in the country rock

COUNTRY ROCK: the rock into which magma is injected to form an intrusion

DIKE: a discordant sheet intrusion

DISCORDANT: having sides (contacts) that are at a substantial angle to the layering in the country rock

LACCOLITH: a concordant, nearly horizontal intrusion that has lifted the country rock above it into a dome-shaped geometry

LAVA: molten rock at or above the surface of the earth

MAGMA: molten rock, still beneath the surface of the earth

SHEET INTRUSION: an intrusion that is tabular, or sheetlike, in shape

SILL: a concordant sheet intrusion

Summary

Igneous rocks form as molten rock cools and solidifies. If this happens at or above the surface of the earth—during a volcanic eruption, for example—the extrusive rock bodies that result are lava flows and ash fall deposits. If it happens beneath the surface of the earth, the resulting rock bodies are called intrusions or plutons. The igneous rocks involved can have a wide range of compositions and textures, and hence a great variety of names. Common ones include basalt and granite; less common ones are monchiquite and lamprophyre. Igneous petrologists study the rock itself, seeking information about how, where, and from what it melted and how it might have been modified prior to its final solidification. Their emphasis is on chemical and mineralogical composition, and they employ phase diagrams and chemical reactions in their work. Information obtained from the body itself—its size, shape, and orientation—is also worth studying. Extrusive rock bodies can also be studied to learn about former volcanoes. A volcano's location, some of the characteristics of its lava, and even details of the topography at the time it was active can be reconstructed long after the volcano has stopped erupting, and even after it has eroded completely away.

Lava flows develop as fluid lava moves downhill, but lava is not a simple fluid. Its behavior is complex. How easily a fluid flows (how "thick" it is) is a function of its strength and its viscosity. A sensitive function of composition, lava viscosities range from those similar to motor oil to those more like asphalt. Within a single flow, strength and viscosity will vary with temperature, gas content, and flow rate—all of which change as the lava is moving. At the edges of flows, cooler lava may form natural levees, confining the flow. Within the flow itself, tubes may develop beneath the surface through which quickly flowing lava can move with little cooling. At the surface, the loss of heat and gases can result in a nearly solid rind that deforms by cracking and breaking.

An ash fall deposit is thickest near the site of the eruption. Its areal extent will be influenced by the winds prevailing at the time of the eruption and the height reached by the ash before its descent. Being blown by the wind, ash may accumulate beneath areas of stagnant air. If still sufficiently hot, particles of ash may weld together as they settle, producing a hard rock that resists the forces of erosion.

Molten rock that has not reached the surface is called magma. As complex as lava, magma moves from regions of higher pressure to areas where the pressure is lower, rather than flowing downhill. During this process, magma may wedge apart solid rock and intrude into the crack it produces. In this way, magma can forge its own subsurface path for many kilometers. Eventually, this forging may bring it to the surface, where an eruption will ensue. An eruption will permit much greater flow through the crack. Flow will be easiest and fastest where the crack is widest. The flowing magma erodes the walls of the crack; the walls near the fastest-flowing magma will erode most quickly. In the narrow parts of the crack, flow will be much less, the magma will be harder to deform, and it will cool and solidify. Thus, the magma conduit, which initially was a fluid-filled crack, transforms during the eruption into a nearly cylindrical form. At any point in this process, the flow of magma may be interrupted, permitting the magma within the conduit system to solidify. The form taken by this frozen magma will be inherited from the walls of the conduits, just as a casting takes its form from a mold.

There are many names for different intrusive forms, and their classification is not very systematic. If the sides of the intrusion are generally parallel to the layering in the rock into which it intrudes (the country rock), it is said to be concordant. If its sides are at a significant angle to that layering, the intrusion is said to be discordant. The names for irregularly shaped discordant intrusions are based on size. Batholiths are huge; stocks are much smaller. Perhaps more important than the sizes which define them, though, are the differences in depths of formation. Batholiths form at great depths,

Volcanic eruptions provide scientists with excellent opportunities to study the features of igneous rock bodies. *(PhotoDisc)*

309

where the temperatures are so high that the country rock behaves in a fairly ductile fashion, whereas many stocks are thought to be remains of the subsurface cylindrical conduits which fed volcanoes.

Commonly, igneous intrusions will form tabular bodies, with one dimension much smaller than the other two. These sheet intrusions are called dikes if they are discordant and sills if they are concordant. These names originated in coal mines, where they were useful in mapping the underground workings. It is now understood that the orientation of a sheet intrusion is more likely to be controlled by the direction of least compression than by attitude of existing layers.

A simple experiment can help to show how the orientation of sheet intrusions is determined by the direction of least compression in the region. The experiment involves a cube made up of several rectangular wooden blocks. In between some of the blocks are uninflated balloons that are all attached to the same source of compressed air, turned off at the beginning of the experiment. Some balloons are horizontal, some are vertical in north-south planes, and some are vertical in east-west planes. If there is pressure on the blocks in an east-west direction, pressure in a north-south direction, and still more pressure surging downward, the balloons that inflate when the compressed air is gradually turned on will resist the pressure the most. Similarly, when molten rock is forced into the crust, it forms dikes or sills depending on which direction is under the least compression. At the surface of the earth, the direction of least compression must be either horizontal or vertical. Sheet intrusions that form near the surface, then, will be emplaced in either a vertical or a horizontal position. Because sedimentary strata are commonly nearly horizontal, a horizontal sheet intruded into such strata is concordant, while a vertical one is discordant. As magma pushes the sides of a sheet intrusion away from each other, stress is concentrated at the tip of the crack. When this stress exceeds the strength of the rock, the rock splits apart and the crack grows longer. The longer crack fills with fluid under pressure, forcing the sides farther apart. Now, with even greater leverage acting on it, the crack tip fails again. The process is repeated as long as there is enough fluid pressure.

The country rock ahead of a sheet intrusion fails in extension. The fracture produced is much like a joint and as such will have many of the surface decorations common to joints. One of these decorations, called plumose structure or twist hackle, occurs when the crack breaks into a number of smaller cracks, each slightly rotated from the parent crack to produce a series of parallel offset cracks. These are called *en echelon* cracks. If they become filled with magma, a set of *en echelon* dikes will result. Continued propagation of the crack and filling with magma will cause the individual segments to coalesce. The resulting intrusion will be a dike with a series of matching offsets along its edges, showing where individual *en echelon* segments existed earlier. Such offsets can be used to infer the direction in which the crack initially grew.

310

As the area over which the pressure acts increases, the force produced by
that pressure increases also. This is the principle behind pneumatic jacks. In
the case of a horizontal sill not too far beneath the surface of the earth, the
force pushing up increases with area, while the resisting forces increase with
the perimeter. Because area grows faster than does perimeter, there may
come a time when the rock lying above the sill will be domed up, producing
a laccolith: a concordant intrusion, with a flat floor and a domed roof. Many
other names have been assigned to intrusions with different geometries.
Lopoliths, sphenoliths, bysmaliths, phacoliths, ductoliths, harpoliths, ak-
moliths, and even cactoliths have been described. Yet, it is not clear whether
such nomenclature has useful general applicability, and the use of these
terms has fallen out of favor. Indeed, even the distinction between dikes and
sills is often no longer made, the phrase "sheet intrusions" being preferred
by many geologists.

Methods of Study

Lava flows and ash fall deposits are usually studied in the field. If the
outcrops are favorable, detailed maps are constructed, often on top of aerial
photographs. These might show variations in thickness, locations, and orientations of lava tubes, levees, and surface fracture patterns. From the
patterns that emerge and by radiometric dating of the deposits, geologists
seek to understand the history of the volcanoes involved. Which vents were
active when? How large was the largest flow? The smallest? What is the
average rate at which lava has been produced by this volcano?

Even if the deposits are buried under thousands of feet of sedimentary
rock, they may still be susceptible to study where they are exposed in canyons
or encountered in wells. Most of the detail will have been obliterated, but
trends in the variation of their thickness can still be used to indicate where
the eruptions that produced them occurred. This study has been conducted
on rocks hundreds of millions of years old, enhancing the understanding of
the earth's tectonic history.

Intrusions are also studied in the field. Measurements are made of the
size and shape of the body being studied. Because much of the intrusion is
usually buried and much of it has been removed by erosion, reconstructing
the original shape may not be easy. Yet, even a thin slice of data through an
intrusion can provide important information. When examined carefully,
many dikes show an asymmetric cross section, something like a long, thin
teardrop. This cross section has been interpreted to indicate gradients in
magma pressure or regional stresses active during emplacement. The edges
of sheet intrusions frequently display offsets and occasionally grooves, which
are thought to indicate the direction in which the crack occupied by the
intrusion initially grew. Such information can be utilized to help reconstruct
the three-dimensional form of the intrusion and to define the sequence of
events that produced it. The transition from sill to laccolith represents a
particularly opportune situation. Estimates of the depth of overburden,

strength, and resistance to bending of the overlying rock, and the pressure and mechanical behavior of the magma may all be derived if sufficient data are available for a substantial number of laccoliths and sills in an area.

Dikes often occur in groups, called dike swarms. By mapping such swarms, geologists may find that they reveal systematic patterns that can be interpreted as images of the stress field in the region. The next step is to use computers and the theory of elasticity to analyze such stress fields, and then try to find causes for the stresses discovered. Because the igneous rock making up the intrusions can often be dated, a stress history of the region may be developed.

To understand field exposures, geologists may construct models representing the intrusions they wish to study. Some of these models are physical, with motor oil or petroleum jelly acting as magma and being intruded under pressure into gelatin, plaster, or clay. Conditions can be controlled, and the experiments can be halted to make measurements, photographs, and even films of the process. Another way to model intrusions is on a computer. Using a system of equations derived from the theory of elasticity, geologists can predict the stresses and displacements in the vicinity of an intrusion with a given shape, containing a magma with known properties, and surrounded by rock with known elastic behavior subjected to a known stress field. By letting each of these variables change, their effects can be studied independently.

Laboratory simulations can suggest which field measurements are most likely to be significant in learning about the conditions at the time of intrusion. Armed with an intuition developed in the lab, the field geologist is better prepared to understand field exposures. Field data, in turn, often pose dilemmas that yield to analysis in the laboratory.

Context

A volcanic eruption is certainly one of nature's most spectacular displays. For reasons ranging from fascination and curiosity to hazard mitigation and self-preservation, people have studied volcanoes for centuries from many different perspectives. One rewarding approach has been to examine the exhumed remains of the deposits and underground plumbing systems of former volcanoes. The sizes, shapes, orientations, and distribution of such igneous rock bodies have been interpreted to reveal important physical characteristics of the molten rocks that produced them, as well as the conditions prevailing at the time of their formation.

By studying the composition and textures of igneous rocks, scientists are able to decipher much about the rocks from which they melted and some of the details about how the melt changed between its initial formation and its final emplacement. Such study, however, usually does not provide much information on the physical processes involved. Some of this information can be obtained by a study of the rock body as a whole, instead of the rock of which it is composed. From the shape of a dike or a sill (a sheet intrusion), it may be possible to determine the direction in which the intrusion initially grew and something about the pressure of the molten rock. From the dimensions of another type of intrusion called a laccolith, it may be possible to make estimates of the fluid pressure of the magma and the original depth of emplacement.

Because they are useful pressure gauges, igneous rock bodies can also provide information about the stresses active at the time of their emplacement. Patterns produced by swarms of dikes have been interpreted in terms of the horizontal stresses acting throughout the region. Such information has direct bearing on questions concerning the mechanism of mountain building, plate tectonics, and the history of the earth in general.

Finally, owing to the fact that igneous rock is frequently more resistant to weathering and erosion than is the sedimentary rock into which it may have been emplaced, igneous rock bodies often form impressive scenic features. Devils Postpile in California, the Henry Mountains of Utah, Ship Rock in New Mexico, and the Palisades Sill across the Hudson from New York City are some of the scenic attractions produced by igenous rock bodies. Appreciation of these and similar features is enhanced when one understands something about their formation.

Bibliography

Billings, Marland P. *Structural Geology.* 3d ed. Englewood Cliffs, N.J.: Prentice-Hall, 1972. Chapter 16, "Intrusive Igneous Rocks," presents classic descriptions of many different intrusive rock bodies. There is little emphasis on the mechanisms of formation or on the interpretation of the bodies described, but each is defined and most are illustrated. Descriptive, with little prior knowledge assumed, this book is suitable for the general reader.

Decker, Robert, and Barbara Decker. *Volcanoes.* San Francisco: W. H. Freeman, 1981. Suitable for the general reader, this book provides useful background for the understanding of igneous rock bodies. Of particular interest are chapter 10, "Roots of Volcanoes," and chapter 8, "Lava, Ash, and Bombs."

Hargraves, R. B., ed. *Physics of Magmatic Processes.* Princeton, N.J.: Princeton University Press, 1980. This book is a fairly technical review of much of the work in progress at the time of its publication. The first seventeen pages of chapter 6, "The Fracture Mechanisms of Magma Transport from the Mantle to the Surface," by Herbert R. Shaw, are very informative and easily understood by the general reader. Most readers will also learn much from chapter 7, "Aspects of Magma Transport," by Frank J. Spera, which describes some of the complexities of magma behavior.

Hunt, Charles B., Paul Averitt, and Ralph L. Miller. *Geology and Geography of the Henry Mountain Region, Utah.* U.S. Geological Survey Professional Paper 228. Washington, D.C.: Government Printing Office, 1953. A classic paper, this 234-page report describes the laccoliths of this region in detail. Profusely illustrated and accompanied by maps and cross sections, the information presented will provide the reader with an excellent sense of what these laccoliths look like and the map patterns they produce. There is little, however, in the way of a convincing discussion concerning how they formed or what inferences may be drawn from their locations or geometries. Suitable for the general reader.

Johnson, Arvid M. *Physical Processes in Geology*. San Francisco: Freeman, Cooper, 1970. This book has strongly influenced the way igneous rock bodies have been studied ever since its publication. Building on an approach developed to study the formation of laccoliths, Johnson leads the reader through discussions of elasticity and viscosity and proceeds to apply some of the results obtained to problems of dike intrusion and the flow of magma. Although differential and integral calculus are needed to follow the derivations, much of the general approach can be appreciated by those with less mathematical training. Suitable for the technically oriented college student.

Suppe, John. *Principles of Structural Geology*. Englewood Cliffs, N.J.: Prentice-Hall, 1985. Chapter 7, "Intrusive and Extrusive Structures," shows how the emphasis in the study of igneous rock bodies has shifted since publication of Billings' book. In this text, there is little in the way of categorizing bodies in terms of their shapes and much more on their interpretation in terms of the physical conditions existing during the time of their emplacement. Although mechanically sound, the math used is not intimidating. Suitable for the general reader.

Williams, Howel, and Alexander R. McBirney. *Volcanology*. San Francisco: Freeman, Cooper, 1979. This book is a general text on volcanoes. Chapter 5, "Lava Flows," and chapter 6, "Airfall and Intrusive Pyroclastic Deposits," are most relevant to igneous rock bodies. The writing seems to put an unnecessary stress on terminology. Suitable for college students with some background in geology.

Otto H. Muller

Cross-References

Igneous Rocks, 315; Joints, 340; Magmas, 383; Pyroclastic Rocks, 512; Stress and Strain, 607; Volcanoes: Recent Eruptions, 675; Volcanoes: Types of Eruption, 685.

Igneous Rocks

> The classification of igneous rocks depends on their texture and composition. "Texture" refers to the grain size of the constituent minerals of the rock and depends on how slowly or quickly a magma cooled to form the igneous rock. "Composition" refers to both chemical and mineralogical features.

Field of study: Petrology

Principal terms

COLOR INDEX: the percentage by volume of dark minerals in a rock; it is used for quick identification of rocks

EXTRUSIVE ROCK: a fine-grained, or glassy, rock which was formed from a magma that cooled on the surface of the earth

FELSIC: characterized by a light-colored mineral such as feldspar or quartz, or a light-colored rock dominantly composed of such minerals

INTRUSIVE ROCK: an igneous rock which was formed from a magma that cooled below the surface of the earth; it is commonly coarse-grained

MAFIC: characterized by a dark-colored mineral such as olivine, pyroxene, amphibole, or biotite, or a rock composed of such minerals

MAGMA: a molten rock material largely composed of silicate ions

MINERAL: a natural substance with a definite chemical composition and an ordered internal arrangement of atoms

MODE: the type and amount of minerals actually observed in a rock

NORM: the type and amount of minerals derived by a set of calculations from a chemical analysis of a rock

SILICA: silicon dioxide, or quartz

Summary

Igneous rocks are classified according to their texture and composition. With regard to texture, there are two groups: fine-grained, or glassy, and coarse-grained. With regard to composition, the rocks can be classed chemically or mineralogically. Chemically, they can be grouped according to their silicon dioxide content or their combination of oxides; mineralogically, they can be grouped according to mode, norm, or color index.

Rocks that are formed from magmas that cool beneath the surface of the earth are called "intrusives." Depending on grain size, intrusives are grouped into plutonics, hypabyssal rocks, and pegmatites. Plutonics are coarse-grained igneous rocks that are formed from magmas that cooled slowly deep beneath the surface of the earth. Hypabyssal rocks are fine-grained, because of a magma's comparatively rapid rate of cooling at a shallow depth. A variety of intrusive which commonly occurs as a tabular body that cuts across other rocks is a pegmatite. A pegmatite is a light-colored and extremely coarse-grained rock.

Igneous rocks which are formed from a magma that has extruded onto the surface of the earth are known as either "extrusives" or "volcanics." In such environments, the magmas cool rapidly, so the minerals are fine-grained. In some cases, a magma may cool so rapidly that minerals may not form at all. Instead, the magmas turn into rocks with no ordered arrangement of atoms, and as such with no crystalline structure. Such an amorphous and unorganized arrangement of atoms characterizes glass. A rock composed of glass is called "obsidian" and generally is black.

Extrusives are subdivided into "pyroclastics" and "lavas." Pyroclastics are formed by the forcible extrusion of highly gas-charged magmas. The top portion of such magma is full of gas and frothy, much like the top part of beer that is poured into a glass. The melt that surrounds the spherical gas bubbles cools into glass shards, yielding a "pumice" rock while, or soon after, the magma is extruded. Glass shards, pumice fragments, early formed crystals, fragmented magma, and rocks that surround the conduit may be ejected into the atmosphere or flow laterally. An accumulation of such fireborne fragmentary material is called a "pyroclastic rock." Commonly, the fragments are flattened and welded together because of heat and overlying weight, which produces a "welded tuff." When very fine-grained glass shards which were explosively ejected into the atmosphere rain down and settle on the ground, they produce volcanic ash, or ash fall tuff. Occasionally, a rising magma may encounter a mass of water which creates steam and causes a powerful phreatomagmatic eruption. In such a case, a ring-shaped cloud of steam with minor solid particles moves swiftly away from a vertical eruption column above the vent. Rocks that are formed from the ring of cloud are called "base surge deposits" and show many structures similar to sedimentary rocks.

Lavas are volcanic rocks formed from magmas that flow gently. "Lava" is also a name for the magma itself. Such magmas do not contain much gas. The gas that may be derived from dewatering vegetation and soil over which the magma moves ascends to the top surface of the cooling magma body, so the top part of a lava rock may contain air bubbles, or vesicles. When there are abundant vesicles in the rock, the lava is called a "scoria." In some lavas, the vesicles may be filled by secondary minerals which precipitated from groundwater. Such rocks are called "amygdaloidal lavas." Some magmas erupt and flow beneath a body of water. The ensuing lavas have a texture that resembles pillows and are therefore called "pillow lavas." Depending on their composition, they are pillow basalts (fine-grained, dark rocks) or spilites (greenish rocks). In the latter, the basalt is altered by interaction with the surrounding body of water.

Simple Classification of Igneous Rocks

	Felsic	*Intermediate*	*Mafic*	*Ultramafic*
Extrusive (volcanic)	rhyolite	dacite/andesite	basalt	
Intrusive (plutonic)	granite	tonalite/diorite	gabbro	peridotite

← increasing silica

increasing iron and magnesium →

Some igneous rocks are porphyritic, in that a few minerals in the rock are coarser-grained than the majority of the minerals. The coarse-grained minerals formed first and probably at greater depths than the fine-grained ones; commonly, the term "porphyritic" is used as a prefix or suffix of a rock name, as in porphyritic lava or lava porphyry.

Although a few igneous rocks are identified only by their textural characteristics, most are classified on the basis of their composition. The composition of an igneous rock depends on the magma from which the rock was derived. A magma is itself derived by the melting of some portion of the top part of the earth. This part of the earth is composed primarily of the following eight major elements: oxygen, silicon, aluminum, iron, calcium, sodium, potassium, and magnesium. Oxygen and silicon together comprise about 75 percent of the top part of the earth. Thus, the combination of silicon and oxygen dominates rocks which are formed from this part. When igneous rocks are chemically analyzed, the results are dominated by seven different oxides, including silicon dioxide. These oxides are used to classify igneous rocks.

The most common parameter used for classifying igneous rocks is the silicon dioxide, or silica, content of a rock. Based on its weight percent of silica, a rock could be placed in one of the following four major classes of igneous rocks: felsic (greater than 66 percent), intermediate (52 to 66 percent), mafic (45 to 52 percent), and ultramafic (less than 45 percent). These four major groups of igneous rocks can also be identified by the percentage of dark minerals, or the color index (CI): Felsics have a CI of less than 30 percent; intermediates, 30 to 60 percent; mafics, 60 to 90 percent; and ultramafics, more than 90 percent. An uncommon group of rocks called "carbonatites" are dominantly composed of carbonate minerals instead of silica minerals.

The combination of the seven oxides in igneous rocks is used as another chemical classification scheme. The exact definition of the types of igneous rock distinguished by such a classification scheme requires using graphs that show the variation of the selected oxides. Such a graph is augmented by trace element and isotope ratios of elements in the rocks. Essentially, however, the method classes igneous rocks into alkaline, peralkaline, and subalkaline rocks. The subalkaline rocks are divisible into tholeiitic and calc-alkali rocks. Generally, the tholeiitic rocks have mafic to ultramafic rock associations as a predominant component. In calc-alkali rocks, felsic to intermediate rock associations are most abundant. Alkaline and subalkaline rocks can form at any place where a deep-seated magma source, or a hot spot, penetrates the surface of the earth, such as in Hawaii. More commonly, though, igneous rock associations are formed at plate boundaries, zones that separate adjacent shifting plates.

Scientists believe that the top part of the earth is compartmentalized into tectonic plates that are in constant motion with respect to one another. The plate boundaries are the geologic settings in which igneous rock associations are formed. Adjacent tectonic plates may move toward each other and converge at their mutual boundary. This geologic setting is called a convergent plate boundary, and the igneous rocks that form in such a setting are in the calc-alkali class. The group of igneous rocks found in the Andes of South America or the mountains of the Caribbean islands are examples of such igneous rocks. In another geologic setting, adjacent plates move away from each other and a rising molten rock material pushes upward at their mutual boundary. This mutual boundary is called a divergent plate boundary, and there are two types of igneous rock association that can form in such a geologic setting. In continental areas, where the plate motion has not succeeded in tearing apart the continent and where there is a continental bulge with a depression at its center (called a rift valley), the igneous rocks at the boundary are of the alkaline class. The East African Rift Valley is an example of a place where alkaline rocks are found. In contrast, where plate motion has succeeded in tearing apart a continent and has created an intervening oceanic floor, the igneous rocks are of the tholeiitic class, particularly the mafic-ultramafic associations.

Another compositional parameter used for classifying igneous rocks is mineralogic composition. The mineral composition is dependent on the chemical composition of the parental magma. A magma is commonly called a silicate melt, because it is composed mainly of silicon and oxygen that are combined to form the silicate ion. This ion has a four-sided configuration, with oxygens at the corners and silicon at the center. The charge on the corner oxygens of a silica tetrahedron is such that the oxygen atoms can bond with equal strength to adjacent tetrahedra. Thus, silica tetrahedra have the capacity to link, or to polymerize. It is the polymerization of the silica tetrahedra that explains why a magma has more resistance to flow, or is more viscous, than liquid water. The degree to which a magma is polymerized determines its viscosity. Felsic magmas are more viscous, so they form volcanic domes; the less viscous, mafic magmas form lava flows. Silicates are a class of minerals that contain the silicate ion in their composition. The

TYPICAL ORE MINERALS ASSOCIATED WITH IGNEOUS ROCKS

Rock Type	Mineral	Metal or Other Commodity Obtained
Felsic—Intermediate		
Granite	Feldspar	Porcelain, scouring powder
	Native gold	Gold
Pegmatite	Cassiterite	Tin
	Beryl	Berryllium, gemstones (emerald; aquamarine
	Tourmaline	Gemstone
	Spodumene	Lithium
	Lepidolite	Lithium
	Scheelite	Tungsten
	Rutile	Titanium
	Apatite	Phosphorus
	Samarskite	Uranium, niobium, tantalium, rare-earth elements
	Columbite, Tantalite	Niobium, tantalium, used in electronics
	Thorianite	Uranium, thorium
	Uraninite	Uranium
	Amazonite (microcline feldspar)	Gemstone
	Rose quartz	Gemstone
	Topaz	Gemstone
	Sphene (titanite)	Titanium, gemstone
	Muscovite mica	Electrical insulation
	Zircon	Zirconium
Rhyolite	Chalcopyrite	Cooper
	Molybdenite	Molybdenum
Mafic—Ultramafic		
Gabbro and Anorthosite	Ilmenite	Titanium
	Labradorite (plagioclase feldspar)	Gemstone
	Chalcopyrite	Copper
	Bornite	Copper
	Pentlandite	Nickel
Peridotite	Chromite	Chromium
	Native plantinum	Platinum
	Sperrylite	Platinum
	Serpentine	Nickel (from weathered soils)

resulting silicates are divisible into groups based on how many of their oxygen atoms are shared among adjacent tetrahedra.

Felsic minerals are composed of oxygen, silicon, aluminum, sodium, potassium, and calcium. They are silicates in which all oxygen atoms are shared among adjacent tetrahedra. Felsic minerals include quartz (silicon dioxide), which is commonly colorless to transparent; feldspars, such as potassic feldspar, sodic feldspar, and calcic feldspar; and feldspathoids, such as nepheline and leucite. In general, the felsic minerals are light-colored, because they may not contain transition metals. Transition metals are elements whose atomic numbers range from 21 to 30 and whose outer electrons can be excited by light to the same energy levels that correspond to the colors of visible light. Thus, felsic igneous rocks that are dominantly composed of felsic minerals are light-colored, or leucocratic, with a color index of less than 30 percent. A felsic rock is formed from the cooling of a felsic magma, a highly polymerized and viscous magma. Typically, a magma is such that it either does or does not have sufficient silicon dioxide to form quartz; in the latter case, feldspathoids form. In other words, feldspathoids and quartz are incompatible minerals and are not to be found in the same igneous rock.

In contrast to felsic minerals, mafic minerals are dominantly composed of magnesium and iron. These minerals have a structure in which a maximum of three oxygens are shared among adjacent silica tetrahedra; the unshared oxygens are bonded with magnesium and iron. The mafic minerals include olivine, green and equidimensional; pyroxene, dark green or black, and stout; amphibole, black and elongate; and biotite, black and tabular. In most cases, the mafic minerals are dark-colored as a result of the presence of transition metals such as iron. A mafic igneous rock which is dominantly composed of these silicates is therefore also dark-colored.

The visible mineralogic composition, or the mode, is used to classify a rock if it consists of minerals of sizes that can be identified and amounts that can be counted; however, some rocks are very fine-grained, or they contain glass, so their mineral composition cannot be estimated even after magnification under a microscope. In such cases, the rocks are analyzed for their major element content, and the minerals that would have formed had the magma cooled slowly are obtained by calculation. The minerals and their amounts determined by calculation give the norm of a rock, and the individual minerals are normative minerals. The normative mineral composition can be used for classification purposes in the same way as the modal composition.

Methods of Study

Division into coarse-grained intrusives and fine-grained, or glassy, extrusives is done by visual examination of the texture of the igneous rock. Further classification of these is undertaken by determining the mode or the norm of the rock and applying a classification scheme.

The mode may be determined by trained persons who can identify the minerals in a rock with the naked eye or after magnifying the minerals ten to thirty times with a magnifying glass, such as a pocket-sized hand lens or a binocular microscope. Better mineral identification is done by scientists after a rock is cut to a small size, mounted on glass, and then ground until a very thin section (0.03 millimeter) of the rock, capable of transmitting light, is prepared. The thin sections are placed on a stage of a transmitted light polarizing microscope. A lens below the stage polarizes light by allowing the transmission only of light which vibrates in one direction—for example, east to west. A lens above the stage allows the passage of light vibrating in a north-south direction. When glass is placed on the stage and the lower polarizer is inserted across the transmission of light, the color of the glass is seen. When the upper polarizer is also inserted, the glass appears dark, because no light is transmitted. The properties of most minerals with respect to the transmission of light (optical properties) are different from those of glass.

Other accessories are used in addition to the polarizing lenses in order to characterize accurately the optical properties of minerals. Magnification by the microscope permits better determination of the physical properties of minerals, such as their shape and cleavage. Cleavage refers to a set of planes along which minerals break, and it is related to the ordered internal arrangement of the atoms of a mineral. The felsic minerals are used to classify the non-ultramafic rocks. A rock sample or thin section of such a rock may be inserted into a dye that stains one of the felsics, usually the alkali feldspars. This method simplifies the counting of minerals for classification.

The norm of a rock is calculated from the major elements obtained by chemical analysis. Modern methods of chemical analysis use emission or absorption of radiant energy that is unique to each atom and therefore can lead to the identification of that atom. There are four such methods: X-ray fluorescence (XRF) spectrometry, electron microprobe analysis (EMP) analysis, instrumental neutron activation analysis (INAA), and atomic absorption spectrometry (AAS).

In the XRF method, radiant energy is focused on a sample of a rock. This energy removes electrons from the lower electron-energy levels of the atoms. The place of the removed electrons may be taken by other electrons, which fall from higher energy levels by emitting radiant energy. Depending on the energy levels from which the electrons fall, X rays of different energy are released from one element, and the spectrum of these is unique to that element. The X rays from many elements are guided to a crystal, which diffracts and disperses them for easy detection by an X-ray counter. The counter triggers electronic signals, which may be either recorded digitally and interfaced to computers or displayed on strip-chart recorders as separate peaks. The peak positions on the chart correspond to the elements in the rock sample. The peak heights are related to the concentration of the elements, the exact amount being determined by a comparison to that of a standard element admixed to the rock sample.

In EMP analysis, electrons are used as the radiant energy source, and this energy is focused on a rock sample that is polished and coated with carbon. It is a nondestructive analytical method that is better suited for determining the compositional variation within a mineral. In INAA, the sample is bombarded by neutrons supplied by either a nuclear reactor or an accelerator. Atoms in a sample acquire neutrons and become heavy and unstable (radioactive) isotopes. The samples are removed and placed in a chamber, where the emitted radiant energy (gamma rays) resulting from the decay of the artificially induced radioactive isotopes is guided to a crystal which separates the sample rays for easy detection by a recording device. This method is suitable for detecting the type and concentration of even those elements which may be found only in trace amounts.

In AAS, one uses the radiant energy emitted from a known element in a cathoderay tube to detect the presence of an element in a sample by noting whether the energy is reduced and by how much. A rock sample is dissolved in a solution and then heated until chemical bonds are broken and the solution contains individual atoms. When radiant energy is made to pass through the atomized sample, the energy will be absorbed by the sample if there are elements in the sample that are identical to those in the cathode. This absorption of energy by the sample causes a reduction in the detection of the source radiant energy, and the reduction corresponds to the amount of the element in the sample.

The XRF method is by far the most widely used technique for analyzing chemical composition. The other techniques have their application in the determination of trace elements that are used to facilitate the classification of igneous rocks by igneous rock association, such as alkaline or subalkaline.

Context

Igneous rocks are one of the three major types of rock, the others being sedimentary and metamorphic rocks. In some regions, igneous rocks are the only types of rock that are found. Inhabitants of such regions have reason to know the varieties of igneous rocks; moreover, many tourists visit sites of igneous rocks, such as the granites of Stone Mountain, Georgia, and of Yosemite National Park, California, and marvel at the beauty of the rocks and the landscape in which they are found.

Humans throughout the centuries have used igneous rocks for a variety of purposes. Prehistoric man used, and certain tribesmen of the Third World still use, obsidian to fashion axes for cutting softer material, and arrowheads or spearheads for hunting. Many kinds of igneous rock, especially plutonic rocks, are used as grinding stones to produce flour for preparing bread and other foods. Monuments are often made of igneous rock. Typically, granite and granodiorite, which are light-colored rocks, are used for tombstones. Other igneous rocks, such as diorite and gabbro, which are black, have been selected for dark-colored monuments, such as the Vietnam Memorial in Washington, D.C. In some cultures, obelisks and places of worship are hewn from solid igneous plutons, irrespective of their hardness. The three-thousand-year-old obelisks at Axum, Ethiopia, were carved out of granodiorite plutons. Many other types of building can be completely or partly constructed with igneous rock. Architects and masons choose the rock's color, type, and dimensions. In general, granite is the igneous rock most often used for expensive structures.

The classification of igneous rocks has its most useful application in the search for economic metal deposits. Ultramafic and mafic rocks are associated with chromium, platinum, palladium, iridium, osmium, rhodium, ruthium, nickel, iron, titanium, and gold deposits. Intermediate and felsic rocks are associated with copper, molybdenum, tin, tungsten, silver, lead, and zinc deposits. Alkaline rocks, particularly the kimberlites, are associated with diamonds. These different igneous associations are found in different geologic environments. Knowing the geologic environment, then, helps both in classifying the rocks and in anticipating the types of metallic deposit that could be sought in such regions.

Bibliography

Best, Myron G. *Igneous and Metamorphic Petrology*. New York: W. H. Freeman, 1982. An easy-to-read book that provides classification as well as description of igneous rocks.

Carmichael, Ian S., Francis J. Turner, and Joan Verhoogen. *Igneous Petrology*. New York: McGraw-Hill, 1974. An excellent book on the classification and description of igneous rocks for college-level students.

Hutchison, Charles S. *Laboratory Handbook of Petrographic Techniques*. New York: John Wiley & Sons, 1974. This book provides a discussion of technique used in the laboratory for identifying minerals and rocks.

Klein, Cornelis, and C. S. Hurlbut, Jr. *Manual of Mineralogy*. 20th ed. New York: John Wiley & Sons, 1985. A useful book for study of minerals. Chapter 12 provides a succinct treatment of igneous rocks.

Prinz, Martin, et al., eds. *Simon & Schuster's Guide to Rocks and Minerals*. New York: Simon & Schuster, 1978. Rocks and minerals are described in this easy-to-read book. Illustrated in color.

Zussman, J., ed. *Physical Methods in Determinative Mineralogy*. 2d ed. London: Academic Press, 1978. An excellent collection of articles on microscopy and instrumental analytical methods. Suitable for advanced students and research geologists.

Habte Giorgis Churnet

Cross-References

Igneous Rock Bodies, 307; Magmas, 383; Minerals: Physical Properties, 415; Pyroclastic Rocks, 512; Rocks: Physical Properties, 550.

Intraplate Volcanism

Intraplate volcanoes are found at localized hot spots that are formed in response to hot columns of mantle material that originate near the earth's core and rise toward the surface. These hot spots initiate volcanoes and volcanic chains located far from the edges of the earth's tectonic plates. Long lines of extinct intraplate volcanoes provide a way of tracing the motion of the earth's lithospheric plates.

Field of study: Volcanology

Principal terms

ASTHENOSPHERE: the shell within the earth beneath the lithosphere that yields plastically to stress; found at depths ranging from 5 kilometers beneath spreading ridges to 100 kilometers under continents

CONVECTION CELLS: movement of materials as a result of density differences caused by heating

CONVERGENT PLATE MARGIN: a plate margin where two crustal plates are moving toward each other

DIVERGENT PLATE MARGIN: a plate margin where two crustal plates are moving away from each other

LITHOSPHERE: the outer, cooler portion of the earth that behaves more or less rigidly; the lithosphere consists of the crust and the uppermost part of the mantle

MANTLE: the thick layer of the interior of the earth between the crust and core

MANTLE PLUME: a localized column of hot mantle material originating near the core-mantle boundary and rising to the base of the crust

PLATE TECTONICS: a theory that interrelates the internal and external processes of the earth and involves the interaction of lithospheric plates

RIFT ZONE: a long, broad trough bounded by faults on each side, formed by stretching the crust; common at divergent plate margins

SUBDUCTION ZONE: the linear belt at the surface of the earth where one lithospheric plate sinks beneath another; common at convergent plate margins

Summary

Planets may lose their internal heat by a number of different mechanisms that transfer heat from the hot, deep interior to the surface. The least efficient form of heat transfer is lithospheric conduction, which involves simple heat migration through solid rock. Rock is a very poor conductor of heat, and the conductive heat loss is extremely slow. Another method of heat transfer in planets is by mantle plumes, in which a rather narrow column of hot material, from deep within the mantle near the core-mantle boundary, rises toward the surface. Plumes of this kind are the most effective means of heat transfer from the deeper regions of a planet. In contrast, mantle convection, in which the plastic interior of a planet undergoes slow convective motion, is the mechanism by which the upper mantle loses heat. These convection cells, within the upper mantle, pull and push on the more brittle lithosphere of the earth. This movement of the mantle is responsible for breaking the lithosphere of the planet into mobile plates separated by narrow zones of upwelling mantle material and zones of downwelling mantle.

The earth's surface is broken into approximately twelve large (and several small) mobile, rigid, lithospheric plates that jostle about. The plates consist of 100- to 200-kilometer-thick slabs of relatively rigid crustal and upper mantle material overlying lower plastic materials of the mantle. Convection cells in the mantle, below the lithosphere, drive plate motions at the surface of the earth. Ninety-five percent of the earth's volcanoes occur along the boundaries of these plates. At the spreading centers between divergent plates, large rift valleys are formed as plate margins are pulled apart. Melting in the mantle beneath the rift zone, caused by the upwelling hot mantle material and the decrease in pressure beneath the rift, gives rise to nearly continuous igneous activity, as basaltic magmas intrude along fractures in the crust to form new ocean crust and erupt onto the floor of the rift valley.

Where plates converge, one side of the convergent zone is pulled down and subducted into the mantle. The zone of convergence is marked by a deep oceanic trench flanked by arcuate chains of islands. As the plate segment descends, it is heated by the natural increase in temperature at greater depth in the earth. When the descending plate reaches a depth where the temperature is high enough to begin melting, part of the plate is assimilated into the mantle, and part is melted to form magma that rises toward the surface. Those magmas that make it to the surface form volcanoes aligned along the junction created by colliding plates. These active volcanic chains form island arcs such as the Japanese Islands, the Philippines, and the Aleutian Islands in the Pacific Ocean and mountain chains such as the Andes of South America and the Cascade Range of the northwestern United States.

In contrast to volcanoes on the edge of plates, some very prominent volcanoes are found within the interior of plates far from any spreading ridge or subduction zone. These intraplate volcanoes are driven by mantle plumes. Mantle plumes are concentrated regions of hot upwelling mantle material that are independent of the plate boundaries and the convection cells associated with them. These plumes give rise to local hot spots at the surface, which are the homes of the earth's largest volcanoes. Mantle plumes are important components of the convection system by which the earth loses heat. Intraplate volcanism shows up as chains of volcanic islands, as seamounts on the ocean floor, and as flood basalts or isolated volcanic fields on the continents. More than one hundred mantle plumes have been active in the past ten million years. They occur in both oceanic and continental settings. The largest concentration of intraplate volcanoes is on the African plate.

Probably the best known of the intraplate volcanic systems is the Hawaiian Island-Emperor Seamount Chain situated within the Pacific Ocean plate. The volcanic activity of Hawaii builds very large volcanoes with gentle sloping sides; these are known as "shield volcanoes." Many of the older volcanoes of this chain have sunk below sea level and are now identified as "seamounts." Other intraplate volcanic trails are common in the Pacific, including the Line Island-Tumoto Chain and the Marshall-Gilbert-Ellice-Samoan Island Chain. Most of the intraplate volcanoes of the Atlantic and Indian Oceans are marked by seamounts and broad ridges or rises. The New England-Corner Seamounts and the Great Meteor Tablemount are sites of past intraplate volcanoes in the North Atlantic. Tristan Da Cuna of the South Atlantic, Reunion Island in the Indian Ocean, and the Tasmantid Seamount east of Tasmania are active volcanic centers associated with chains of extinct volcanoes on the ocean floor.

On the North American continent, intraplate volcanoes include the Snake River-Yellowstone Volcanic Province and the Columbia Plateau flood basalts. The one-billion-year-old basalts of the Lake Superior region represent a mantle plume that was responsible for erupting in excess of 400,000 cubic kilometers of flood basalts. In Africa, the isolated mountain ranges in the central Sahara, such as the Tibetsi, Hoggar, and Jebel Marra, are major intraplate volcanic complexes in which some of the volcanoes reach more than three thousand meters in height. In contrast to the Pacific plate plumes, the African plumes have remained active in the same locality for thirty million years.

The source of a mantle plume at the core-mantle boundary remains stationary over hundreds of millions of years. Therefore, as the earth's lithospheric plates move over the plumes, the plumes generate hot spot trails or chains of volcanic islands. Unlike the volcanic chains of the island arcs, which are active all along the arc, hot spot trails are only active above the mantle plume. The volcano located over a mantle plume becomes extinct as the moving plate moves it away from the source, and a new volcano forms over the plume. Thus, the volcanic rocks of the volcanic chain become progressively older with increasing distance from the hot spot. The age of the volcanic rocks of the chain can be used to trace absolute plate motions.

Hot spot trails are less obvious on the continents. A mantle plume that rises beneath the much thicker and insulative continental crust may give rise to magma that cools within the crust, known as a "pluton," or it may cause a regional bulge in the crust with accompanying stretching, thinning, and fracturing of the crust. The release of pressure at depth gives rise to great quantities of magma, which rises to pour out of great long fissures onto the surface as lavas. If eruption rates are high, fluid lava does not build a cone but spreads across the surface to bury the landscape under a series of massive lava flows. These "flood basalts" are the most extensive type of volcanic activity on the earth's continents. The one- to two-kilometer-thick Columbia River Plateau basalt covering an area of 200,000 square kilometers of southern Washington, northern Oregon, and western Idaho is an example of a flood basalt that erupted on the North American continent from six million to seventeen million years ago. This same mantle plume may be responsible for the basalts of the adjacent Snake River Plains and for the current activity of the Yellowstone region. The lavas of Yellowstone National Park, Wyoming, are rhyolitic rather than basaltic; nevertheless, Yellowstone is thought to represent a hot spot that incorporates much more continental rock in its magma than that of the Columbia River Plateau basalts.

By definition, intraplate volcanism is not associated with plate boundaries, yet the two cannot be completely separated. Some mantle plumes may fortuitously reach the surface at a plate boundary, and some hot spots under continents may evolve through time to initiate rifting. Many onetime continental, intraplate volcanic centers sitting above a mantle plume continue to develop along a characteristic pattern. The expansion of the crust over the mantle plume stretches the lithosphere to the point that normal faulting creates a down-dropped trough or rift zone. Magma, erupting along the faults, flows onto the floor of the rifts. Eventually, the faulted arch is thinned to the point that its crest founders. The floor of the rift drops below sea level and floods with water to become an embryonic ocean. As the rift widens with time, the once adjacent pieces of the continent are slowly pushed apart by the intrusion of more magma between them, and a new divergent plate margin is born. Often such rift zones begin with three radially divergent rift zones as a "triple junction." Such a triple junction can be seen where the Red Sea, the Gulf of Aden, and the East African Rift zone meet. Two of the arms of the junction (the Red Sea and the Gulf of Aden) have opened to form narrow arms of the ocean, while the third arm (East African Rift zone) has not yet collapsed enough for flooding. The East African Rift zone, which is the site of recent and current volcanic activity, was the site of earlier extensive flood basalt eruptions. In this way, a mantle plume that gave rise to intraplate volcanism has continued to develop until it initiated a spreading center that now defines the edges of tectonic plates.

Vast basalt accumulations, such as the Parana flood basalts on the eastern edge of South America and the Etendeka flood basalts of the western edge of Africa, began as continental, intraplate volcanic centers, but they are associated with the separation of those continents by the opening of the southern Atlantic Ocean 120 million to 130 million years ago. In like manner, the Greenland flood basalts and the Hebridean basalts of northern Scotland and Ireland once formed a single volcanic province that was split by the separation of the northern Atlantic Ocean sixty million years ago.

Iceland is the only part of the northern Mid-Atlantic Ridge to grow above sea level. Because of the large volume of basalt, it is suggested that Iceland is a hot spot that coincides with the ridge, and its volcanoes are of dual origin-mantle plume and spreading center. The diverse composition of lavas—an apparent mixture of ridge and hot spot compositions—would seem to support the theory of the composite origin of Iceland. Another hot spot that coincides with spreading centers includes the Rio Grande Rise and the Walvis Ridge of the south Atlantic Ocean, which has left symmetrical trails across the ocean floor on either side of the spreading center at Tristan da Cunha.

Intraplate volcanism is the source of the geysers and other thermal phenomena of the Yellowstone region. *(National Park Service)*

The great size of the volcanoes on Mars is attributed to the presence of mantle plumes and the absence of plate tectonics. Olympus, the largest volcano in the solar system at nearly 600 kilometers in diameter and 20 kilometers in height, owes its great size to continuous outpouring of lava at one spot. This long-term coupling of a magma source at depth has allowed Martian volcanoes to build over long periods of time, as long as the mantle plume was active—probably hundreds of millions of years. In contrast, the largest volcanoes on Earth, the Hawaiian Islands, do not grow to such large dimensions because they are cut off from their source as the Pacific plate continues to move over the hot spot. Although early models attempted to fit Venus to a plate tectonic theory, there is no strong evidence for large-scale crustal spreading or subduction. The distribution and shape of large volcanic features on the surface of the planet are more consistent with mantle plume-driven volcanism similar to that of Mars and with intraplate volcanism on Earth, as are the active volcanoes on Io, a satellite of Jupiter.

Methods of Study

The recognition of plate tectonics as a major explanation for the large-scale structure of the earth's surface largely grew out of attempts to understand the morphology and structure of the ocean floor. The ocean floor was the last and greatest unexplored terrain of the earth's surface, primarily because of its inaccessibility. With the advent of technology that would allow the depths of the ocean to be sounded following World War II, the physiography and geology of the ocean floor was exposed. What was once assumed to be a rather featureless abyssal plain was found to contain myriad features, including fracture zones, midoceanic mountain ranges, and seamounts. As the theory of plate tectonics grew, it became necessary to explain the isolated volcanoes, groups of islands and island chains that are not explained by plate tectonics and yet are obvious in the Pacific Ocean. J. Tuzo Wilson, an important figure in the development of the plate tectonics theory, first advanced the concept of hot spots in the early 1960's, and the theory was amplified by Jason Morgan in the early 1970's. Radiometric age-dating of the islands and seamounts of the Hawaiian-Emperor Chain was not completed until the 1980's.

Mineralogical and chemical studies have been employed to compare and contrast basalts of the oceanic islands to those of the midoceanic ridges and to continental volcanic materials. The plume-related volcanic rocks are much more diverse in composition than the spreading-center basalts, which suggests that the mantle is quite heterogeneous. Experimental crystallization studies and mantle xenoliths provide information regarding the composition of the mantle at the source where the magmas form. Phase studies indicate that intraplate magmas form at greater depths (60 to 100 kilometers) than those of the midoceanic ridges (20 to 40 kilometers). Geophysical data indicate that little or no melting actually occurs in the lithosphere immediately beneath the oceanic island volcanoes;

this supports the idea that magmas actually come from deeper in the mantle. The development of techniques for seismic tomography, which allows a three-dimensional imaging of the location of seismic-wave velocity throughout the mantle, is providing a clearer picture of the distribution of temperatures within the mantle. It is now possible to image upwelling currents and descending slabs within the mantle as three-dimensional models.

Relative positions of the islands and radiometric studies of the ages of basalts on the various islands and seamounts of the Hawaiian-Emperor Chain provide a record of the Pacific plate motion over the last seventy million years. The big island of Hawaii is currently active. Volcanic rocks of Maui are dated as from zero to 1.3 million years old; rocks of Oahu are 2.3 million to 3.3 million years old; and those of Kauai are 3.8 million to 5.6 million years old. Midway Island, 2,400 kilometers to the northwest, is 27.2 million years old. The entire Hawaiian-Emperor Chain is 6,000 kilometers long and more than 70 million years old. The Pacific plate is drifting to the northwest at approximately 8.8 centimeters per year. A bend in the chain 700 kilometers northwest of Midway Island, which marks a change in the direction of plate motion, is dated at 43 million years. Other chains of the Pacific Ocean basin exhibit a progression of ages similar to that of the Hawaiian chain, as well as a similar shift in direction of motion. This date, associated with the bend, correlates to some tectonic events in western North America and to the time that the Indian subcontinent collided with Asia.

Context

An understanding of intraplate volcanism is important to understanding the internal mechanisms that continue to shape the ever-changing surface of the earth. Unlike igneous activity associated with continental margins, intraplate volcanism plays a minor role in the origin of tangible economic resources. The diamond-bearing kimberite pipes of South Africa may be associated with prerift arching over a mantle plume; however, for the most part, intraplate volcanoes are not associated with significant ore deposition. The volcanoes may be important as a local source aggregate and building materials; however, their economic potential is chiefly a function of secondary factors such as location, scenery, and habitat. The chief economic value of the oceanic intraplate volcanoes is tied to their location as way stations in the vast Pacific Ocean. The islands provide a source of fresh water and a habitat for terrestrial plant and animal life. The shallow waters surrounding the islands provide favorable conditions for a diverse marine ecology. Fringing reefs built by coral and other marine organisms in shallow water on the coast of volcanic islands survive as atolls enclosing shallow lagoons after subsidence of extinct volcanoes. Some coral islands rooted on foundered extinct volcanoes become nesting grounds for seabirds. The phosphate-rich guano reacts with limestone to form phosphate rock, a much sought after component used in fertilizer manufacture.

The lava beds of the Columbia River basalts are important groundwater reservoirs in the northwestern United States. Geysers and hot springs of the Yellowstone area are famous for their scenic qualities. Areas of active or recent volcanism are important as potential sources of hydrothermal energy.

Intraplate volcanoes are subject to the same volcanic hazards as any other volcanically active region. Earthquakes, landslides, and lava flows are potential hazards to inhabitants of such areas. Although intraplate volcanoes are less explosive than volcanoes associated with convergent plate margins, volcanic activity on the big island of Hawaii is, nevertheless, responsible for the destruction of roads, homes, businesses, and government buildings. On the positive side, recent eruptions are responsible for adding new land to the island.

Bibliography

Burke, K. C., and J. Tuzo Wilson. "Hot Spots on the Earth's Surface." *Scientific American*, August, 1979. A description of "hot spots" by one of the first proponents of the concept. The authors describe the occurrence and probable mechanisms of hot-spot origin and its relationship to continental rifting.

Decker, R. W., and B. B. Decker. *Mountains of Fire: The Nature of Volcanoes.* New York: Cambridge University Press, 1991. An extremely lucid description of Earth's volcanoes and their relationship to plate tectonics.

Francis, P. W. *Volcanoes.* Harmondsworth, England: Pelican Books, 1976. Written for the general reader, this book provides a clear description of volcanic phenomenon. Makes good use of eyewitness accounts of major volcanic events, ranging from that of Vesuvius in A.D. 79 to more recent activity.

Frankel, Charles. *Volcanoes of the Solar System.* New York: Cambridge University Press, 1996. Describes volcanism on Earth and proceeds to discuss the Moon, Mercury, Mars, Venus, Io, and the icy satellites of Jupiter, Saturn, and Uranus. The text is suitable for the general reader, but it contains enough detail for introductory earth science students.

Tilling, R. I., C. Heliker, and T. L. Wright. *Eruptions of Hawaiian Volcanoes: Past, Present, and Future.* Washington, D.C.: U.S. Geological Survey, 1987. Offers a comprehensive description of the world's most famous and most intensively studied hot spot.

Wessel, Paul. "Sizes and Ages of Seamounts Using Remote Sensing: Implications for Intraplate Volcanism." *Science* 277, no. 5327 (August 8, 1997): 802-806.

René De Hon

Cross-References

Earth's Mantle, 195; Hot Spots and Volcanic Island Chains, 297; The Ocean Ridge System, 464.

Island Arcs

Island arcs are arc-shaped chains of volcanic islands formed by the collision of two oceanic plates. They are the sites of most of the world's explosive volcanic eruptions and large earthquakes. Tsunami and ash clouds generated by these events can affect people around the globe.

Field of study: Tectonics

Principal terms

ANDESITE: a light-colored volcanic rock rich in sodium and calcium feldspar, with some darker minerals

BASALT: a dark-colored igneous rock containing minerals such as feldspar and pyroxene, high in iron and magnesium

BENIOFF ZONE: the dipping zone of earthquake foci found below island arcs, named after Hugo Benioff, the seismologist who first defined it

EARTHQUAKE FOCUS: the region in the earth that marks the starting site of an earthquake

GRANITE: a light-colored igneous rock containing feldspar, quartz, and small amounts of darker minerals

GRAVITY ANOMALIES: differences between observed gravity readings and expected values after accounting for known irregularities

LITHOSPHERE: the rigid outer shell of the earth, composed of a number of plates

SUBDUCTION: the process by which a lithospheric plate containing oceanic crust is pushed under another plate

TSUNAMI: a seismic sea wave generated by vertical movement of the ocean floor, caused by an earthquake or volcanic eruption

Summary

An island arc is a long, arcuate chain of volcanic islands with an ocean on the convex (or outer) side of the arc. Paralleling the arc lies a long, narrow trench with steeply sloping sides that descend far below the normal ocean floor. A map of the world shows that many island arcs occur in and around the Pacific Ocean, such as the Aleutians, Japan, Tonga, Indonesia, New Zealand, and the Marianas. The West Indies are an island arc bordering the Atlantic Ocean. The associated trenches contain the deepest places on earth. The Marianas trench near the island of Guam reaches a maximum depth of 10,924 meters. This is farther below sea level than Mount Everest (at 8,848 meters) is above sea level. These island arc features, though merely topographic ones, demonstrate that island arcs and deep-sea trenches are parts of the same earth structure. There are six basic features common to island arc-trench systems: chains of volcanoes, deep ocean trenches, earthquake belts, a shallow sea behind the island arc, large negative gravity anomalies, and rock deformation in later geologic time. Some features are better displayed in one arc than another, but all are present.

All island arcs consist of an arc-shaped chain of volcanoes and volcanic islands. Many volcanoes are currently active or have been active in the recent geologic past. Some scientists group island arcs into two types: island arcs composed of volcanic islands located on oceanic crust (such as the Aleutians, Kurils, Marianas, and West Indies) and chains of volcanoes on small pieces of continental crust (such as Japan, Indonesia, New Zealand, and the Philippines). The main difference is that the continental-type arc is older, has had a more complex geologic history, and thus represents a later stage in the evolution of island arcs. The volcanoes of both types produce andesitic magma. Andesite is a light-colored, fine-grained igneous rock composed primarily of sodium- and calcium-rich feldspar. In composition, andesite lies midway between quartz-rich granites and iron- and magnesium-rich basalts. The andesitic magma also contains large amounts of gases that cause extremely explosive and destructive eruptions. Examples of island arc volcanic eruptions are Krakatoa, Indonesia, in 1883; Mount Pelée, Martinique, in 1902; and La Soufrière, St. Vincent, in 1902.

On the ocean side of all island arcs lie deep-sea trenches, long, narrow features that parallel the island chains. They have steep, sloping sides extending to great depths. Minor differences occur. Some trenches, such as the Marianas and the Kuril, have V-shaped cross sections and are rock-floored to their bottom. Others, such as the Puerto Rico and southwest Japan trenches, have a flat bottom. Detailed studies have shown that these flat-bottomed trenches are sediment-filled and that the underlying rock floor is also V-shaped.

Island arcs are active seismic regions and the sites of many of the world's largest and deepest earthquakes. The region in which an earthquake begins is called its focus. The foci of earthquakes in arc regions lie along a narrow, well-defined zone that dips from near the trench below the island arc. The number of earthquakes generally decreases with depth, with some foci reaching 600-700 kilometers below sea level. This dipping seismic zone is called the Benioff zone, after the seismologist Hugo Benioff, who first defined it.

Behind the island arc lies a shallow marginal sea; examples are the Sea of Japan, the Philippine Sea, and the Caribbean Sea. Below some marginal seas, the crust is partly continental but becomes oceanic toward the arc. Below other seas, the crust is entirely oceanic. The composition of the oceanic crust beneath the marginal seas is more like andesite than the basalt of the normal ocean floor.

Also common to arc-trench regions are large negative gravity anomalies. Geophysicists have found that the value of gravity over the earth's surface varies by slight amounts. Most of the variations can be accounted for and result from irregularities in altitude and topography. After observed gravity readings are corrected, however, variations called gravity anomalies still remain. These are caused by differences in rock density from place to place below the earth's surface. A negative anomaly shows that a greater volume of lighter (less dense) rocks is present in one area than in surrounding ones. Large negative anomalies are associated with the deep-sea trench and imply the presence of a great volume of low-density rocks at depth.

Deformation of rocks in the recent geologic history of an island arc is common. Some rocks have been folded, others metamorphosed. Areas of local uplift and subsidence are found that may also be related to shallow earthquakes and to faulting. These features are more easily seen and studied on the island arcs located on continental crust, such as Japan or New Zealand. Deformation, however, is present in all island arcs.

Earth scientists have long sought to explain the origin of island arcs. The remains of ancient marine volcanic islands and volcano-derived sediments are found in the core of the present-day Appalachian Mountains. These rocks and large amounts of continental sediments were compressed,

A satellite view of the Hawaiian Islands, an example of an island arc. *(National Aeronautics and Space Administration)*

faulted, and folded to form the ancient Appalachians. Yet, the relation of the volcanic islands to the mountain-building process was unclear. The development of the concept of plate tectonics has provided an explanation.

The outer portion of the earth is composed of a number of rigid lithospheric plates. Driven by forces in the mantle, the plates are pushed and shoved about the face of the earth. A lithospheric plate may contain continental crust, oceanic crust, or (more commonly) both. Island arcs form when the ocean portion of two colliding plates forces one plate under the other. The continuing collision of the plates may eventually result in the creation of a new mountain range, such as the ancient Appalachians. The process in which oceanic lithosphere is pushed under another plate is called subduction. Subduction of the plate causes the geological and geophysical features observed in the island arc system. (It should be noted that oceanic plates can also be subducted beneath continental plate margins. The resulting features are similar to those found in island arcs except that the andesite volcanoes form along the edge of the overriding continental margin. The Andes and Cascade mountains are the island arc equivalent for subduction beneath a continent. Mount St. Helens is an andesite volcano.)

As the two plates converge, one bends and is pushed under the other. The line of initial subduction is marked by a deep ocean trench. Subduction is not a smooth process. Friction between the subducting plate and the overriding plate and between the downgoing plate and the mantle tries to prevent movement. When frictional forces are overcome, an earthquake occurs. The location of earthquake foci outlines the subducting lithospheric plate. As subduction continues, earthquakes occur at greater depths. The lack of earthquakes below 700 kilometers suggests that this is the maximum depth that the plate can reach before it becomes part of the mantle. The downgoing oceanic plate drags along any deep-sea sediments that have been deposited on it or in the trench area. Both the plate and the sediments are heated, primarily by friction and by the surrounding hotter mantle. At about a depth of 100 kilometers, partial melting occurs, giving a magma rich in sodium, calcium, and silica. This magma mixes with the iron- and magnesium-rich mantle, creating a magma less dense than is the surrounding mantle. Forcing its way upward through zones of weakness in the overlying plate, the magma generates andesite volcanoes. The gravity anomaly associated with the trench is caused by the light crustal rocks of the subducting plate being held (or pushed) down by the overriding plate. The increased volume of less dense rocks produces a large negative anomaly.

Also common to island arc systems are recently deformed rocks. The overriding plate does not slip smoothly over the subducting plate. Rocks in the leading edge of the plate are compressed (pushed together), causing faulting, folding, and uplift. Pieces of the subducting plate can be broken off and folded into the island arc. Heat from the mantle causes metamorphism in the overlying rocks. Behind the island arc, shallow faulting caused by tension (pulling apart) creates earthquakes.

The marginal sea between the island arc and the continent is called the back-arc basin. The presence of these basins is not totally understood. Some, such as the Aleutian and the Philippine basins, were formed from pieces of preexisting ocean. Others have features suggesting that they were once continental crust that has been turned into oceanic crust. Still other basins appear to have been created by interarc spreading, like that seen along mid-ocean ridges. The origin of back-arc basins is a topic of active geologic research.

Methods of Study

In studying island arcs, scientists use a wide range of geological and geophysical techniques. The geological methods generally study the accessible portions of island arcs and include mapping and sample collection and analysis. The geophysical methods study the deep features of island arcs using earthquake and explosion seismology, gravity surveys, and heat-flow measurements. Computers aid in the analysis of data and in the generation of island arc models. Earth scientists studying certain aspects or features of island arcs select a combination of tools most appropriate to their region of interest.

Geologic mapping requires direct access to the rocks forming island arcs. The geologist surveys a region, recording the type of rock found and its extent and the orientation of observed faults and folds. Rock samples are collected for later study. These may be supplemented by drilling to sample rocks below the surface. The field data are transferred to a topographic map and, with the aid of aerial photographs, a geologic map is drawn. Aerial and satellite photographs have become increasingly helpful in the mapping of regions covered by vegetation. The traces of faults and the effects of changing bedrock can be reflected in surface features visible from high altitudes.

The rock samples collected are subjected to chemical and mineralogical analyses. The presence of trace elements or certain minerals can provide clues to the source region of a rock's components or to the thermal history of the rock since it was formed. Minerals containing radioactive elements, such as potassium 40 and rubidium 87, can be used to obtain the age of the rock units. Microscopic analysis of the rocks yields information on their thermal and deformational history. Direct collection of rock samples is limited to the exposed portions of island arcs. Dredging is used for sample collection in shallow ocean regions such as the back-arc basins. The use of submersibles, such as the *Alvin*, operated by Woods Hole Oceanographic Institution, has allowed scientists to photograph and collect rocks and other data from the ocean floor far below sea level, enabling them to extend the study of island arcs. These methods, however, do not reach the regions far below the earth's surface.

Indirect methods are used by geophysicists to study the deeper island arc regions. The two most widely used methods are earthquake and explosion seismology. Earthquake seismology studies the seismic waves generated by earthquakes and provides an average velocity structure of the crust and mantle. The distribution of earthquakes in arc regions, particularly Japan, led to the discovery of the dipping seismic zone beneath the arc. Scientists also study plate movements to learn the mechanism causing earthquakes.

Explosion seismology uses seismic waves generated by controlled explosions to study the detailed crustal and lithospheric structure of island arcs. Within the ocean regions, this technique has revealed the steep topography of the deep-sea trenches and the deformed rock layers near the base of the overriding plate.

Measurements of the earth's gravity field can be obtained on land and at sea. The data are corrected for irregularities in altitude and terrain. Remaining differences relate to density variations deep in the crust. The negative anomalies in the trench areas reflect the great thickness of crustal rocks at the plate boundary. In other areas, such as Japan, gravity data suggest a thicker crust or a less dense mantle below the back-arc basin (Sea of Japan) than below the Pacific Ocean.

Heat-flow measurements also reflect regional features. On the average, heat flow is the same over both continental and oceanic regions as a result of a deep, common source. In Japan, one of the most thoroughly studied island arc areas, a region of high heat flow coincides with the distribution of volcanoes and hot springs. A second high below the Sea of Japan suggests that the mantle is hotter than average, perhaps as a result of an interarc spreading center. A zone of low heat flow occurs on the Pacific side of the arc.

Earth scientists seek to unravel the history of island arcs to understand their formation. Computers are used to form models of island arcs so that theories of arc formation can be tested. By modifying the model to fit the observed geological and geophysical data, scientists can increase their overall understanding of the island arc system.

Context

Formed by the subduction of one oceanic plate beneath another, island arcs are active volcanic and seismic regions. Understanding island arcs aids earth scientists in unraveling the processes by which geologic features are formed. Knowledge of island arcs is also important to the nonscientist for two reasons: volcanoes and earthquakes. Though of greatest importance to people living in island arc regions, these active geologic zones can have far-reaching effects.

Subduction zone volcanism, in island arcs and in continental margins, accounts for 88 percent of known fatal eruptions. The high gas content of andesitic magma often results in abrupt and highly explosive eruptions, such as the eruption of Mount Pelèe, Martinique, in 1902, which destroyed the town of St. Pierre and killed most of its inhabitants. The effects of eruptions can be far-reaching. In 1883, Krakatoa, Indonesia, erupted, sending volcanic ash over a 700,000-square-kilometer area. Fine dust from the explosion rose high in the atmosphere, where it stayed for several years, lowering the earth's average annual temperature a few degrees by partially blocking solar radiation. The explosion also generated a tsunami (a seismic sea wave), which washed away 165 villages in the Sunda Strait and was recorded as far away as the English Channel. In addition to having climatic effects, the presence of large amounts of ash and dust today would pose a hazard to aircraft navigation.

Many of the world's greatest earthquakes, such as those in Tokyo in 1923 and Alaska in 1964, originate in island arc regions. The danger posed is especially great in highly populated arcs, such as Japan or southern California. Ground movements caused by earthquakes and related landslides damage rigid structures—buildings, bridges, and pipelines—which disrupts basic health and safety services as well as economic production. More than 40 percent of the deaths in Tokyo in 1923 resulted from uncontrollable fires indirectly caused by the earthquake. Vertical movement of the sea floor associated with an earthquake can also generate tsunamis. Tsunamis caused by the 1964 Alaskan earthquake heavily damaged towns in the Gulf of Alaska. Tsunamis from the same earthquake also killed twelve people in Crescent City, California.

Understanding island arc processes can lead to better evaluation of the earthquake and volcanic hazard potential for nearby citizens. Identification and evaluation of events that are harbingers of earthquakes and volcanic eruptions may eventually lead to the prediction of both. The far-reaching effects of earthquakes and volcanic eruptions cannot be predicted. Tsunamis and ash clouds, however, may take several hours to reach distant cities. The Seismic Sea Wave Warning System, established in 1946 to reduce the danger from Pacific tsunamis, issues international warnings when tsunami risk from an earthquake is high. A volcanic watch service has been set up by the Federal Aviation Administration and the National Oceanic and Atmospheric Administration. Using satellite imagery and ground stations, the program will provide warnings to aircraft when dangerous ash clouds from volcanic eruptions are present.

Bibliography

Bolt, Bruce A. *Earthquakes*. New York: W. H. Freeman, 1988. A popular book on the many aspects of earthquakes. Chapters cover distribution of earthquakes, tsunamis, earthquake prediction, and hazard protection planning. Illustrated, with bibliography and index. Suitable for the general reader.

Decker, Robert, and Barbara Decker. *Volcanoes and the Earth's Interior*. San Francisco: W. H. Freeman, 1982. A series of articles on studies in volcanism. The first article, "The Subduction of the Lithosphere," discusses the relationship between subduction zones and island arcs. Illustrated, with bibliography. Suitable for the general reader.

King, Philip B. *The Evolution of North America*. Rev. ed. Princeton, N.J.: Princeton University Press, 1977. A revised edition of the classic book on the geology of North America. An excellent discussion of features of an island arc-trench system (the West Indies) appears on pages 84-90. Requires some knowledge of geology.

Lambert, David. *The Field Guide to Geology*. New York: Facts on File, 1988. A concise introduction to basic geologic terms and concepts. Chapter 2 explores the structure of the lithosphere. Chapter 3 discusses volcanoes and igneous rocks. A well-illustrated book suitable for high school students or beginning students at any level. Index and bibliography.

Miller, Russell. *Continents in Collision*. Alexandria, Va.: Time-Life Books, 1983. An extensively illustrated introduction to plate tectonics. Traces the historical development of the theory, with color illustrations and photographs. Index and bibliography. Suitable for the general reader.

National Research Council. *Explosive Volcanism: Inception, Evolution, and Hazards*. Washington, D.C.: National Academy Press, 1984. A study on all types explosive volcanism, prepared for the National Academy of Sciences. Includes a series of background reports on various aspects of explosive volcanism, including Mount St. Helens and Kilauea. Recommends plans for future research and emergency planning in the United States. Reports are technical but suitable for the knowledgeable reader.

Press, Frank, and Raymond Siever. *Earth*. 2d ed. San Francisco: W. H. Freeman, 1978. A book for the beginning reader in geology. Of interest are chapters 15, 17, and 19 for an overall view of earthquakes, volcanism, and plate tectonics. Illustrated and supplemented with numerous marginal notes. Chapter bibliographies and glossary.

Talwani, Manik, and Walter C. Pitman III, eds. *Island Arcs, Deep Sea Trenches, and Back-Arc Basins*. Washington, D.C.: American Geophysical Union, 1977. Series of technical papers from a symposium on island arcs, held in 1976. Papers cover a wide range of active research areas of the island arc-trench-back-arc system. Suitable for the college student or knowledgeable reader. Extensive references.

Walker, Bryce S. *Earthquake*. Alexandria, Va.: Time-Life Books, 1982. Extensively illustrated book on earthquakes, their causes, and their effects on humans. Chapter 1 covers the cause and effects of the 1964 Alaskan earthquake. Traces the historical development of scientific attempts to understand earthquakes and to predict them. Index and bibliography. Suitable for the general reader.

Pamela R. Justice

Cross-References

Hot Spots and Volcanic Island Chains, 297: Igneous Rocks, 315; Plate Tectonics, 505; Subduction and Orogeny, 615.

Joints

Joints form when rocks undergo brittle failure, usually during expansion. Their orientations, physical features, and patterns of occurrence can be used to help infer the physical conditions present at the time of failure. Because joints are important conduits for fluids, particularly oil and water, that move beneath the surface, understanding their formation and occurrence has economic benefits.

Field of study: Structural geology

Principal terms

CHEMICAL WEATHERING: changes in rocks produced by reactions with fluids near the surface of the earth

COLUMNAR JOINTING: the formation of columns, often with hexagonal cross sections, as joints grow inward from the outer surfaces of cooling igneous rock bodies

CONJUGATE SHEAR SETS: two sets of joints that make angles with each other of something close to 60 degrees and 120 degrees

EXFOLIATION: the splitting off of curving sheets from the outside of a body of rock; also called sheeting

EXTENSION: expansion, or stretching apart, of rocks

FRACTOGRAPHY: the study of fracture surfaces to determine the propagation history of the crack

JOINT: a fracture in a rock across which there has been no substantial slip parallel to the fracture

Summary

Joints are the ubiquitous cracks found in nearly every outcrop. They are unquestionably the most common structure at the surface of the earth. They vary in size from microscopic fractures visible only within an individual grain in a rock to fractures kilometers in length, some of which are responsible for the magnificent scenery of Arches National Park. If one smashes a rock with a hammer, one produces joints. Joints also form as molten rocks solidify and cool, as weathering alters the volume of the outer layers of a rock, and even as erosion removes overlying layers, reducing the weight on the layers below and permitting them to expand and crack.

There is no appreciable slip across a joint. (Failure surfaces accommodating large amounts of slip are faults.) Often, the same mineral grain can be observed on both sides of a joint, neatly cut in two but otherwise not disturbed. This indicates failure in extension; such joints are sometimes called extension joints. The sides of the fracture moved away from each other but did not slide past each other at all. The forces producing such joints were literally pulling the rock apart. An engineer might call these tensile forces. Geologists work with rocks that are nearly always in a state of compression, however, and true tension is uncommon. Therefore, geologists usually call such forces the forces of least compression. The direction of least compression is the direction in which extension occurs. That is, as the crack opens and the sides move away from each other, they will move in the direction of the least compressive force. Consequently, the plane of an extension joint will be perpendicular to the direction of the least compressive force.

Such extension may be produced in a variety of ways. Most obvious are mechanisms involving large-scale deformation of the rock. The folding of a unit of rock causes extension on the outside of the fold. Often, extension joints are found fanning around the "nose" of a fold. The injection of molten rock into cooler rock, and its subsequent cooling, can fracture the cooler rock, producing joints. Extension is also produced during weathering when, because of chemical reactions at the surface of a rock, the surface layer expands more than the interior does. This expanded surface pulls away from the rest of the rock much as an onionskin pulls off the outside of an onion. This process is called exfoliation or sheeting. One classic example of this is Half Dome, in Yosemite National Park.

Contraction, too, can cause joints. The familiar cracks that form in dried-up mud puddles are an example: Moist mud contracts as it dries, and cracks in the surface result when the forces involved in this contraction overcome the cohesive strength of the mud. As rock cools, such as when a molten, igneous rock body solidifies and then cools further, it contracts. The cooling and contraction are greatest at the surface of such a body. The polygonal patterns of cracks that form on such surfaces are very similar to those seen in dried mud. As the hot rock continues to cool, these cracks extend into the interior of the body. This may result in spectacular columns, such as those seen at Devils Postpile. The process is called columnar jointing.

The most common way joints form, however, is probably when rocks that equilibrated at depth are brought to the surface, either by mountain-building forces or when erosion removes the overlying rocks. This means of forming joints was proposed by Neville Price in his 1966 book *Fault and Joint Development in Brittle and Semi-Brittle Rock*. Although the model cannot be aplied directly at any particular location because the deformation history, local topography, and other factors vary too much from place to place, it is instructive to consider the process in general terms.

How can vertical uplift produce horizontal extension? Consider large suspension bridges such as the Verrazano Narrows in New York City or the Golden Gate Bridge in San Francisco. The vertical towers supporting these bridges diverge from one another by about a hundredth of a degree, because of the curvature of the earth. The tops of these towers are farther apart than their bases, so a rope that exactly reached between the bases would have to stretch a bit if it were raised to their tops. If it were unable to stretch, it would break.

Would uplift of 5 kilometers be sufficient to produce joints in a typical rock? If the earth's circumference is 40,074 kilometers, the circumference of a circle lying 5 kilometers beneath the surface would be 40,043 kilometers. If that circle were brought to the surface, it would have to be stretched by 31 kilometers, or by 0.078 percent. This amount of stretch might not seem like much, as a block of rock 100 meters long would need to extend only 7.8 centimeters. But if one tried to stretch that block of rock by attaching a gigantic pulling apparatus on it, experimental data show that the block would break before stretching that much. In addition to this geometric extension, the rock would generally cool as it came up to the surface, contracting and making more joints in the process. Because the state of compression at depth would likely be different from that at the surface, the changes that occur during uplift might encourage or inhibit joint formation, depending on local conditions. Still, if all the joints currently at the surface of the earth are considered, it appears as if the majority of extension joints may form by uplift. The fact that joints form during uplift and erosion does not mean, however, that they are necessarily unrelated to the structural history of the rocks in which they occur. Deformed rocks often contain stored-up energy, much like the energy stored in a spring, which was caused by the deformation. This energy, usually called residual stress, can influence the development of joints. Thus, joints that form hundreds of millions of years after a rock was initially deformed will often occur in patterns and orientations clearly related to that deformation.

Some joints are not formed strictly by extension. These joints develop as a series of cracks, called shear joints, which break the rock into diamond-shaped pieces. These pieces slide slightly past one another, accommodating the deformation. Careful examination of these joints may show some slight offset of grains across the joint, but the displacement across any one joint is small. The cumulative effect across hundreds of joints, however, can be considerable. Often, these joints occur in two parallel sets, with angles of about 60 degrees and 120 degrees between them. Such sets are called conjugate shear sets. Shear joints are not perpendicular to the direction of least compression, but it is possible to determine the direction of compression from the orientation of the conjugate shear sets. The direction of intermediate compression is indicated by the line of intersection of the joints. The direction of least compression bisects the obtuse angle between the sets, and the direction of maximum compression bisects the acute angle. In terms of a diamond-shaped piece, the direction of least compression is the short way across the diamond, and the direction of greatest compression is the long way across the diamond. It is not uncommon for conjugate shear sets to occur in conjunction with extension joints. In this case, the extension joints will be parallel to the long axes of the diamonds.

Consider the joints that might be associated with a fold that forms in a horizontal layer of rock not too far beneath the surface. The fold will form as the layer buckles in response to forces acting along it—in the north-south direction, for example—which is similar to the way a playing card flexes when one squeezes the edges between one's fingers. Early in the deformation process, least compression in the east-west direction results in extension

The magnificent scenery of Arches National Park is produced by joints. *(Utah Travel Council)*

joints running north-south and shear joints running northeast and northwest at a 30-degree angle. Eventually a buckle develops, folding the layer and producing extension fractures in the east-west direction. Much later, erosion and uplift may bring parts of this layer to the surface. The direction of the least compression at that time, which may have no relation at all to the forces that originally produced the fold, will control the vertical extension joints that develop because of this uplift. Finally, weathering and the vagaries of the topography at the time the rock is exposed to weathering control the exfoliation joints that will follow the shape of the exposed surface.

Fracture "decorations"—patterns—on the joint surfaces can yield useful information about the speed of fracture growth and the direction in which it grew. This field of study is called fractography and has been developed by ceramic engineers concerned with reconstructing the brittle failure of glass and ceramic objects in order to improve their design. It can be directly applied to the study of joint surfaces.

When a fracture begins to grow, it starts with a low velocity but accelerates quickly. While it is moving slowly, the front of the fracture is usually a smooth curve, and the decoration it leaves on the fracture surface may be perfectly smooth, called the mirror region, or slightly frosted in appearance, in which case it is called mist hackle. If the crack grows intermittently, arrest lines may result. These curves show where the crack front was at different times when it temporarily stopped growing. As it increases in speed, the fracture front divides into a number of fingerlike projections. These commonly move a bit beyond the initial plane of the fracture as the fracture continues to grow. The result is a pattern on the surface of the fracture that has long been called plumose structure by geologists but is known as twist hackle by fractographers. It looks very much like a feather. The directions in which the fracture grew are shown by the directions of each slightly offset, curving element. Many of these features can be seen on building stones, flagstones, and slate floor tiles.

Joints provide conduits for the movement of fluids beneath the surface. Just as cracks in a pot permit water to leak through the pot, joints in the bedrock greatly enhance the rate at which water, oil, natural gas, and other fluids move through it. Near the surface, water is the fluid most likely to move through joints. As it does so, it is likely to attack the rock on both sides of the joint, chemically weathering it. This process enlarges the joint, increasing the flow of water through it, which in turn causes it to be weathered further, and the process continues. When the jointed rocks are limestone, the result may be elaborate systems of caverns, such as Carlsbad Caverns and other famous caves. Maps of such caves clearly demonstrate that joints controlled their development. In areas underlain by less soluble rock, joints may provide access to groundwater resources. By studying joint patterns displayed on geologic maps, aerial photographs, or satellite images, hydrologists are sometimes able to see where the natural underground flow of water may be greatest, and they can exploit this knowledge in their search for water. Similarly, petroleum geologists seek conditions where joints may facilitate the movement of oil and gas toward potential well sites. Because the rocks of interest to them are often much deeper than those with useful water resources, petroleum geologists may be forced to guess the location of joints at depth. Although the surface traces of joints seen on maps can help, it is often necessary to apply an understanding of how and why joints form in order to predict where they may be at depth. In some cases, artificial joints are produced by pumping fluids under very high pressure into the rocks.

Methods of Study

Joints can be studied at a variety of different scales. Sometimes a regional picture is sought, in which case aerial photographs or topographic maps can be used to see where joint surfaces intersect the surface of the earth. These lines are called the traces of the joints. The advantage of this technique is that large areas can be studied easily, with no need to be in the field. The disadvantages include the fact that vertical and other very steep joints may be well represented, but horizontal or gently dipping joints may not be visible. Knowledge that such biases exist enable useful results to be derived from such studies; however, care must be exercised. The significance—or even existence—of physical causes for the lines and lineaments perceived in such studies is often open to question.

When more three-dimensional information is sought, the orientations of joint planes can be measured in the field and plotted on maps. Approaches vary from visually estimating which joints on an outcrop are significant and then measuring only a representative sample of them (the selection method) to establishing a grid of stations so that every joint within a circle of a certain radius is measured (the inventory method). If the study involves a large number of joints, statistical techniques are employed, such as plotting the data on a graphic device called a stereonet and contouring the results.

If the details of the propagation history of the joints are important, exposed joint surfaces are studied and mapped. Careful measurement of joint openings and offsets (if they exist) can give an estimate of strain. Traces on the outcrop surface can be studied to determine the extent and significance of joint interactions. Finally, the forces that had the greatest influence in the joint development can be isolated, and their magnitudes can be estimated.

Context

Joints are fractures in solid rock. Unlike faults, which are also fractures, there is little or no slip across joints. In many ways, joints are like cracks in a pane of glass. If a baseball hits a window hard enough to crack it, but not hard enough to go through it, it may produce a set of fractures, radiating from the point of impact, which look a bit like a spider's web. That pattern strongly suggests an impact and is not what would be expected from a more

345

gradual process, such as the settling of a house. Similarly, the orientations of some joints can be used to infer the directions of the stresses and forces in the outer part of the earth during the time they formed. This information may help in predicting where the walls and ceilings of mines might collapse and where earthquakes will occur in the future.

Many joints are controlled by stresses locked into the rock long before it fractures. The tempered glass in the rear window of a car always shatters the same way, regardless of how it is broken; the characteristic net of cracks results from the high stresses locked into that type of glass when it was manufactured. The patterns produced by such joints in rocks can provide useful insights into deformations that occurred millions of years ago.

Because they are planes of weakness, joints often control how rocks weather and erode. Cliffs, caves, and mountaintops often owe their shapes to joints. When rocks fail and landslides or rockslides result, joints usually influence where the failure occurs. Fluids utilize joints in order to move through otherwise impermeable rock. In cold climates, near the surface of the earth, water can freeze inside joints, wedging them apart. Water moving through joints frequently reacts chemically with the rock it contacts, weathering the rock and weakening it. This process can be a factor in the formation of additional joints.

At greater depths, joints provide conduits through which water—and also oil and gas—can migrate. Scientists seeking water resources use aerial photographs, satellite images, and topographic maps to locate areas where joints are prevalent. Petroleum geologists study complicated models of deformation deep within the earth in order to guess where naturally produced jointing may increase the flow of oil and gas. If necessary, engineers can increase fluid flow rates, producing joints by pumping fluids into the earth under high pressure.

The surface of the earth would look much different, weather and erode at much lower rates, and have much slower movement of fluids within it, if it were not for the millions and millions of joints that fracture it so pervasively.

Bibliography

Billings, Marland P. *Structural Geology*. 3d ed. Englewood Cliffs, N.J.: Prentice-Hall, 1972. Chapter 7, "Joints" (33 pages), gives a general description of joints and how they are studied. Includes four photographs and a discussion of brittle failure. Suitable for college students.

Davis, George H. *Structural Geology of Rocks and Regions*. New York: John Wiley & Sons, 1984. Chapter 10, "Joints" (28 pages), covers the subject from a generally field-oriented perspective. The discussion of field methods is clear and comprehensive. Includes a section on joint-related structures, including veins and stylolitic joints. Largely descriptive, easily read, and suitable for anyone interested in joints and the study of joints.

Johnson, Arvid M. *Physical Processes in Geology.* San Francisco: Freeman, Cooper, 1970. Chapter 10, "Formation of Sheet Structure in Granite," shows how a particular variety of joints has been studied using a quantitative approach. A technical approach that reveals how interesting some of the questions raised by jointing become when studied in detail. Appropriate for technically oriented college students.

Marshak, Stephen, and Gautam Mitra, eds. *Basic Methods of Structural Geology.* Englewood Cliffs, N.J.: Prentice-Hall, 1988. Chapter 12, "Analysis of Fracture Array Geometry," by Arthur Goldstein and Stephen Marshak (18 pages), gives a good description of joint surface morphology, an overview of the field methods that have been used in studying them, and a summary of graphing and statistical techniques. The treatments are brief, but references are complete. Seems to have been designed for the reader who already knows something about joints but wishes to learn more about how to study them. College level.

Price, Neville J. *Fault and Joint Development in Brittle and Semi-Brittle Rock.* London: Pergamon Press, 1966. This 176-page book is a classic in its field. Most of the work done on joints since its publication has been done by people who have read this book, so its influence is great. Suffers, however, from having been written before modern experimental methods for studying rock deformation had been fully developed and from its use of the British system of measurement. Contains only four photographs, of somewhat limited usefulness, and other figures are schematic, though clear. Not overly technical, with a minimum of mathematical derivations, this book gives a fair summary of what was know about joints at the time it was written. Suitable for college-level students.

Suppe, John, ed. *Principles of Structural Geology.* Englewood Cliffs, N.J.: Prentice-Hall, 1985. Chapter 6, "Joints," presents an excellent overview of the classification, appearance, and formation of joints. The sixteen photographs show all manner of joints at a variety of scales, and some of the fourteen line drawings provide good examples of how joint studies have been used in interpreting structures and the state of stress in the crust. At times, the discussion becomes a bit technical, and several references are made to chapter 4, "Fracture and Brittle Behavior," which is more technical yet. A useful digression concerns the state of stress in the upper and lower crust and in the mantle. College-level reading.

Otto H. Muller

Cross-References

Folds, 248; Igneous Rock Bodies, 307; Karst Topography, 348; Stress and Strain, 607; Weathering and Erosion, 701.

Karst Topography

Karst topography is a landform produced by the dissolving action of surface and groundwaters on the underlying bedrock of a region. The landforms produced are unique because they represent internal or underground drainage, forming such features as sinking streams, sinkholes, caves, natural bridges, and springs.

Field of study: Geomorphology

Principal terms

CARBONIC ACID: a weak acid formed by mixing water and carbon dioxide; it is important in the dissolving of the most common karst rock, limestone

CAVE: an underground chamber open to the surface; in karst topography, caves have been dissolved out and act (or once acted) as underground conduit flow routes for water

KARST: the international term for landforms dissolved from rock

LIMESTONE: a common sedimentary rock containing the mineral calcite; the calcite originated from fossil shells of marine plants and animals

NATURAL BRIDGE: a bridge over an abandoned or active watercourse; in karst topography, it may be a short cave or a remnant of an old, long cave

SINKHOLE: a hole or depression in the landscape produced by dissolving bedrock; sinkholes can range in size from a few meters across and deep to kilometers wide and hundreds of meters deep

SINKING STREAM: a stream or river that loses part or all of its water to pathways dissolved underground in the bedrock

SPRING: a place where groundwater reappears on the earth's surface; in karst topography, a spring represents the discharge point of a cave

Summary

Karst topography is a unique landscape produced by the dissolving of the bedrock of a region, with the consequent development of underground drainage. Most common bedrock materials, such as granite, sandstone, and shale, are resistant to the process of dissolving (also known as dissolution), and landscapes are carved into these bedrock materials by the mechanical action of water, wind, and ice. Some bedrock materials, such as limestone, dolomite, gypsum, and rock salt, dissolve relatively easily. Gypsum and rock salt are soluble in plain water and in most landscapes are chemically destroyed very quickly. Karst landscapes on these two rock types persist only in dryer climates, such as in the American Southwest. Gypsum and rock salt are also mechanically weak rocks and are destroyed rapidly by mechanical erosive activities. Limestone, and its close cousin dolomite, are mechanically strong rocks; therefore, they resist normal erosive activity. Yet, calcite, the mineral of which limestone is composed, is especially vulnerable to dissolving by water that is slightly acidic. Under normal conditions, the source of such acidity is from carbonic acid, a natural combination of carbon dixoide and water. Rain-water absorbs carbon dioxide from the atmosphere and is naturally slightly acidic (modern pollution has accentuated this natural tendency to produce acid rain). In the soil, organic activity can increase the amount of carbon dioxide to levels well above those found in the atmosphere, and water moving through such soils can become very acidic. When rainwater or soil water charged with carbonic acid meets limestone bedrock, it will slowly dissolve the limestone. Limestone, mechanically resistant, supports the development of karst topography well. Dolomite reacts more slowly to acid waters, and generally the karst features found on dolomites are subdued and take longer to form than those on limestones. Discussion of karst topography is therefore, in most situations, a discussion of the dissolving of limestone.

The rate at which limestone dissolves depends on the amount of water in the environment and the amount of carbon dioxide available. Atmospheric levels of carbon dioxide are similar worldwide, but the amount of carbon dioxide in the soil varies greatly. Organic activity controls the amount of carbon dioxide available, and for that reason, the most acid groundwater tends to be found in the warm, wet zones of the tropics, where organic activity is high. Dry, cold climates have erosion rates as little as a few millimeters per one thousand years, whereas warm, wet climates can have rates up to 150 millimeters per one thousand years. These erosion rates are averages of the rate of surface lowering for a region. Over the course of hundreds of thousands of years, significant landforms can be developed.

The dissolving action of carbon dioxide-charged water produces a number of etching patterns on exposed limestone bedrock, from microscopic features to large trough structures more than a meter deep and many meters long. Karst topography develops when the water penetrates down pores, cracks, and openings into the limestone. Once inside the rock, the water is capable, over long periods of time, of dissolving voids and passageways in the rock. These openings in the rock become integrated into underground flow networks similar to stream patterns on the earth's surface. The result is a series of underground passageways called caves, which collect water from sinkholes and sinking streams on the earth's surface and transmit the water back to the earth's surface at springs. While the flow pattern of caves is often similar to the pattern of surface streams, caves are tubes in bedrock and may migrate up and down, as well as from side to side, in a manner that surface streams cannot. As the cave system grows and matures, it can capture larger volumes of surface water and enlarge. Eventually it will not only carry material in solution but also mechanically transport sediment underground. On the land surface, this underground or internal drainage produces sinking streams and sinkholes. In mature systems, the sinkholes may be kilometers across and hundreds of meters deep, containing many sinking streams. Depending on the exact nature of the limestone, the climate, and the presence of mountains, a wide variety of karst landscapes may appear. The landscape may be as simple as rolling hills or a flat plain with mostly normal surface drainage and only a few scattered sinkholes, sinking streams, and caves. Other landscapes, however, may have no significant surface drainage and have sinkholes covering the land surface, producing a sinkhole plain. One of the best examples of a sinkhole plain is in the Mammoth Cave area of Kentucky. Under extreme conditions, usually associated with the tropics, the landscape is so altered by dissolution that the sinkholes deepen faster than they widen, producing a landscape of tall limestone towers and pinnacles standing above a flat plain. This landform is called tower karst and is best known from southern China, where thousands of towers several hundred meters high dot the landscape.

The ever-downward erosion by groundwater can produce cave systems at lower elevations, causing earlier, higher cave systems to become abandoned. These abandoned caves may persist for hundreds of thousands of years, developing complex mineral displays of stalactites, stalagmites, and other deposits. Animals and people may use the cave for shelter, leaving behind important fossil and cultural material. Continuing erosion of the land surface will eventually breach into underlying cave systems to produce new entrances and truncated cave fragments called natural bridges. Sinkholes become natural traps, collecting unwary animals as well as plant remains, soil, and pollen. The transport of material into the subsurface allows the material to be preserved from destruction in the earth's surface environment.

Like all landscapes, karst topography can go through a series of developmental stages. Karst development begins when the soluble rock is first exposed at the earth's surface. In some cases, groundwater reaches the soluble rock before it is ever exposed, and cave development can begin before the rock is present on the surface. In either case, the progressive development of a karst landscape by chemical and mechanical erosion will remove the layer of soluble rock responsible for the karst processes. In time, karst processes become less important in the landscape as the soluble bedrock disappears, and the landscape will begin to revert to a more typical, mechanically produced form.

In the United States, limestones form important areas of karst topography in the Virginias, Pennsylvania, Indiana, Kentucky, Tennessee, Alabama, Florida, Missouri, Arkansas, and Texas. Many other states have minor amounts of limestone karst, such as New York, Minnesota, Iowa, Colorado,

The Wind Cave system of South Dakota is located in an area of karst topography. *(Wind Cave National Park)*

and New Mexico. Some states have major cave systems located in minor areas of karst, such as Carlsbad Caverns in New Mexico and the Jewel and Wind Caves in South Dakota. Gypsum karst is locally important in Kansas, Oklahoma, and New Mexico, while dolomite karst is found in Missouri. Marble, limestone altered by heat and pressure, forms small karst areas in California, New England, and Oregon. Rock salt karst is very rare in the United States but can be found in Spain and the Middle East. Almost every state and every country has some form of karst development.

The unique features of karst topography can be mimicked by other landscapes in special cases. Such cases are called pseudokarst, because they imitate karst topography but are produced by other phenomena. For example, volcanic areas produce eruptive craters and lava tubes that look like sinkholes and caves from karst areas. In deserts, streams can dry up and appear to be sinking, and winds can produce excavated hollows in the sand. Glaciers can produce depressions in the ground that look like sinkholes. In fine-grained clay deposits—which can be seen in the Badlands of South Dakota—caves, bridges, and sinkholes are produced by mechanical flushing of the clay particles, a process called suffusion. Wave activity on rocky coastlines can carve sea caves in a variety of rock materials.

Karst topography is unique because of the chemical manner in which the bedrock is destroyed and because of the internal or underground drainage that results. Most karst is developed in limestone because of this rock's abundance, slow chemical dissolution rates, and great mechanical strength.

In certain areas of the world, karst topography is the dominant landscape for thousands of square kilometers. Cave systems in excess of 500 kilometers of surveyed passage exist, and caves have been followed as deep as 1,700 meters beneath the surface.

Methods of Study

Karst topography is most easily recognized from topographic maps, which show the earth's elevations as a series of contour lines. Areas of karst topography will show internal drainage, with sinking streams, large sink-holes, and springs. Well-known caves will be listed on the map. Field examination of the area is necessary to prove that the bedrock involved has undergone chemical attack and has dissolved. Geologists determine the rate at which the landscape has formed by analyzing rock samples in the laboratory under controlled conditions to see how it dissolves under a variety of temperatures and acid concentrations. Measurements of soil carbon dioxide, regional rainfall and temperature, and amount of soluble rock present allow the rate of landscape formation to be established. Sinking streams and springs are measured to establish a water balance; how much water is entering the system and how much is leaving it. To understand the underground flow paths, a geologist will explore and map a cave system; in many cases, however, the cave system is blocked by collapse, sediment, or excess water, and the explorer is stopped. To define the water's underground flow paths completely requires stream tracing, in which a known quantity of dye is placed in a sinking stream; springs in the area are monitored to see where (and how much) of the dye returns to the surface. Special dyes are used that are safe, detectable in tiny quantities, and resistant to loss or destruction in the groundwater system. The dye is detected by automatic water samplers and a fluorometer, a device that can detect dyes even in parts per billion concentrations. The speed at which the dye transits the cave system and the degree to which it has become diluted can reveal the nature and configuration of the unexplored portion of the cave system.

The successful delineation of the cave systems of a karst area, along with data about the nature of the rock and climate, can show geologists what is happening in the present. In many cases, however, the geologist is also interested in the past development of the karst area in order to detect trends that allow prediction of the karst area in the future: Will the caves get bigger? Will sinkholes collapse? Will there be flooding? These are questions important to land use. Geologists often go to the abandoned caves to learn about the history of a karst landscape. Inside the caves, they find sediments that were washed in from the surface when the abandoned caves were active. The size and composition of those sediments reveal how much water was entering the cave and from what source areas. Animal fossils and human cultural remains can explain how the climate may have changed.

A special technique involving stalagmites, calcite mineral deposits that grow up from the floors of caves, can provide hard numerical data on past conditions. The calcite that makes up the stalagmites is deposited in layers, much like the growth rings in a tree trunk. Uranium, a radioactive element, is often present in trace amounts in the drip water that falls from the cave roof to make the stalagmite. As the calcite crystallizes to make the layering in the stalagmite, it traps small amounts of uranium with it. Through time, the radioactive uranium begins to decay into the element thorium. The older the stalagmite, the more calcite layers it has. The bottom layers of calcite will have both uranium and thorium in them: the uranium from the initial deposition and the thorium from decay of some of the uranium. Higher up in the stalagmite, each calcite layer is younger, and therefore the amount of thorium should be less. Since the rate of decay of uranium is well known, the ratio of uranium to thorium is a measure of how old the calcite layer is. The more thorium relative to uranium in the layer, the older it is. When these sorts of measurements were first done, a startling discovery was made. The age of many stalagmites, especially those from northern North America, did not get uniformly younger from the base of the stalagmite to the top. Instead, there were sudden jumps in age. The bottommost portion of the stalagmite might show a progression in age from older to younger layers, but then there would be an abrupt jump to a much younger age. Geologists reasoned that the sudden jumps in age meant that at certain times in the history of the stalagmite, it was not growing; the drip water feeding the stalagmite had become shut off. This was clearly a signal of climatic conditions on the surface of the earth. The earth was either too dry or too cold to allow liquid water into the cave ceiling, where it could drip and make the stalagmite. When the times of stalagmite growth and non-growth were examined, it was found that for northern North American caves, the nongrowth occurred when the region had been covered by ice sheets during ice ages. The technique has since been refined to allow geologists to use the rate of stalagmite growth in caves to determine many aspects of the duration and intensity of climatic changes on the earth's surface above the caves. The stalagmite data, when joined to the sediment and fossil data, have allowed the history of many karst areas to be determined with a higher degree of accuracy than is possible in most nonkarst areas.

Context

About 25 percent of the world's population either lives on or obtains its water from areas of karst topography. Because of the internal, underground nature of landscape development and water flow, many environmental problems occur in karst areas that do not happen on other landscapes. The ability of limestone to react with acids means that soils in limestone karst areas are rarely acidic and are therefore often highly productive agriculturally.

Abandoned caves in karst areas preserve a wealth of information about past landscape history, extinct animals, and past human culture. Some of the most famous paleontological and archaeological sites in the world have been situated in or near caves. Cultural uses of caves have ranged from homes and churches to prisons and air-raid shelters. The earliest known human artworks have been found in caves of the Pyrenees Mountains of France and Spain, cave paintings produced more than thirty thousand years ago.

Karst areas are confronted with an unusual array of environmental problems. Sinkholes are convenient sites for waste disposal, yet they feed directly into the underground water supply. Unlike traditional underground water reservoirs or aquifers, water transmission underground in karst areas is through tubes in the bedrock, caves. There is no filtration in a cave system, as there is in an aquifer of porous sandstone. In addition, caves can transmit water long distances in very short periods of time. Cave streams often cross underneath surface divides to release groundwater at springs in a completely different valley from the one in which the water first sank underground. These unique aspects of groundwater flow in karst areas make groundwater pollution and contamination a serious problem. The pollution can travel rapidly with minimal alteration over long distances to unexpected locations.

In some regions, agricultural, industrial, and domestic use of water has resulted in the lowering of the water table in karst areas. Lowering of the water table can lead to sinkhole collapse with large-scale disruption on the land surface. In South Africa, Pennsylvania, Missouri, and most dramatically in Florida, excessive groundwater pumping has resulted in the loss of property and life. The collapse can be sudden or gradual, and special engineering techniques are required to restore the landscape to its original function.

Land use in areas of karst topography requires a complete understanding of karst processes and of the nature of the landscapes those processes produce. The list of engineering failures in karst areas is long: landfills that polluted a city's water supply, sewage lagoons that drained underground overnight, dams that collapsed, and reservoirs that never filled. Karst topography, because of caves, is tied directly to the early development of humans. The cryptic nature of the underground environment continues to lure amateur explorer and scientist alike.

Bibliography

Beck, B. F., and W. L. Wilson, eds. *Karst Hydrogeology: Engineering and Environmental Applications*. Boston: A. A. Balkema, 1987. This book is the proceedings volume of the Second Multidisciplinary Conference on Sinkholes and the Environmental Impacts of Karst and contains sixty-three papers on a variety of topics concerning human activity and karst, with an emphasis on the high rate of mistakes made as a result of ignorance of karst processes. Accessible to the nonspecialist.

Carleton, Cathleen L. "Karst Assessment." *Science Teacher* 62, no. 6 (September, 1995): 68-74.

Jennings, J. N. *Karst Geomorphology*. New York: Basil Blackwell, 1985. A college-level text that emphasizes karst processes overall, as opposed to only their role in cave formation. The text requires some knowledge of the basic sciences. It is interesting to read, as it uses many Australian examples and contains the lifetime knowledge of one of the world's leaders in karst research.

Middleton, John, and Anthony C. Waltham. *The Underground Atlas: A Gazetteer of the World's Cave Regions*. New York: St. Martin's Press, 1987. An easy-to-read guide to the famous long and deep caves of the world, this atlas covers the continents in a general review and then describes the cave resources of the world's countries, from Afghanistan to Zimbabwe. Simple maps are presented to portray the largest and deepest caves. The appendices supply a listing of record-holding caves in terms of length, depth, and volume, as well as data about caving organizations and sources of additional reading.

Moore, George W., and G. Nicholas Sullivan. *Speleology: The Study of Caves*. Teaneck, N.J.: Zephyrus Press, 1978. This text is designed to give a scientific introduction to caves and karst for nonscientists. It explores the chemistry, atmospherics, mineralogy, biology, and cultural uses of caves. Oriented to the American reader, examples are primarily from North America. Contains an appendix that lists caves open to the public throughout the United States.

Palmer, Arthur N. *A Geological Guide to Mammoth Cave National Park*. Teaneck, N.J.: Zephyrus Press, 1981. Mammoth Cave is the longest explored cave in the world, with more than 500 kilometers of surveyed passage. This very readable book explains subtle aspects of karst geology in a way understandable to the layperson but at the same time impressive to the specialist. Demonstrates how surface karst topography relates to the internal drainage of caves.

Sasowsky, I. D. *Cumulative Index for the National Speleological Society Bulletin, Volumes 1 Through 45, and Occasional Papers of the N.S.S., Numbers 1 Through 4*. Huntsville, Ala.: National Speleological Society, 1986. The Bulletin of the National Speleological Society is carried by many libraries and contains papers on the science of speleology and karst processes. This index to that journal allows quick access to published papers on specific topics, by topic, author, or geographical area.

Waltham, Anthony C. *Caves*. New York: Crown, 1974. An easy-to-read popular account of caves, how they form, what they contain, and how they are explored. The author is both an internationally recognized cave explorer and a top scientist in the area of karst topography. The text contains a good index and a bibliography and uses worldwide examples to explain the variety of features found above and below ground in karst areas.

White, William B. *Geomorphology and Hydrology of Karst Terrains*. New York: Oxford University Press, 1988. An advanced college textbook written by one of the leaders in karst research in North America, this volume contains a well-written and comprehensive discussion of karst topography.

John E. Mylroie

Cross-References

Drainage Basins, 141; Reefs, 518; Sedimentary Rocks, 568; Water-Rock Interactions, 692.

Lakes

Lakes are geologically short-lived features, and sediments deposited in lakes (called lacustrine sediments) constitute only a tiny part of the sedimentary rocks of the earth's crust. Nevertheless, lake sediments are the most important sources of information about past climates. Several important economic resources, including oil shales, diatomaceous earth, salt, some limestones, and some coals, originate in lakes.

Field of study: Sedimentology

Principal terms

ALLOGENIC SEDIMENT: sediment that originates outside the place where it is finally deposited; sand silt and clay carried by a stream into a lake are examples

BIOGENIC SEDIMENT: sediment that originates from living organisms

CLASTIC SEDIMENTS: sediments composed of durable minerals that resist weathering

CLAY: a size term referring to any mineral particle less than 2 micrometers in diameter

CLAY MINERALS: a mineral group that consists of structures arranged in sandwichlike layers, usually sheets of aluminum hydroxides and silica, along with some potassium, sodium, or calcium ions

ENDOGENIC SEDIMENT: sediment produced within the water column of the body in which it is deposited; for example, calcite precipitated in a lake in summer

MINERAL: a solid with a constant chemical composition and a well-defined crystal structure

MINERALOID: a solid substance with a constant chemical composition but without a well-ordered crystal structure

PLANKTON: plant and animal organisms, most of which are microscopic, that live within the water column

SESTON: a general term that encompasses all types of suspended lake sediment, including minerals, mineraloids, plankton, and organic detritus

Summary

Several geologic mechanisms can provide the closed basins that are needed to impound water and produce lakes. The most important of these mechanisms include glaciers, landslides, volcanoes, rivers, subsidence, and tectonic processes.

Continental glaciers formed thousands of lakes by the damming of stream valleys with moraine materials. Glaciers also scoured depressions in softer bedrock, and these later filled with water to form lakes. Depressions called kettles formed when buried ice blocks melted. Mountain glaciers also produce numerous small, high alpine lakes by plucking away bedrock. The bowl-shaped depressions that occur as a result of this plucking are called cirques; lakes that occupy these cirques are called tarns. Sometimes, a mountain glacier moves down a valley and carves a series of depressions along the valley that, from above, look like a row of beads along a string. When these depressions later fill with water, the lakes are called paternoster lakes, the name coming from their similarity to beads on a rosary.

Landslides sometimes form natural dams across stream valleys. Large lakes then fill behind the dam. Volcanoes may produce lava flows that dam stream valleys and produce lakes. A volcanic explosion crater may fill with water and make a lake. After an eruption, the area around the eruption vent may collapse to form a depression called a caldera. Some calderas, such as Crater Lake in Oregon, fill with water. Rivers produce lakes along their valleys when a tight loop of a meandering channel finally is eroded through and leaves behind an oxbow lake, isolated from the main channel. Sediment may accumulate at the mouth of a stream, and the resulting delta may build and bridge across irregularities in the shoreline to create a brackish coastal lake.

Natural subsidence creates closed basins in areas underlain by soluble limestones or evaporite deposits. As the underlying limestone is dissolved away, the earth above collapses to form a cavity (sinkhole), which later fills with water. Finally, large-scale (tectonic) downwarping of the earth's crust produces some very large lakes. Large basins result when the crust warps or sinks downward in response to deep forces. The subsidence produces very large closed basins that can hold water. A few immense lakes owe their origins to tectonic downwarping.

With few exceptions, most lakes exist in relatively small depressions and serve as the catch basins for sediment from the entire watershed around them. The natural process of sedimentation ensures that most lakes fill with sediment before very long periods of geologic time have passed. Lakes with areas of only a few square kilometers or less will fill within a few tens of thousands of years. Very large lakes, the inland seas, may endure for more than ten million years. Human-made lakes and reservoirs have unusually high sediment-fill rates in comparison with most natural lakes; they fill with sediment within a few decades to a few centuries.

Lake sediments come from four sources: allogenic clastic materials that are washed in from the surrounding watershed; endogenic chemical precipitates that are produced from dissolved substances in the lake waters; endogenic biogenic organic materials, produced by plants and animals living in the lake; and airborne substances, such as dust and pollen, transported to the lake in the atmosphere.

Allogenic clastic materials are mostly minerals; they are produced when rocks and soils in the drainage basin are weathered by mechanical and chemical processes to yield small particles. These particles are moved downslope by gravity and running water to enter streams, which then transport them to the lake. Clastic materials also enter the lake via waves, which erode the materials from the shoreline, and via landslides that directly enter the lake. In winter, ice formed on the lake can expand and push its way a few centimeters to a meter or so onto the shore. There, the ice may pick up large particles, such as gravel and cobbles. When spring thaw comes, waves can remove that ice, together with its enclosed particles, and float it out onto the lake. The process by which the large particles are transported out on the lake is called ice-rafting. As the ice melts, the large clastic particles drop to the bottom; they are termed dropstones when found in lake sediments. A landslide into a lake or a flood on a stream that feeds into the lake can produce water heavily laden with sediment. The sediment-laden water is more dense than is clean water and therefore can rush down and across the lake bottom at speeds sufficient to carry even coarse sand far out into the lake. These types of deposits are called turbidite deposits.

Endogenic chemical precipitates in freshwater lakes commonly consist of carbonate minerals (calcite, aragonite, or dolomite) and mineraloids that consist of oxides and hydroxides of iron, manganese, and aluminum. In some saline and brine lakes, the main sediments may be carbonates, together with sulfates such as gypsum (hydrated calcium sulfate), thenardite

Oregon's spectacular Crater Lake fills a volcanic caldera. *(National Park Service)*

(sodium sulfate), epsomite (hydrated magnesium sulfate) or with chlorides such as halite (sodium chloride) or more complex salts. Of the endogenic precipitates, calcite is the most abundant. Its precipitation represents a balance between the composition of the atmosphere and that of the lake water. The mechanism that triggers the precipitation of calcite is usually an algal bloom, often of diatoms. Diatoms are distinctive microscopic algae that produce a frustule (a kind of shell) made of silica glass that is highly resistant to weathering. When seen under a high-powered microscope, diatom frustules appear to be artwork—beautiful and highly ornate saucer- and pen-shaped works of glass. A tiny spot of lake sediment may contain millions.

A lake's sediment may contain from less than 1 percent to more than 90 percent organic materials, depending upon the type of lake. Most organic matter in lake sediments is produced within the lake by plankton and consists of compounds such as carbohydrates, proteins, oils, and waxes that are made up of organic carbon, hydrogen, nitrogen, and oxygen, with a little phosphorus. Plankton, with an approximate bulk composition of 36 percent carbon, 7 percent hydrogen, 50 percent oxygen, 6 percent nitrogen, and 1 percent phosphorus (by weight), includes microscopic plants (phytoplankton) and microscopic animals (zooplankton) that live in the water column. Lakes that are very high in nutrients (eutrophic lakes) commonly have heavy blooms of algae, which contribute much organic matter to the bottom sediment. Terrestrial (land-derived) organic material such as leaves, bark, and twigs form a minor part of the organic matter found in most lakes. Terrestrial organic material is higher in carbon and lower in hydrogen, nitrogen, and phosphorus than is planktonic organic matter.

Airborne substances constitute only a tiny fraction of lake sediment. The most important material is pollen and spores. Pollen usually constitutes less than 1 percent of the total sediments, but that tiny amount is a very useful component for learning about the recent climates of the earth. Pollen is some of the most durable of all natural materials. It survives attack by air, water, and even strong acids and bases. Therefore, it remains in the sediment through geologic time. As pollen accumulates in the bottom sediment, the lake serves as a kind of tape recorder for the vegetation that exists around it at a given time. By taking a long core of the bottom sediment from certain types of lakes, a geologist may look at the pollen changes that have occurred through time and reconstruct the history of the climate and vegetation in an area.

Volcanic ash thrown into the atmosphere during eruptions enters lakes and forms a discrete layer of ash on the lake bottom. When Mount St. Helens erupted in 1980, it deposited several centimeters of ash in lakes more than 160 kilometers east of the volcano. Geologists have used layers of ash in lakes to reconstruct the history of volcanic eruptions in some areas. Although dust storms contribute sediment to lakes, such storms are usually too infrequent in most areas to contribute significant amounts.

Lake waters are driven into circulation by the agents of temperature and wind. Most freshwater lakes in temperate climates circulate completely twice each year; they are termed dimictic lakes. Circulation exerts a profound influence on water chemistry of the lake and the amount and type of sediment present within the water column. During summer stratification, the lake is thermally stratified into three zones. The upper layer of warm water (epilimnion) floats above the denser cold water and prevents wind-driven circulation from penetrating much below the epilimnion. The epilimnion is usually in circulation, is rich in oxygen (from algal photosynthesis and diffusion from the atmosphere), and is well lighted. This layer is where summer blooms of green and blue-green algae occur and calcite precipitation begins. The middle layer (thermocline) is a transition zone in which the water cools downward at a rate of greater than 1 degree Celsius per meter. The bottom layer (hypolimnion) is cold, dark, stagnant, and usually poor in oxygen. There, bacteria decompose the bottom sediment and release phosphorus, manganese, iron, silica, and other constituents into the hypolimnion.

Sediment deposited in summer includes a large amount of organic matter, clastic materials washed in during summer rainstorms, and endogenic carbonate minerals produced within the lake. The most common carbonate mineral is calcite (calcium carbonate). The regular deposition of calcite in the summer is an example of cyclic sedimentation, a sedimentary event that occurs at regular time intervals. This event occurs yearly in the summer season and takes place in the upper 2 or 3 meters of water. On satellite photos, it is even possible to see the summer events as whitings on large lakes, such as Lake Michigan.

As the sediment falls through the water column in summer, it passes through the thermocline, into the hypolimnion, and onto the lake bottom. As it sits on the bottom during the summer months, bacteria, particularly anaerobic bacteria (those that thrive in oxygen-poor environments), begin to decompose the organic matter. As this occurs, the dissolved carbon dioxide (CO_2) increases in the hypolimnion. If enough CO_2 is produced, the hypolimnion becomes slightly acidic, and calcite and other carbonates that fell to the bottom begin to dissolve. The acidic conditions also release dissolved phosphorus, calcium, iron, and manganese into the hypolimnion, as well as some trace metals. Clastic minerals such as quartz, feldspar, and clay minerals are not affected in such brief seasonal processes, but some silica from biogenic material such as diatom frustules can dissolve and enrich the hypolimnion in silica. As summer progresses, the hypolimnion becomes more and more enriched in dissolved metals and nutrients.

Autumn circulation begins when the water temperature cools and the density of the epilimnion increases until it reaches the same temperature and density as the deep water. Thereafter, there is no stratification to prevent the wind from circulating the entire lake. When this happens, the cold, stagnant hypolimnion, now rich in dissolved substances, is swept into circulation with the rest of the lake water. The dissolved materials from the hypolimnion are mixed into a well-oxygenated water column. Iron and manganese that formerly were present in dissolved form now oxidize to form tiny solid particles of manganese oxides, iron oxides, and hydroxides. The sediment therefore becomes enriched in iron, manganese, or both during the autumn overturn, the amount of enrichment depending upon the amount of dissolved iron and manganese that accumulated during summer in the hypolimnion. Dissolved silica is also swept from the hypolimnion into the entire water column. In the upper water column, where sunlight and dissolved silica become present in great abundance, diatom blooms occur. The diatoms convert the dissolved silica into solid opaline frustules.

As circulation proceeds, the currents may sweep over the lake bottom and actually resuspend a centimeter or more of sediment from the bottom and margins of the lake. The amount of resuspension that occurs each year in freshwater lakes is primarily the result of the shape of the lake basin. A lake that has a large surface area and is very shallow permits wind to keep the lake in constant circulation over long periods of the year.

As winter stratification comes, an ice cover forms over the lake and prevents any wind-induced circulation. Because the circulation is what keeps the lake sediment in suspension, most sediment quickly falls to the bottom; sedimentation then is minimal through the rest of winter. If light can penetrate the ice and snow, some algae and diatoms can utilize this weak light, present in the layer of water just below the ice, to reproduce. Their settling remains contribute small amounts of organic matter and diatom frustules. At the lake bottom, the most dense water (that at 4 degrees Celsius) accumulates. As in summer, some dissolved nutrients and metals can build up in this deep layer, but because the bacteria that are active in releasing these substances from the sediment are refrigerated, they work slowly, and not as much dissolved material builds up in the bottom waters.

When spring circulation begins, the ice at the surface melts, and the lake again goes into wind-driven circulation. Oxidation of iron and manganese occurs (as in autumn), although the amounts of dissolved materials available are likely to be less in spring. Once again, nutrients such as phosphorus and silica are circulated out of the dark bottom waters and become available to produce blooms of phytoplankton. Spring rains often hasten the melting, and runoff from rain and snowmelt in the drainage basin washes clastic materials into the lake. The period of spring thaw is likely to be the time of year when the maximum amount of new allogenic (externally derived) sediment enters the lake.

Spring diatom blooms continue until summer stratification prevents further replenishment of silica to the epilimnion. Thereafter, the diatoms are succeeded by summer blooms of green algae, closely followed by blooms of blue-green algae. Silica is usually the limiting nutrient for diatoms; phosphorus is the limiting nutrient for green and blue-green algae.

After sediments are buried, changes occur; this process of change after burial is termed diagenesis. Physical changes include compaction and dewatering. Bacteria decompose much organic matter and produce gases such as methane, hydrogen sulfide, and carbon dioxide. The "rotten-egg" odor of black lake sediments, often noticed on boat anchors, is the odor of hydrogen sulfide. After long periods of time, minerals such as quartz or calcite slowly fill the pores remaining after compaction.

One of the first diagenetic minerals to form is pyrite (iron sulfide, or FeS_2). Much pyrite occurs in microscopic spherical bodies that look like raspberries; these particles, called framboids, are probably formed by bacteria in areas with low oxygen within a few weeks. In fact, the black color of some lake muds and oozes results as much from iron sulfides as from organic matter. Other diagenetic changes include the conversion of mineraloid particles containing phosphorus into phosphate minerals such as vivianite and apatite. Manganese oxides may be converted into manganese carbonates (rhodochrosite) and, in a few rare cases, freshwater manganese oxide nodules may form.

Methods of Study

Scientists who study lakes (limnologists) must study all the natural sciences—physics, chemistry, biology, and geology—because lakes are complex systems that include biological communities, changing water chemistry, geological processes, and interaction between water, sunlight, and the atmosphere.

Modern lake sediments are collected from the water column in sediment traps (cylinders and funnels into which the suspended sediment settles over periods of days or weeks) or by filtering large quantities of lake water. Living material is often sampled with a plankton net. Older sediments that have accumulated on the bottom are collected with dredges and by coring, which involves pushing a sharpened hollow tube (usually about 2.5 centimeters in diameter) downward into the sediment. A special tip at the bottom of the tube prevents the soft sediment from falling out as the core is withdrawn from the lake bottom. Cores are valuable because they preserve the sediment in the order in which it was deposited, from oldest at the bottom to youngest at the top. Once the sample is collected, it is often frozen and taken to the laboratory. There, pollen and organisms may be examined by microscopy, minerals may be determined by X-ray diffraction, and chemical analyses may be made.

Varves are thin laminae that are deposited by cyclic processes. In freshwater lakes, each varve represents a year's deposit; it consists of a couplet with a dark layer of organic matter deposited in winter and a light-colored layer of calcite deposited in summer. Varves are deposited in lakes where annual circulations cannot resuspend bottom sediment and therefore cannot mix it to destroy the annual lamination. Some lakes that are small and very deep may produce varved sediments; Elk Lake in Minnesota is an example. In other lakes, the accumulation of dissolved salts on the bottom eventually produces a dense layer (monimolimnion), which prevents disturbance of the bottom by circulation in the overlying fresher waters. Soap Lake in Washington State is an example. Because each varve couplet represents a year, a geologist may core the sediments from a varved lake and count the couplets to determine the age of the sediment in any part of the core. The pollen, the chemistry, the diatoms, and other constituents may then be carefully examined to deduce what the lake was like during a given time period. The study is much like solving a mystery from a variety of clues. Eventually, the history of climate changes of the area can be known from the study of lake varves.

Context

Lake sediments (particularly varved sediments) are among the most important sources of information about how climates have changed over the past ten thousand to fifteen thousand years. The study of lake varves can show what the temperatures were and how much rainfall occurred. Scientists can thus use varves to look for periodicity, or repeating patterns of climate fluctuation. For example, the eleven-year period of the sunspot cycle produces a cyclic "fingerprint" in lake sediments, which show temperature and rainfall variations in accord with that cycle. Other longer-period variations have been noted, but the mechanisms that cause them have yet to be well understood. Clearly, such knowledge of the past is important for understanding present climate trends such as global warming and droughts.

Varved lakes also give scientists an opportunity to monitor environmental changes that have occurred in the earth as a result of human activities. For example, an increase in lead content in the upper layers of sediments in many lakes marks the time when leaded gasoline and automobiles became abundant. Careful measurements of radioactivity in lake sediments can detect the layers deposited during the years of nuclear weapons testing in the 1950's. Studies of varves can show whether lakes are becoming acidified in the present and if and when they had ever become acidic in prehistoric times through natural causes. Increased sedimentation often shows the effects of construction and mining. The alteration of vegetation is reflected within the pollen record. Therefore, lake sediments serve as a storehouse for vast amounts of information about the recent past of the planet.

Lake sediments sometimes yield valuable commercial deposits. Saline lakes such as the Great Salt Lake, Utah, yield a variety of evaporite salts, including the common table salt, halite. Commercial borate deposits that include borax (hydrated sodium borate) and other boron-containing minerals are found in lake sediments in California and also in Kashmir, Tibet, India, the Soviet Union, and Iran. Soda ash, used in the manufacture of glass, baking soda, and many chemicals, is the mineral trona (a complex salt of sodium carbonates and bicarbonates and water), which is mined in Wyoming from the lake deposits of the Green River formation.

After long burial under suitable geological conditions, the abundant hydrogen in lake plankton yields an organic matter, kerogen, that is conducive to the formation of petroleum. The world's oil shales mostly originate in lakes; the Green River formation of Utah and Wyoming is the largest and best known. Oil is produced when the volatile organic matter from these shales is heated and distilled. The Green River shales may contain over a trillion barrels of oil that can be recovered in this way. Clearly, these lake sediments constitute a major source of future fuel and petrochemicals.

Many coal beds are nonmarine and were deposited in lakes associated with swamps, marshes, and delta plains. Lakes also yield deposits of diatomaceous earth, which is simply a deposit of diatom frustules. The tiny pieces of broken diatom frustules pierce the outside of insects and kill them as they try to crawl through it; therefore, diatomaceous earth is used as a pesticide in shipment of seeds. It is also used as a filter medium and absorption medium in the manufacture of explosives.

Bibliography

Bailey, Ronald. *Rivers and Lakes.* New York: Time-Life Books, 1985. A book on lacustrine (lake) and fluvial (river) environments, suitable for general readers. Part of the Time-Life Planet Earth series.

Håkanson, Lars, and M. Jansson. *Principles of Lake Sedimentology.* New York: Springer-Verlag, 1983. Though this book is a reference for professionals in the field of lake sedimentology, parts of it may be understood by high school students. Most books on limnology focus on lake water; this reference is one of the few to focus on lake sediments in detail. Provides methods of sampling and discusses the influence of lake type and shape on the sediments formed in the lake, the circulation of lake waters, the chemistry of sediments, and the pollution of lakes.

Hutchinson, G. Evelyn. *A Treatise on Limnology.* 3 vols. New York: John Wiley & Sons, 1957-1975. A highly comprehensive reference about lakes. The set derives its information from worldwide sources and is well indexed by subject as well as by specific lakes and their geographic locations. Volume 1 (in two parts: *Geography and Physics of Lakes* and *Chemistry of Lakes*) discusses the geologic formation of lakes, lake types, the interaction of sunlight with lake waters, water color, heat distribution, and water circulation. The chemistry of lake waters is discussed in detail. Volume 2, *Introduction to Lake Biology and Limnoplankton*, covers plankton and the factors that influence their growth. Volume 3, *Limnological Botany*, covers larger aquatic plants (macrophytes) and attached algae. Most of the treatise is readable by college undergraduates and high school seniors, but a few parts will be well understood only by specialists.

Lerman, Abraham, ed. *Lakes: Chemistry, Geology, Physics.* New York: Springer-Verlag, 1978. This book fills a gap in the Hutchinson treatise by focusing on the geologic processes of sedimentation in freshwater and brine lakes. The book is actually a compilation of chapters, each written by specialists. Particular attention is given to carbonate sediments, clastic and endogenic minerals, human influence on natural lakes, and organic compounds. The chapters on lake sediments and carbonate sedimentation are well illustrated. Most parts of these chapters are accessible to college undergraduates and high school students. Other chapters will be well understood only by specialists.

Stumm, Werner, ed. *Chemical Processes at the Particle-Water Interface.* New York: John Wiley & Sons, 1987. A highly technical reference book, designed for specialists and graduate students. Chapters are authored by a variety of experts, who focus on the chemical interactions that occur between sediment particles and the surrounding lake waters, the process of clotting of particles, and the role of particle surfaces in removing trace metals from water. Chapter 12, by Laura Sigg, focuses specifically on lake sediments and may be understood by undergraduates who have had rigorous courses in introductory chemistry and geology.

Wetzel, R. G., ed. *Limnology.* 2d ed. Philadelphia: Saunders, 1983. A very well written textbook typical of those used by undergraduates and graduates in their first limnology courses. Covers physical, biological, and chemical aspects of lakes. High school algebra, chemistry, physics, and biology courses will be necessary prerequisites to understand most of the text.

Edward B. Nuhfer

Cross-References

Calderas, 43; Coastal Processes and Beaches, 71; Drainage Basins, 141; Sedimentary Rocks, 568.

Lava Flow

The Hawaiian words pahoehoe and aa are used all over the world to designate two very common types of lava flow. The outward appearance of lava flows tells much about their eruption temperatures, chemical compositions, and viscosities. The surfaces of lava flows vary from smooth and glassy to rough and clinkery, depending on the viscosity of the lava and whether it is emplaced on land or under water.

Field of study: Volcanology

Principal terms

AA: a Hawaiian term (pronounced "ah-ah") that has been adopted for lava flows with rough, clinkery surfaces

AUTOCLASTIC BRECCIA: the clinkery or blocky rubble that forms on some lava flows

BLOCK LAVA: lava flows whose surfaces are composed of large, angular blocks; these blocks are generally larger than those of aa flows and have smooth, not jagged, faces

BRECCIA: a general term for any deposit composed mainly of coarse volcanic rock fragments

PAHOEHOE: a Hawaiian term (pronounced "pa-hoy-hoy") that is used in reference to lava flows with smooth, ropy surfaces

PILLOW LAVA: a type of bulbous, glassy-skinned lava that forms only when basaltic lava flows erupt under water

Summary

Aa and *pahoehoe* are Hawaiian words that describe the surface textures of lava flows. Aa is a rough, clinkery variety of lava, and pahoehoe has a smooth, ropy surface. Although aa and pahoehoe are the most common types of lava flow on land, other forms occur if lava is erupted under water or if it is extremely viscous. An explanation of why lava flows assume different forms requires a basic understanding of the physical properties of magma.

Magma is molten rock that originates by partial melting of either the crust or the mantle of the earth. "Lava" is a general term for magma that has erupted onto the earth's surface. Lava flows are bodies of magma that have been emplaced onto the surface as coherent, flowing masses of lava. The viscosity of lava, its resistance to flowage, largely controls the external appearance of the lava. In turn, the viscosity of lava is controlled mainly by its temperature and chemical composition. Most magmas are dominantly composed of the elements silicon and oxygen, which form strong chemical bonds with each other in the magma; together, these elements are referred to as silica. The silica content of magma exerts a major influence on magma viscosity. Similar to adding flour to batter, increased silica content means higher magma viscosity.

Although geologists recognize hundreds of kinds of volcanic rocks, there are three fundamental types of magma. Basalt magma is produced in the earth's upper mantle and is poorer in silica (about 50 percent by weight) and dissolved gases than other types of magma. It erupts at the highest temperatures of all lava, usually about 1,100-1,200 degrees Celsius. Because of its low silica content and high temperature, basalt lava has relatively low viscosity, or resistance to flowage. With the consistency of honey or peanut butter (although considerably denser, at around 2.7 grams per cubic centimeter), basalt lava is relatively fluid and usually erupts as thin, sheetlike lava flows. Andesite magma erupts at slightly lower temperatures than does basalt (around 1,000 degrees Celsius) and is richer in dissolved gases and silica (about 60 percent by weight). Andesite magma has a viscosity much greater than that of basalt. Andesite lava viscosity is similar to that of cold putty or frozen caulking compound, although its density of about 2.6 grams per cubic centimeter is considerably higher. Because of their relatively high viscosity, andesite lava flows are thick and tonguelike in shape and do not travel as far from the vent as do fluid basalt lava flows. Rhyolite, the third fundamental type of magma, is formed by melting of the continental crust. It is very rich in silica (greater than 70 percent by weight) and often contains high amounts of dissolved gases. Thus, it tends to erupt explosively and to

Volcanic eruptions can produce spectacular lava flows miles in length. *(PhotoDisc)*

form voluminous deposits of frothy pumice. The viscosity of molten rhyolite lava is extremely high: even at its white-hot eruption temperature of around 900 degrees Celsius, rhyolite behaves more like solid material than a liquid. It fractures when struck with a hammer, and its viscosity is sometimes similar to that of glacial ice. Thus, rhyolite magma moves very slowly and oozes onto the earth's surface as thick, pasty lava flows or steep-sided volcanic domes. The latter are steep piles of lava that grow directly on top of volcanic vents because the lava is too viscous to flow away from its source. Because the crystallization of rhyolite is very sluggish, its lava flows and domes are often composed of obsidian, a type of silica-rich volcanic glass.

Pahoehoe seldom forms on lava flows other than basalt, which is the hottest and most fluid type of magma. The smooth surface texture on pahoehoe lava flows is formed during quenching of the flow surface against the earth's atmosphere. A several-centimeter-thick crust of brittle lava is formed, commonly with a thin skin of glass. The smooth, brittle crust forms a barrier to rising gas bubbles (vesicles), and there is often a frothy zone of round vesicles directly underneath the crust, which rides atop the hot, fluid interior of the lava flow. The lava crust is an insulating barrier that prevents the flow of heat from the interior of the lava flow, which can remain hot for weeks, months, or even years, depending on its thickness. During flowage of the molten interior, the smooth crust is commonly deformed into folds and ridges, as when a carpet is shoved against a wall. The folds and ridges assume many forms, as they are repeatedly stretched and refolded onto themselves. Several varieties of pahoehoe have been recognized. Entrail pahoehoe is commonly formed when lava tubes are breached, spilling their contents as entwined, elongate bulbs of glassy-skinned lava. Shelly pahoehoe forms only near volcanic vents, where centimeter-thick lava crusts are stacked atop one another and are separated by cavernous gas pockets.

Pahoehoe lava flows are able to travel tens of kilometers from their vents because rocks are poor conductors of heat. As the top and sides of the lava flow solidify, an insulating barrier is formed around the interior of the flow, which remains hot and fluid. Like blood within a system of blood vessels, fluid lava moves within circular channels known as lava tubes, and the advancing nose of the lava flow is fed by this network. When eruption of lava ceases, some of the tubes drain out and their roofs may collapse in places, leaving accessible caves with flow marks on the walls and lava dripstones hanging from the ceilings.

Pillow lava is the subaqueous counterpart of pahoehoe and forms only when fluid basalt lava is erupted under water. Pillow lava has been observed forming around the Hawaiian Islands, and beneath its blanket of sediment, most of the sea floor is composed of this type of basalt lava. When basalt lava erupts under water, large quantities of steam are formed, sapping heat from the lava surface. As a result, a glassy skin quickly forms. Back-pressure from the still-fluid lava inside the lava flow eventually bursts the brittle skin, which is shattered into plates, chunks, and small shards of glass. New bulbs of lava ooze out of the cracks in a sort of budding process. Continuation of the process builds entwined masses of bulbous, glassy-skinned lava that look like toothpaste extrusions. The lava is mingled with layers of glassy fragments from the brittle skin of the advancing lava flow. Because it looks like a stack of pillows, the bulbous, glassy lava is appropriately called pillow lava.

Aa is a type of lava with a rough, clinkery surface. It is typical of basalt and andesite lava flows that are somewhat fluid but are too viscous to have pahoehoe surfaces. Jagged, spiney blocks of lava are formed during crumbling of the viscous mass. The spiney fragments continually rub against one another, eventually forming a thick outer envelope of debris that grades into the hotter, more fluid interior of the lava flow. The clinkery rubble rides on the flow until it eventually tumbles down the nose of the moving flow front: Similar to an advancing tractor tread, the fluid interior of the lava flow overrides its own debris. Eventually, the interiors of aa flows may become so viscous that they can no longer flow as liquids and instead begin to shear along horizontal fractures (platy joints) that form near their bases. The shear planes are similar to those that develop near the bases of glaciers, and the process is much like spreading out a deck of cards across a table top.

Single lava flows can change from pahoehoe into aa, but the reverse of this process has never been observed. When lava is first emitted, it is hottest and most fluid. As the lava cools, loses its gases, and slowly crystallizes, its viscosity is irreversibly increased. The originally smooth lava crust becomes thicker, and slabs of it begin to grind against one another. Eventually, smooth crust can no longer develop because the outer portion of the lava flow has become too brittle. In addition to cooling and the resulting increase in viscosity, the pahoehoe-to-aa transition has also been observed to occur when lava flows undergo high rates of internal shear. This can occur, for example, when a pahoehoe flow travels over steep terrain such as a cliff face, sometimes continuing as an aa flow at the base of the cliff. Under such circumstances, the lava must flow faster and internal shearing is therefore increased. During the pahoehoe-to-aa transition, not only does the brittle lava crust become thicker, but the interior of the flow also becomes more sluggish, as shown by the generally slower rates of movement of active aa flows (meters to hundreds of meters per hour), as compared to pahoehoe flows (hundreds of meters to tens of kilometers per hour). Pahoehoe flows have been clocked at speeds up to about 60 kilometers an hour in open channels and in lava tubes.

A third type of surface is formed on lava flows with extremely high viscosities. In some andesite and most rhyolite lava flows, fragmentation of the lava is very thorough because of the high viscosity of the mass. Great volumes of large, angular blocks are formed, each block having relatively smooth (not sharp, clinkery) faces and sharp edges between the faces. Called block lavas, the flows are thick, crumbling masses of fine-grained lava or obsidian that can barely move away from the vent. The noses of block-lava

A subterranean tube formed by lava flow. *(D. R. Clark/Craters of the Moon National Monument)*

flows are steep (30-35-degree) embankments of coarse rubble, and the tops of the flows are pocked with irregular depressions, lava spines, and automobile-sized lava chunks. The jagged rubble of block and aa lava flows is called friction breccia, crumble breccia, or autoclastic ("self-fragmented") breccia. "Breccia" is a general size term that refers to any deposit that is composed of coarse volcanic rock fragments (greater than 64 millimeters).

When the output of block lava is relatively low and magma viscosity is high, steep-sided volcanic domes may form over the vent. Most domes grow from within, by internal expansion, rather than by the additon of surface flows of lava. As a result, the thick carapace (shield or shell) of blocky rubble and lava spines is continually shoved aside, and the debris tumbles down the margins of the growing dome. Volcanic domes are very unstable and tend to collapse unpredictably into piles of blocky rubble. During collapse, ground-hugging avalanches of hot debris are sometimes jetted outward from the dome. Hence, volcanic domes are among the most dangerous of volcanic phenomena. Lava domes have repeatedly formed in the crater of Mount St. Helens since its catastrophic eruption of 1980, and the explosion of a Mount Pelée lava dome destroyed the city of St. Pierre on the Caribbean island of Martinique in 1904.

Methods of Study

Lava flows are a major component of most volcanoes. Together with chemical analyses of lava, the physical features of ancient lava flows have revealed much about how lava flows are emplaced. From laboratory experiments and theoretical calculations based on the chemistry of lava, the viscosity of virtually any type of magma can be derived.

Observations on active lava flows have revealed much about how lava moves. The temperatures of active lava flows can be estimated in several ways. The incandescent color of the lava gives a general indication of temperature, in much the same way that blacksmiths judge the temperature of forged steel: dull red is about 600 degrees Celsius, orange is about 900 degrees Celsius, and golden yellow is about 1,100 degrees Celsius. A more precise method uses an optical pyrometer to measure the wavelength of visible and near-infrared radiation, which varies uniformly as a function of temperature. The instrument is essentially a telescope in which a wire filament is mounted. The filament is heated by increasing an electric current until its color temperature matches that of the object being viewed; the temperature of the filament (and hence, the lava) can then be calculated. Both the visual method and the optical pyrometer can give inaccurate results because of atmospheric effects and because only the surfaces of objects are measured (the interiors of lava flows are of course much hotter than their surfaces). Another method that is less subject to error but considerably more hazardous in use is the thermocouple, a pair of wires of different composition welded together at both ends. When one end of the circuit is immersed in hot material (a difficult procedure in viscous lava), an electrical current is generated, its strength depending on the temperature difference between both ends of the circuit. An ammeter near the cold end of the circuit reads the electrical current, from which the temperature at the hot end of the circuit can be calculated.

Viscosities of active lava flows can be estimated by the use of penetrometers: When a known amount of force is applied to a steel rod of known diameter, the penetration of the rod into the lava will depend on the viscosity of the lava. Again, this is a superficial measurement that may not give an accurate estimate of the viscosity of the fluid interior of the lava. The chemical compositions of lavas are now routinely analyzed, and theoretical viscosities can then be calculated from the chemical data. Similarly, if the chemical composition of a lava is known, its crystallization temperature can be accurately estimated, and such a value is usually close to the eruption temperature of the lava. Field measurements are often used together with theoretical calculations in order to arrive at the best understanding of the flow behavior and physical properties of lava.

Context

If the thin blanket of ocean sediments is ignored, about three-fourths of the earth's surface is covered by subaerial and subaqueous lava flows. With the advent of space exploration, scientists have learned that other planets are also the hosts of volcanoes and that lava flows are common on their surfaces. The study of lava-flow features on earth, so far the only place where detailed observations can be made, is therefore of great benefit to planetary geologists. Hundreds of thousands of geochemical and minera-logical analyses have been collected on terrestrial and lunar igneous rocks during the past century, many of these from lava flows. Using the geophysi-cal and geochemical data from magma and lava, much has been learned about the internal constitution and melting behavior of the earth and its moon.

Although lava flows are among the least dangerous of volcanic phenom-ena, they are nevertheless the most common kind of material to erupt from many volcanoes and are potentially destructive of property. Particularly around the Pacific Rim, a substantial proportion of the world's population resides near active or potentially active volcanoes, and knowledge of the behavior of lava flows is therefore of great practical benefit.

Bibliography

Cas, Ray A. F., and J. V. Wright. *Volcanic Successions: Modern and Ancient.* Winchester, Mass.: Allen & Unwin, 1987. This 530-page text is written for geologists at the college level. Presents very thorough coverage of volcanic materials at an advanced level; much of the information on lava flows is readily digested by persons with some college science background. Contains many black-and-white photographs of volcanic features.

Decker, Robert, and Barbara Decker. *Volcanoes.* New York: W. H. Free-man, 1981. This 240-page paperback is geared to the general public and contains much useful information and interesting anecdotes. Easy to use, it includes a glossary, selected references for each chapter, an index, and appendices on the world's individual volcanoes and volcano information centers. Highly recommended as a starting point for anyone wanting to know more about volcanoes.

Francis, Peter. *Volcanoes.* New York: Penguin Books, 1981. This 370-page paperback is an excellent introduction to volcanoes. Written at about the level of beginning college students, it would be profitable for people with a good high school science background. Well illustrated and has an index but no reference list. Contains a chapter on extrater-restrial volcanism.

Green, Jack, and N. M. Short, eds. *Volcanic Landforms and Surface Features: A Photographic Atlas and Glossary.* New York: Springer-Verlag, 1971. This 522-page, black-and-white photographic atlas shows lava flows of di-verse compositions and from worldwide localities. The main purpose of this book is to provide examples not describe phenomena or give backround; hence it is best used in conjunction with other, more detailed descriptive references.

Kennish, Michael J. "Morphology and Distribution of Lava Flows on Mid-Ocean Ridges: A Review." *Earth Science Reviews* (May, 1998): 63-91.

Macdonald, Gordon A. *Volcanoes.* Englewood Cliffs, N.J.: Prentice-Hall, 1972. This college-level textbook is an excellent source of descriptive information concerning all aspects of volcanism. Its chapter on lava flows is one of the best available. The 510 pages include an index, a reference list, and a useful appendix, "Active Volcanoes of the World." Profusely illustrated with black-and-white photographs and line drawings.

Williams, Howell, and A. R. McBirney. *Volcanology.* San Francisco: Freeman, Cooper, 1979. This 400-page college textbook has the complete descriptive coverage of the book by Macdonald but is more quantitative. Geared to geologists. It nevertheless contains much descriptive information that can be understood by nonscientists. Chapters on the physical nature of magma and on lava flows are excellent, and the text is liberally illustrated with graphs, drawings, and black-and-white photographs. A subject index and reference list are given.

William R. Hackett

Cross-References

Magmas, 383; The Oceanic Crust, 471; Shield Volcanoes, 577; Volcanoes: Types of Eruption, 685.

The Lithosphere

Within the lithosphere, earthquakes occur, volcanoes erupt, mountains are built, and new oceans are formed. An understanding of the lithosphere's structure is needed in the search for oil and gas, for the prediction of earthquakes, and for the verification of a nuclear test ban treaty.

Field of study: Geophysics

Principal terms

ASTHENOSPHERE: the partially molten weak zone in the mantle directly below the lithosphere

BASALT: a dark-colored igneous rock containing minerals, such as feldspar and pyroxene, high in iron and magnesium

CRUST: the rocky, outer "skin" of the earth, made up of the continents and ocean floor

GRANITE: a light-colored igneous rock containing feldspar, quartz, and small amounts of darker minerals

MANTLE: the thick, middle layer of the earth between the crust and the core

MOHOROVIčIĆ discontinuity (Moho): the boundary between the crust and the mantle, named after the Yugoslavian seismologist Andrija Mohorovičić, who discovered it in 1909

PERIDOTITE: an igneous rock made up of iron- and magnesium-rich olivine, with some pyroxene but lacking feldspar

REFLECTED WAVE: a wave that is bounced off the interface between two materials of differing wave speeds

REFRACTED WAVE: a wave that is transmitted through the interface between two materials of differing wave speeds, causing a change in the direction of travel

Summary

The lithosphere is the rigid outer shell of the earth. It extends to a depth of 100 kilometers and is broken into about ten major lithospheric plates. These plates "float" upon an underlying zone of weakness called the asthenosphere. The phenomenon is somewhat like blocks of ice floating in a lake: As lake currents push the ice blocks around the lake, so do currents in the asthenosphere push the lithospheric plates. The plates carry continents and oceans with them as they form a continually changing jigsaw puzzle on the face of the earth.

The word "lithosphere" is derived from the Greek *lithos*, meaning stone. Historically, the lithosphere was considered to be the solid crust of the earth, as distinguished from the atmosphere and the hydrosphere. The words "crust" and "lithosphere" were used interchangeably to mean the unmoving, rocky portions of the earth's surface. Advances in the understanding of the structure of the earth's interior, resulting mostly from seismology, have forced the redefinition of old terms. "Crust" presently refers to the rocky, outer "skin" of the earth, containing the continents and ocean floor. "Lithosphere" is a more comprehensive term that includes the crust within a thicker, rigid unit of the earth's outer shell. To appreciate the reason for this redefinition, it is necessary to learn about the nature of the earth's interior.

Except for the upper 3 or 4 kilometers, the earth's interior is inaccessible to humans. Therefore, indirect methods, such as studying earthquakes and explosions, are used to learn about the inside of the earth. Earthquakes and explosions, both conventional and nuclear, generate two types of energy waves: compressional (P) waves and shear (S) waves. P waves travel faster than do S waves and are generally the first waves to arrive at an observation station. The speed of a wave, however, depends on the rock through which it travels. When seismic waves encounter a boundary between two different rocks, some energy is reflected back, and some is transmitted across the boundary. If the rock properties are very different, the transmitted waves travel at a different speed and their travel path is bent, or refracted. This phenomenon can be illustrated by placing a pencil in a glass of water. Light in water travels at a speed different from that of light in air, so light is refracted, or bent, as it travels from water to air. Thus, the pencil appears to be bent. P waves and S waves are reflected and refracted as they travel through the earth. Waves following different paths travel at different speeds.

Since 1900, seismologists have studied P and S waves arriving at different locations from the same earthquake. They discovered three distinct layers in the earth: the crust, the mantle, and the core. The boundaries separating these layers show abrupt changes in both P- and S-wave speeds. These changes in wave speeds provide information about the earth's interior. Scientists studying the theory of traveling elastic waves, such as earthquake waves, related the speed of waves to the physical properties of the material through which they travel. It was found that S waves do not travel through liquids. From this finding, scientists concluded that the earth's core had a liquid outer region and a solid inner region. Other scientists measured the P- and S-wave speeds of many different rocks and provided clues to the kind of rocks found inside the earth.

The quantity and quality of seismological and related information have grown rapidly since the end of World War II. There has been an increase in the number of seismological observatories and in the quality of seismographs, the instruments that record the arrivals of seismic waves. The methods of explosion seismology, developed for use in the search for oil and gas, have also been applied to the lithosphere. Experimental rock studies have been undertaken at higher and higher temperatures and pressures. In addition, the development of high-speed computers has enabled scientists to handle the vast amounts of data being generated and to test more complex models of the earth's properties with the observed seismic data. The combined use of explosion seismology, earthquake studies, and the experimental and theoretical studies of rocks has provided a very detailed picture of the seismic structure of the crust and upper mantle.

The continental crust averages 30-40 kilometers thick and is divided into two main seismic layers. One layer, the upper two-thirds of the crust, has P- and S-wave speeds corresponding to those of granitic rocks. The speeds increase slightly in the bottom third of the continent, corresponding to rocks of basaltic composition. The average oceanic crust is 11 kilometers thick and is of basaltic composition. Beneath both continental and oceanic crust, the P- and S-wave speeds increase sharply. This boundary between the crust and mantle is called the Mohorovičić discontinuity, or Moho. The Moho marks a compositional change to a dense, ultramafic rock called peridotite.

At an average depth of 100 kilometers, the S-wave speed decreases abruptly. It remains low for about 100-150 kilometers. This region is called the low velocity zone (LVZ). Laboratory experiments have shown that seismic-wave speeds, particularly those of S waves, decrease in rocks containing some liquid. The LVZ in the mantle indicates a zone of partial melting, perhaps 1-10 percent melt. The presence of the melt reduces the overall strength of the rock, giving the region its name, "asthenosphere," from the Greek *asthenes*, meaning "without strength."

The partially molten asthenosphere is very mobile, allowing the more rigid lithosphere above it to move about the earth's surface. The boundary between the lithosphere and the asthenosphere does not mark a change in composition; it marks a change in the physical properties of the rocks. The lithosphere defines this region of crust and mantle from the mantle region below by its seismic-wave speeds and its physical properties.

Seismic-wave speeds and earthquake distribution provide information about the lithospheric plates and the boundaries between them. Like the earth's crust, lithospheric plates are not the same everywhere. For example, the Pacific plate contains primarily oceanic crust, the Eurasian plate is mostly continental, and the North American plate contains both continental and oceanic crust. The lithosphere is thinnest at spreading centers, or regions where two plates are moving away from each other, such as the Mid-Atlantic Ridge and the East Pacific Rise. Here, the asthenosphere is close to the surface and the melt portion pushes upward, separating the plates and creating new lithosphere. Shallow earthquakes occur as the new crust is cracked apart. In areas such as western South America or southern Alaska, two plates are coming together, with the oceanic lithosphere being thrust under the continental plate. Earthquakes occur as deep as 700 kilometers as one plate slides under the other. Along the California coast, two plates slide past each other along faults that cut through the lithosphere. Earthquakes are common, and the faults can move several meters at a time. Where two continental plates, India and Eurasia, have collided, the crust is highly faulted and 65 kilometers thick. Earthquakes in and near the Himalaya are numerous, often occurring along deep fault zones.

Although earthquakes are most common along plate boundaries, they can also occur within lithospheric plates. Some earthquakes are related to newly forming boundaries. The Red Sea is believed to be a recently formed spreading center pushing the Arabian Peninsula and Africa apart. Some earthquakes result from the movement along ancient geologic faults buried within the crust. The causes of some earthquakes, however, such as the one in 1886 in Charleston, South Carolina, remain unknown.

Structural details within plate regions cannot be determined by earthquake studies alone. P and S waves generated by explosions are reflected and refracted by layers within the lithospheric plates. Regional studies show the upper lithosphere to be highly variable. In mountainous regions, such as the Appalachians or the Rocky Mountains, the continental crust is thicker than average and shows much layering. In the midcontinent and the Gulf of Mexico regions, the crust consists of thick layers of sediments and sedimentary rocks. Oil companies, combining the data from many controlled explosions, discovered petroleum and natural gas within these layers from the changes in P- and S-wave speeds. Other regional seismic studies have found ancient geological features deep within the crust. Similarities in the seismic structure between these and other known features can uncover potential sites of much-needed natural resources. The discovery of the oil fields of northern Alaska was prompted by the area's structural similarity to the Gulf of Mexico, a known source of oil and gas.

The seismic structure of the lower lithosphere is less well known. Early studies show that it is also highly variable and that crustal structures are often related to features deep in the lithosphere. Much work, however, remains in unraveling the details of the lithosphere.

Methods of Study

Scientists use a number of seismic techniques to study the lithosphere. They use P and S waves generated by earthquakes that travel through the earth (body waves) and along the earth's surface (surface waves). Reflection and refraction seismology uses seismic waves generated by explosions to study the continental and oceanic lithosphere. Data from experimental studies of rocks are used to relate seismic speeds to specific kinds of rocks. Computers help analyze the vast amounts of seismic data and are used to develop models to aid in the understanding of the earth.

The use of P and S waves from earthquakes is the oldest method of studying earth structure. The time at which P and S body waves, reflected and refracted by the layers in the earth, arrive at different distances from the same earthquake is related to the average speed at which the waves travel. The arrival time of surface waves also depends on the layer speeds. Using seismic waves from many earthquakes, seismologists can determine the seismic structure of the lithosphere.

In regions with numerous earthquakes, seismologists record P and S waves using many portable seismographs, instruments that record seismic-wave arrivals. The scientists can then determine a more detailed regional structure. Earthquakes, however, do not occur regularly everywhere on the earth. Until an average regional structure is known, it will be difficult to determine the precise location and time of an earthquake.

Explosions as a source of seismic waves to study crustal structure have been developed and used extensively by the oil industry. With an explosive source, its location and time of detonation can be precisely controlled. Two basic techniques using artificial sources are reflection seismology and refraction seismology. Refraction seismology studies the arrivals of waves that are refracted, or bent, by the layers in the crust. The scientist determines an average velocity structure for an area by recording the time the first waves arrive at receivers located varying distances from the explosion. To determine deep structure, the distance between the explosion and the receivers must be very large. Reflection seismology allows a deeper look into the crust by studying reflections from many different layers. The seismic-wave receivers do not need to be placed as far from the source as they must in refraction studies. The reflection technique combines the results from many explosions, producing a picture of the earth's layers. This method is used extensively in the search for oil and gas. The techniques of reflection and refraction seismology have been applied to the lithosphere. Long reflection and refraction profiles have been acquired over geologically interesting but little-understood regions.

Seismic waves are vibrations traveling around and through the earth. Because of friction, these vibrations eventually stop, and seismic waves no longer travel. Earthquakes and explosions generate waves that vibrate at many frequencies. The earth slows each frequency differently. As a seismic wave travels through different rocks, the shape of its vibrations recorded on a seismograph is related to the properties of the rocks through which it travels. The analysis of seismic waveforms has shown differences between waves generated by earthquakes and by explosions.

To understand the lithosphere, it is necessary to know about rocks. Using a hydraulic press, scientists squeeze rocks in the laboratory to pressures and heat them to temperatures present deep within the earth. They then measure the rock's physical properties at these conditions. Experimentally measured P and S speeds are compared to wave speeds determined from earthquakes and explosions to infer the kind of rocks and the conditions that exist within the earth. The complexity of the lithosphere, however, does not allow simple answers.

To aid the scientists in their studies, computers are used to develop models—simplified representations—of the earth. By making changes in the model, the scientist can study changes in computed seismic properties and compare them to the observed earth properties. Changes in the model are made to resemble the earth more closely. In modeling the lithosphere, scientists incorporate data from a wide range of sources, such as earthquake studies, experimental rock studies, and geologic maps. The computer allows the earth scientist to test more complex models in an effort to provide a better understanding of the lithosphere.

Context

Understanding the seismic structure of the lithosphere helps in understanding nature. The movement of the lithospheric plates about the earth creates mountain ranges, causes earthquakes, and devours ocean basins. Because much of the earth is inaccessible, seismic waves, generated by earthquakes and explosions, are used to look deep within the earth to provide a picture of the earth's structure. For the earth scientist, increased knowledge of the seismic structure of the lithosphere helps in unraveling the processes by which geologic features are formed.

Increased knowledge of the lithosphere is important to the average person for three reasons: First, earthquakes are caused by movements between and within the lithospheric plates. Every year, lives are lost and millions of dollars in damage occur because of earthquakes and earthquake-related phenomena. Detailed knowledge of the lithosphere helps scientists understand where and how earthquakes occur. This information can lead to regional assessment of the potential for earthquakes and earthquake-related damage. Knowledge of the earthquake potential of a region can result in the improvement of local building codes and the evaluation of existing emergency preparedness plans. Earthquake-hazard assessment can also aid in prediction by determining the probability of future earthquake occurrence. Some success in long-term predictions has been seen in Japan and China. Eventually, the increased understanding of the lithosphere may lead to the short-term prediction of earthquakes.

Second, detailed knowledge of lithospheric structure will lead to the discovery of potential sites of needed natural resources, such as oil, gas, and coal; metals, such as iron, aluminum, copper, and zinc; and nonmetal resources, such as stone, gravel, clay, and salt. Scientists are beginning to unravel the relationship of tectonic features to the formation of many mineral deposits. Detailed knowledge of the structure of the lithosphere from seismic studies can uncover deeply buried features that may provide new sources for critically needed resources.

Finally, in the interest of preserving life on earth and the earth itself, better knowledge of the lithosphere can lead to a nuclear test ban treaty. Scientists require detailed information on the seismic structure of the lithosphere to locate and identify earthquakes and nuclear explosions. More structural information will also lead to better identification of the differences between these two types of seismic wave sources. An accurate and reliable means of distinguishing between earthquakes and nuclear explosions is critical for the verification of any nuclear test ban treaty.

Bibliography

Bakun, William A., et al. "Seismology." *Reviews of Geophysics* 25 (July, 1987): 1131-1214. A series of articles summarizing research in seismology in the United States from 1983 to 1986. Reviews recent findings and unresolved problems in all areas of seismology. Articles are somewhat technical but suitable for the informed reader. Extensive bibliographies.

Bolt, Bruce A. *Earthquakes.* New York: W. H. Freeman, 1988. A popular, illustrated book on the many features of earthquakes. Chapter topics include the use of earthquake waves to study the earth's interior and earthquake prediction. A bibliography and index are included. Suitable for the layperson.

Bullen, K. E., and B. A. Bolt. *An Introduction to the Theory of Seismology.* 4th ed. New York: Cambridge University Press, 1985. Introductory sections of most chapters provide historical and nonmathematical insight into the subject, suitable for the general reader. Contains a selected bibliography, references, and an index. (Designed as a text for the advanced student with a mathematics background.)

Lavier, Luc L., and Michael S. Steckler. "The Effect of Sedimentary Cover on the Flexural Strength of Continental Lithosphere." *Nature* 389, no. 6650 (October 2, 1997): 476-480.

Mutter, John C. "Seismic Images of Plate Boundaries." *Scientific American* 254 (February, 1986): 66-75. An article on the application of explosion seismology to the study of plate boundaries. Summarizes the method of seismic reflection profiling. Shows results of studies across different plate boundaries. Well illustrated. Suitable for the general reader.

Pitman, Walter C. "Plate Tectonics." In *McGraw-Hill Encyclopedia of the Geological Sciences.* New York: McGraw-Hill, 1978. A brief summary of plate tectonics, discussing evidence for the theory and an explanation of causes of present-day features. Cross-referenced, illustrated, with bibliography. Suitable for the general reader.

Press, Frank, and Raymond Siever. *Earth.* 4th ed. San Francisco: W. H. Freeman, 1986. A book for the beginning reader in geology. Of interest are chapter 17, "Seismology and the Earth's Interior," and chapter 19, "Global Plate Tectonics: The Unifying Model," for an overall understanding of the importance of the lithosphere. Illustrated and supplemented with numerous marginal notes. Chapter bibliographies and glossary.

Smith, Peter J., ed. *The Earth.* New York: Macmillan, 1986. A well-illustrated, comprehensive guide to the earth sciences for the general reader. Chapter 3, "Internal Structure," describes historical development of the current view of the earth's lithosphere. Chapters 1, 2, and 5 provide related material. Includes glossary of terms.

Thomson, Ker C. "Seismology." In *McGraw-Hill Encyclopedia of the Geological Sciences.* New York: McGraw-Hill, 1978. A brief summary of the principles of seismology. Discusses methods of determining earth structure and of detecting nuclear explosions and describes related research. Cross-referenced and illustrated, with bibliography. Suitable for the interested general reader.

Pamela R. Justice

Cross-References

Earthquakes, 148; Earth's Crust, 179; Earth's Mantle, 195; Earth's Structure, 224; Oil and Gas: Distribution, 481; Plate Tectonics, 505.

Magmas

Magma is a naturally occurring molten rock material that originates ultimately within the earth's mantle but also within the crust and migrates in the subsurface via both intrusion and extrusion. Recognition of the origin of magmas is essential in understanding the complex relationship between volcanism and regional deformation processes, or plate tectonics.

Field of study: Petrology

Principal terms

ASSIMILATION: the incorporation and digestion of solid or fluid foreign material

ASTHENOSPHERE: that portion of the earth that comprises the part of the upper mantle where isostatic adjustments occur, magma may be generated, and seismic waves are strongly attenuated

CONTAMINATION: the process whereby the chemical composition of a magma is altered as a result of assimilation of inclusions or the surrounding rock being intruded

CRYSTALLIZATION: the formation and growth of a crystalline solid from a liquid or gas

DIFFERENTIATION: the process of developing more than one rock type from a common magma

EXTRUSION: the emission of magma or lava and the rock so formed onto the earth's surface

INTRUSION: the process of emplacement of magma and the rock so formed in preexisting rock

ISOTOPE: one of two or more species of the same chemical element where the element has the same number of protons but different number of neutrons and, therefore, different mass

LITHOSPHERE: that portion of the earth that comprises the crust and part of the upper mantle where deformation at geologic rates occurs

MAGMATIC EVOLUTION: the continuing change in the chemical composition of a magma during its ascent

Summary

Magma is a naturally occurring mobile rock material consisting of a liquid silicate phase, with or without a solid phase of suspended crystals, and entrained gases. Largely a molten heterogeneous material, magma originates within the earth's upper mantle or lower crust. The earth's crust in part is largely a product of magmatic processes within the mantle that began at least 3.8 billion years ago and are evidenced today in certain areas such as the sea floor and in some continental regions. Magmas encompass a considerable range in composition and almost invariably contain one or more solidus phases (feldspar, pyroxene, olivine, and so on) that are enclosed in a silicate melt whose properties, notably viscosity and temperature, are related to its composition. Extrusion is the process whereby magma cools at the surface and forms extrusive rocks. Volcanic (surficial) rocks are derived from magmas that erupted as lava from volcanic vents that have solidified. In contrast, intrusion is the process whereby magma cools at depth (deep-seated) and forms intrusive rocks. Plutonic rocks, which are derived within the crust—notably below volcanic regions where temporary reservoirs of magma exist—are good examples of intrusive rocks. It is inferred that such bodies solidify completely without ever reaching the surface.

The earth is subdivided into the crust, mantle, and core. Based on geophysical evidence, the earth consists of solid material, with the exception of the outer core. The crust extends to a depth of 30-50 kilometers under most continental regions and less than 10-12 kilometers under most oceanic regions. The mantle is below the crust and is characterized by an abrupt increase in seismic velocities. The transition, referred to as the Mohorovičić discontinuity, from the crust to the mantle represents an important change in bulk composition and mineralogy from essentially an andesite-basaltic continental crust and basaltic oceanic crust to a mantle of peridotite. The upper layers of the earth include the asthenosphere and portions of the lower lithosphere. The asthenosphere includes that portion of the upper mantle or layer of weakness that comprises the part of the upper mantle where isostatic adjustments occur, magma may be generated, and seismic waves are strongly attenuated. Overlying the asthenosphere is the lithosphere, which includes the crust and that portion of the upper mantle or layer of strength where deformation at geologic rates occurs. The thickness of the lithosphere is on the order of 100 kilometers. The boundary between the lithosphere and asthenosphere is the low velocity zone, at depths ranging from about 100 to 200 kilometers.

The solid mantle is composed principally of the rock referred to as peridotite. Peridotite is a general term for a coarse-grained, dense plutonic rock composed chiefly of olivine with or without other mafic mineral constituents. These other constituents include pyroxenes but also amphiboles and micas with little or no feldspar. The ultimate source of all magmas is a result of partial melting of the solid preexisting rock, notably the mantle. Pressure and temperature conditions in the asthenosphere are such that a small increase in temperature or decrease in pressure can induce partial melting of peridotite within ranges of about 10-30 percent. Melting is concentrated within the uppermost 150-250 kilometers of the mantle.

Partial melting can be induced in several ways, most clearly by the increase in temperature. Melting is, however, rarely complete. Most melts probably coalesce to form discrete magma bodies that migrate away from the source area, leaving some residue behind; thus, the resulting chemical composition of the magma differs from that of the source material. Partial melting of the originally solid crust or mantle allows for the diverse variety of igneous rock types observed. Only a very small fraction of igneous rocks are representative of primary magmas. Primary variation occurs when magmas of different composition are created at the source, which accounts for the variation in igneous rocks, including differences in composition and type of materials being melted in the source region, degree of partial melting, and conditions under which melting has taken place.

Although the diversity in igneous rocks can be explained in terms of these primary variations, secondary variations (where different magmas subsequently evolved from a common parent) are also very important, as they reflect the magmatic evolution of the igneous rocks. These secondary processes include magmatic differentiation, whereby more than one rock type is developed from a common magma. The incorporation and digestion of solid or fluid foreign material during magma ascent, referred to as assimilation, can also result in the development of a different rock type if, in fact, the chemical composition of a magma is significantly altered. The altering of magma chemistry via assimilation is referred to as the process of contamination. Mixing of magma types can also change the overall chemistry such that once solidified, an intermediate rock type is developed. When a magma chemistry is altered simultaneously by solidifying or crystallizing within the cooler portions of the magma body while also melting portions of the surrounding rocks encompassing the body (the process of zone melting), the evolved rock type can also be changed. Within plutonic rocks, evidence of differentiation, sometimes of contamination, and rarely of magma mixing has been found. With volcanic rocks, however, these processes are more difficult to detect.

Magma generation is closely associated with tectonic processes. Basaltic magmas, for example, are generated by direct partial melting of mantle peridotite. This magma type is most abundant beneath active oceanic ridges, which are the sites where lithosphere plates are being formed. In fact, the oceanic crust is basically basaltic in composition and composed of volcanic rocks solidified from magma erupted on the ocean floor. These rocks are underlain by basaltic dikes, which reflect fissures or conduits for molten magma. These dikes are in turn underlain by gabbros and cumulate layers, which are evolved rocks that have formed when large amounts of crystals settled out of a cooling magma.

Andesitic magmas are generated above subduction zones, where lithospheric plates converge. Subduction zones are long, narrow belts where one lithospheric plate descends beneath another. In this case, the melting of peridotite within the mantle is complicated by the presence of wet sediments and basalts that comprise the descending oceanic or lithospheric plate. These magmas thus tend to be more silica-rich and are responsible for the formation of new continental crust, which grows by progressive accumulation of andesitic volcanic rocks above and dioritic intrusive rocks below.

Granitic magmas to a large extent are generated above subduction zones. The processes at work here include both partial melting at the base of the andesitic crust and differentiation, the process of developing more than one rock type from a common magma of andesitic or even basaltic magmas that have been modified by some mixing with crustal material.

True primary magmas are of basaltic composition, as they can only be generated by partial melting of peridotite within the mantle. In contrast, andesites and granites can be formed either by differentiation of basaltic magmas or by direct partial melting within the mantle or crust. In fact,

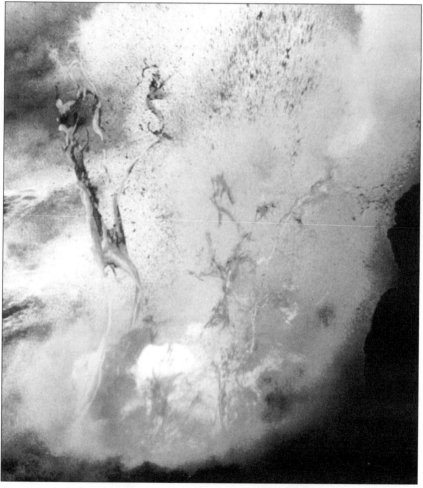

Volcanic eruptions expel magma in the form of lava. *(PhotoDisc)*